COMING
OF AGE
IN THE
MILKY
WAY

Books by Timothy Ferris

The Red Limit
Galaxies
SpaceShots
The Practice of Journalism
(with Bruce Porter)
Coming of Age in the Milky Way

COMING OF AGE IN THE MILKY WAY

TIMOTHY FERRIS

ANCHOR BOOKS
DOUBLEDAY
NEW YORK LONDON TORONTO SYDNEY AUCKLAND

AN ANCHOR BOOK
PUBLISHED BY DOUBLEDAY
a division of Bantam Doubleday Dell Publishing Group, Inc.
1540 Broadway, New York, New York 10036

ANCHOR BOOKS, DOUBLEDAY, and the portrayal of an anchor
are trademarks of Doubleday, a division of Bantam Doubleday
Dell Publishing Group, Inc.

Coming of Age in the Milky Way was originally published in hardcover
by William Morrow and Company, Inc., in 1988.
The Anchor Books edition is published by arrangement with
William Morrow and Company, Inc.

Library of Congress Cataloging-in-Publication Data
Ferris, Timothy.
Coming of age in the milky way/Timothy Ferris.
1st Anchor books ed.
  P.   cm.
  Originally published: New York: Morrow, c1988.
  Includes index.
  ISBN 0-385-26326-0
  1. Science—History.   2. Space and time.   3. Cosmology.
I. Title.
Q125.F425   1989   89-32185
509—dc20   CIP

For Carolyn

If I could write the beauty of your eyes
And in fresh numbers number all your graces,
The age to come would say, "This poet lies—
Such heavenly touches ne'er touched earthly faces."

—SHAKESPEARE

PREFACE

AND

ACKNOWLEDGMENTS

How oft we sigh
When histories charm to think that histories lie!
—Thomas Moore

This book purports to tell the story of how, through the workings of science, our species has arrived at its current estimation of the dimensions of cosmic space and time. The subject is grand, and it goes without saying that the book is unequal to it. Of the limitations and liabilities of this work I would hope to defend only those resulting from its brevity. Obedient to the dictum of Callimachus that "a big book is equal to a big evil," I have striven for economy, but economy has its price.

First, it has of course meant leaving out many things. In a general survey of science it would be ludicrous, for instance, to discuss quantum mechanics without making reference to Erwin Schrödinger, who was one of the principal architects of that innovative and fruitful discipline. My justification is that this is not a general survey. It is a book with one tale to tell—that of the awakening of the human species to the spatiotemporal dimensions of the universe—and owes its loyalties to that theme alone.

In addition to encouraging sins of omission, compression tends to foreshorten history, making it seem more coherent and purposeful than it really is, or was. The real history of science is a maze, in which most paths lead to dead ends and all are littered with the broken crockery of error and misconception. Yet in this book all that is underrepresented, while disproportionate emphasis is devoted to the ideas and observations that have in retrospect proved most salient. A book that assigned a full measure of devotion to the mistakes of science would, however, be almost unreadable: Plowing through it would be like reading a collection of mystery stories of which only one or two came to any satisfactory resolution, while in most the detective switched careers before the identity of the culprit could be ascertained or the butler was irrelevantly run over by a bus.

Similarly, in recounting the long-term development of enduring conceptions one tends to assign missions to people that they did not have, or did not know that they had, at the time. Thus Maxwell becomes the father of unified theory, Fraunhofer a founder of astrophysics, and Einstein the theorist who anticipated the expansion of the universe, though there is no evidence that any of these men ever got up in the morning with the intention of doing any such thing. As Thomas Carlyle wrote, "No hammer in the Horologe of Time peals through the universe when there is a change from Era to Era. Men understand not what is among their hands."[1] But history, as they say, is comprehended backward though it must be lived forward, and when we examine our predecessors we bring our own lamps.

Economy also implies simplification. This book is intended for general readers. It keeps mathematics and jargon to a minimum—such technical terms as seemed unavoidable are explained in the text and the glossary—and in so doing sometimes warps the very concepts it seeks to explain. Where the distortion is excessive or ill-advised the fault is of course entirely my own, but much of it results from a change of perspective: Relativity and quantum mechanics and cosmology look different to a lay observer than to a practicing scientist, just as the experience of making an Atlantic crossing on a cruise ship is different for a passenger than for a stoker in the boiler room. On the other hand, I have tried in general not to oversimplify, preferring that a subtle idea remained subtle in the retelling rather than hammering it so flat as to make it appear trivial or self-evident.

Much the same applies when it comes to ambiguities and disagreements over the facts of our intellectual heritage and their interpretation. The history of science is full of disputations about such questions as just why Galileo was persecuted by the Roman Catholic Church or whether Einstein had the Michelson-Morley experiment in mind when he composed his special theory of relativity. Having tiptoed through more than a few of these minefields, I am full of admiration for scholars who choose to habituate them. Nevertheless, I have devoted little space to detailing the contrasting arguments they have set forth. If the resulting narrative is unambiguous it is also skewed, and can claim to be accurate only insofar as I may have succeeded in supporting or inventing a point of view that may itself prove to be accurate. Here endeth the confession, with the plea that economy is a jealous god.

A word about numerical style. Exponential numbers are employed, in which the exponents express powers of ten; thus 10^3 equals a one followed by three zeros, or 1,000, and 10^{-3} equals 0.001. By the word "billion" is meant the American billion, equal to 1,000,000,000 or 10^9.

* * *

Coming of Age in the Milky Way was written in New York, Los Angeles, and San Francisco over a period of twelve years, from 1976 through 1988. As one might expect, in the course of so long a project I have incurred more debts of gratitude than I can properly retire. I should like, however, to express my thanks for aid and criticism provided by William Alexander, Sherry Arden, Hans Bethe, Nancy Brackett, Ken Broede, Robert Brucato, Lisa Drew, Ann Druyan, David Falk, Andrew Fraknoi, Murray Gell-Mann, Owen Gingerich, J. Richard Gott III, Stephen Jay Gould, Alan Guth, Stephen Hawking, He Xiang Tao, Karen Hitzig, Larry Hughes, Res Jost, Kathy Lowry, Owen Laster, Irwin Lieb, Dennis Meredith, Arthur Miller, Bruce Murray, Lynda Obst, Heinz Pagels, Abraham Pais, Thomas Powers, Carl Sagan, Allan Sandage, David Schramm, Dennis Sciama, Frank Shu, Erica Spellman, Gustav Tammann, Jack Thibeau, Kip S. Thorne, Michael Turner, Nick Warner, Steven Weinberg, John Archibald Wheeler, Houston Wood, and Harry Woolf.

I am grateful to Prairie Prince for the line illustrations that accompany the text and for the starfield paintings employed in the endpapers.

For research aid and secretarial assistance at various stages along the way I am indebted to Eustice Clarke, Dave Fredrick, Russ Gollard, Michele Harrah, Sandra Loh, and Camille Wanat, and to the exertions of librarians too numerous to mention at the American Institute of Physics, Brooklyn College of the City University of New York, CERN, Caltech, the Federal Polytechnic Institute in Zurich, Fermilab, Harvard University, the Massachusetts Institute of Technology, the Mount Wilson and Las Campanas Observatories, New York University, Princeton University, the University of Southern California, the University of California at Berkeley, and the public libraries of New York City, Los Angeles, Chicago, Boston, and Miami.

I am happy to acknowledge the support provided by research grants from the University of California at Berkeley, the Division of Social Sciences of the University of Southern California, and the John Simon Guggenheim Memorial Foundation.

My thanks go as well to my mother, Jean Baird Ferris, for her lively conversation and steadfast encouragement, her tireless proffering of intriguing clippings and articles, and for having taught me, as a boy, to love and to live by books.

Finally I should like to express my deep gratitude to my wife and family, for their loving and generous forbearance through the long years of long hours that writing this book consumed.

—T.F.
Berkeley, California

CONTENTS

CONTENTS

One thing I have learned in a long life: that all our science, measured against reality, is primitive and childlike—and yet it is the most precious thing we have.

—Albert Einstein

The wind was flapping a temple flag, and two monks were having an argument about it. One said the flag was moving, the other that the wind was moving; and they could come to no agreement on the matter. They argued back and forth. Eno the Patriarch said, "It is not that the wind is moving; it is not that the flag is moving; it is that your honorable minds are moving."

—Platform Sutra

PART ONE

SPACE

The self shines in space through knowing.
—The Upanishads

1

THE DOME
OF HEAVEN

> You may have heard the music of Man but
> not the music of Earth. You may have heard
> the music of Earth but not the music of Heaven.
> —Chuang Tzu

> Had we never seen the stars, and the sun, and
> the heaven, none of the words which we have
> spoken about the universe would ever have
> been uttered. But now the sight of day and
> night, and the months and the revolutions of
> the years, have created number, and have given
> us a conception of time, and the power of
> enquiring about the nature of the universe;
> and from this source we have derived philos-
> ophy, than which no greater good ever was
> or will be given by the gods to mortal man.
> —Plato

The skies of our ancestors hung low overhead. When
the ancient Sumerian, Chinese, and Korean astronomers trudged
up the steps of their squat stone ziggurats to study the stars, they
had reason to assume that they obtained a better view that way,

not, as we would say today, because they had surmounted a little
dust and turbulent air, but because they had got themselves appreciably closer to the stars. The Egyptians regarded the sky as a
kind of tent canopy, supported by the mountains that demarked
the four corners of the earth, and as the mountains were not all
that high, neither, presumably, were the heavens; the gigantic
Egyptian constellations hovered close over humankind, as proximate as a mother bending to kiss a sleeping child. The Greek sun
was so nearby that Icarus had achieved an altitude of only a few
thousand feet when its heat melted the wax in his wings, sending
the poor boy plunging into the uncaring Aegean. Nor were the
Greek stars significantly more distant; when Phaethon lost control
of the sun it veered into the stars as suddenly as a swerving chariot
striking a signpost, then promptly rebounded to earth (toasting the
Ethiopians black on its way down).

   But if our forebears had little notion of the depths of space,
they were reasonably well acquainted with the two-dimensional
motions of the stars and planets against the sky, and it was by
studying these motions that they were led, eventually, to consider
the third dimension as well. Since the days of the ancient Sumerians
and probably before, there had been students of the night sky
willing to devote their evening hours to the lonely business of
squinting and straining to take sightings over aligned rocks or along
wooden quadrants or simply across their fingers and thumbs, pa-
tiently keeping records of what they saw. It was a lot of trouble.
Why did they bother?

   Part of the motive may have had to do with the inchoate
longing, mysterious but persistent then as now, to express a sense
of human involvement with the stars. As Copernicus noted, reverence for the stars runs so deep in human consciousness that it
is embedded in the language itself. "What is nobler than the heavens," he wrote, "the heavens which contain all noble things? Their
very names make this clear: *Caelum* (heavens) by naming that which
is beautifully carved; and *Mundus* (world), purity and elegance."[1]
Even Socrates, though personally indifferent toward astronomy,
conceded that the soul "is purified and kindled afresh" by studying
the sky.

   There were obvious practical incentives as well. Navigation,
for one: Mariners could estimate their latitude by measuring the
elevation of the pole star, and could tell time by the positions of

the stars, and these advantages were sufficiently appreciated that seafaring peoples codified them in poetry and mythology long before the advent of the written word. When Homer says that the Bear never bathes, he is passing along the seafarer's knowledge that Ursa Major, the constellation that contains the Big Dipper, is circumpolar at Mediterranean latitudes—that is, that it never sinks beneath the ocean horizon.

Another practical motive was timekeeping. Early farmers learned to make a clock and a calendar of the moving sky, and consulted almanacs etched in wood or stone for astronomical guidance in deciding when to plant and harvest their crops. Hesiod, one of the first poets whose words were written down, emerges from the preliterate era full of advice on how to read the sky for clues to the seasons:

> When great Orion rises, set your slaves
> To winnowing Demeter's holy grain
> Upon the windy, well-worn threshing floor. . . .
> Then give your slaves a rest; unyoke your team.
> But when Orion and the Dog Star move
> Into the mid-sky, and Arcturus sees
> The rosy-fingered Dawn, then Perseus, pluck
> The clustered grapes, and bring your harvest home. . . .
> When great Orion sink, the time has come
> To plough; and fittingly, the old year dies.[2]

The hunter-gatherers who preceded the farmers also used the sky as a calendar. As a Cahuilla Indian in California told a researcher in the 1920s:

> The old men used to study the stars very carefully and in this way could tell when each season began. They would meet in the ceremonial house and argue about the time certain stars would appear, and would often gamble about it. This was a very important matter, for upon the appearance of certain stars depended the season of the crops. After several nights of careful watching, when a certain star finally appeared, the old men would rush out, cry and shout, and often dance. In the spring, this gaiety was especially pronounced, for . . . they could now find certain plants in the mountains. They never went to the mountains until they saw a certain star, for they knew they would not find food there previously.[3]

Stonehenge is one of thousands of old time-reckoning machines the moving parts of which were all in the sky. The Great Pyramid at Giza was aligned to the pole star, and it was possible to read the seasons from the position of the pyramid's shadow. The Mayans of ancient Yucatan inscribed stone monuments with formulae useful in predicting solar eclipses and the heliacal rising of Venus (i.e., its appearance westward of the sun, as a "morning star"). The stone medicine wheels of the Plains Indians of North America ticked off the rising points of brighter stars, informing their nomadic architects when the date had come to migrate to seasonal grazing lands. The twenty-eight poles of Cheyenne and Sioux medicine lodges are said to have been used to mark the days of the lunar month: "In setting up the sun dance lodge," said Black Elk, a priest of the Oglala Sioux, "we are really making the universe in a likeness."[4]

Political power presumably played a role in early efforts to identify periodic motions in the sky, inasmuch as what a man can predict he can pretend to control. Command of the calendar gave priests an edge in the hardball politics of the Mayans, and Christopher Columbus managed to cow the Indians of Hispaniola into providing food for his hungry crew by warning that the moon otherwise would "rise angry and inflamed to indicate the evil that God would inflict on them." Writes Columbus's son Ferdinand, in his journal entry for the night of February 29, 1504:

> At the rising of the moon the eclipse began, and the higher the moon rose the more the eclipse increased. The Indians observed it, and were so frightened that with cries and lamentations they ran from every side to the ships, carrying provisions, and begged the Admiral by all means to intercede for them with God that he might not make them feel the effects of his wrath, and promised for the future, diligently to bring all he had need of. . . . From that time forward they always took care to provide us with all that was necessary, ever praising the God of the Christians.[5]

But the better acquainted the prehistoric astronomers became with the periodic motions they found in the night sky, the more complicated those motions proved to be. It was one thing to learn the simple periodicities—that the moon completes a circuit of the zodiacal constellations every 28 days, the sun in 365¼ days, the visible planets (from the Greek *planetes*, for "wanderers") at intervals ranging from 88 days for fleet-footed Mercury to 29½ years for plodding Saturn. It was another and more baffling matter to learn that the planets occasionally stop in their tracks and move backward—in "retrograde"—and that their paths are tilted relative to one another, like a set of ill-stacked dishes, and that the north celestial pole of the earth precesses, wobbling in a slow circle in the sky that takes fully 26,000 years to complete.*

The problem in deciphering these complexities, unrecognized at the time, was that the earth from which we view the planets is itself a planet in motion. It is because the earth orbits the sun while rotating on its tilted axis that there is a night-by-night shift in the time when any given star rises and sets at a given latitude. The earth's precessional wobble slowly alters the position of the north celestial pole. Retrograde motion results from the combined wanderings of the earth and the other planets; we overtake the outer planets like a runner on an inside track, and this makes each appear first to advance, then to balk and retreat across the sky as the earth passes them. Furthermore, since their orbits are tilted relative to one another, the planets meander north and south as well as east and west.

These complications, though they must have seemed a curse, were in the long run a blessing to the development of cosmology, the study of the universe at large. Had the celestial motions been simple, it might have been possible to explain them solely in terms of the simple, poetic tales that characterized the early cosmologies. Instead, they proved to be so intricate and subtle that they could not be predicted accurately without eventually coming to terms with the physical reality of how and where the sun, moon, and planets actually move, in real, three-dimensional space. The truth is beautiful, but the beautiful is not necessarily true: However aesthetically pleasing it may have been for the Sumerians to imagine that the stars and planets swim back from west to east each day

*This phenomenon, called the precession of the equinoxes, was known to the ancient Greeks and may have been discovered even earlier. Georgio de Santillana, in his book *Hamlet's Mill*, identifies it with the ancient myth of Amlodhi (later Hamlet), the owner of a giant salt grinder that sank to the bottom of the sea while being transported by ship. The mill has ground on ever since, creating a whirlpool that slowly twists the heavens. Whether or not it describes precession, the myth of Hamlet's mill certainly endures; I first heard it at the age of nine, in a rural schoolyard in Florida, from a little girl who was explaining why the ocean is salty.

Apparent path of Mars in the sky

Orbit of Mars

Orbit of Earth

Retrograde motion of Mars occurs when Earth overtakes the more slowly moving outer planet, making Mars appear to move backward in the sky.

via a subterranean river beneath a flat earth, such a conception was quite useless when it came to determining when Mars would go into retrograde or the moon occult Jupiter.

Consequently the idea slowly took hold that an adequate model of the universe not only should be internally consistent, like a song or a poem, but should also make accurate predictions that could be tested against the data of observation. The ascendency of this thesis marked the beginning of the end of our cosmological childhood. Like other rites of passage into adulthood, however, the effort to construct an accurate model of the universe was a bittersweet endeavor that called for hard work and uncertainty and deferred gratification, and its devotees initially were few.

One was Eudoxus. He enters the pages of history on a summer day in about 385 B.C., when he got off the boat from his home town of Cnidus in Asia Minor, left his meager baggage in cheap lodgings near the docks, and walked five miles down the dusty road to Plato's Academy in the northwestern suburbs of Athens. The Academy was a beautiful spot, set in a sacred stand of olive trees, the original "groves of academe," near Colonus, blind Oedipus' sanctuary, where the leaves of the white poplars turned shimmering silver in the wind and the nightingales sang day and night. Plato's mentor Socrates had favored the groves of academe, which even Aristophanes the slanderer of Socrates described lovingly as "all fragrant with woodbine and peaceful content."⁶

Beauty itself was the principal subject of study at the Academy, albeit beauty of a more abstract sort. LET NONE BUT GEOMETERS ENTER HERE, read the motto inscribed above the door, and great was the general enchantment with the elegance of geometrical forms. Geometry (geo-metry, "the measurement of the earth") had begun as a practical affair, the method employed by the Egyptian rope-stretchers in the annual surveys by which they reestablished the boundaries of farmlands flooded by the Nile. But in the hands of Plato and his pupils, geometry had been elevated to the status approaching that of a theology. For Plato, abstract geometrical forms *were* the universe, and physical objects but their imperfect shadows. As he was more interested in perfection than imperfection, Plato wrote encomiums to the stars but seldom went out at night to study them.

He backed this view with an imposing personal authority. Plato was not only smart, but rich—an aristocrat, one of the "guardians" of Greek society, descended on his mother's side from Solon the lawmaker and on his father's from the first kings of Athens—and physically impressive; *Plato*, meaning "broad-shouldered," was a nickname bestowed upon him by his gymnastics coach when as a youth he wrestled in the Isthmian Games. Eudoxus, we may assume, was suitably impressed. He was, however, a geometer in his own right—he was to help lay the foundations of Euclidean geometry and to define the "golden rectangle," an elegant proportion that turns up everywhere from the Parthenon to the paintings of Mondrian—and, unlike Plato, he combined his abstract mathematical reasonings with a passion for the physical facts. When he made his way to Egypt (a pilgrimage to the seat of geometrical wisdom that many Greek thinkers undertook, though Plato seems never quite to have got around to it), Eudoxus not only conducted research in geometry but applied it to the stars, building an astronomical observatory on the banks of the Nile and there mapping the sky. The observatory, though primitive, evinced his conviction that a theory of the universe must answer to the verdict, not only of timeless contemplation, but of the ceaselessly moving sky.

When the mature Eudoxus returned to the Academy, now as a renowned scholar with his own retinue of students, he set to work crafting a model of the cosmos that was meant to be both Platonically pleasing and empirically defensible. It envisioned the universe as composed of concentric spheres surrounding the earth, itself a sphere.* This in itself would have gratified Plato, who esteemed the sphere as "the most perfect" of the geometric solids, in that it has the minimum possible surface area relative to the volume of space it encloses. But the Eudoxian universe was also intended to better fit the observed phenomena, and this aspiration mandated complexity. To the simple, spherical cosmos that had been proposed by Parmenides a century earlier, Eudoxus added more spheres. The new spheres dragged and tugged at those of the sun, moon and planets, altering their paths and velocities, and by adjusting their rates of rotation and the inclination of their axes Eudoxus found that he could, more or less, account for retrograde motion and other intricacies of celestial motion. It took a total of twenty-

---

*By Eudoxus' day, all educated Greeks accepted that the earth was spherical, on the strength of such evidence as the shape of the shadow it casts on the moon during lunar eclipses.

seven spheres to do the job. This was more than Plato would have preferred, but it answered somewhat more closely to the data than had the preceding models. The hegemony of pure, abstract beauty had begun its slow retreat before the sullen but insistent onslaught of the material world.

But, ultimately, even so complex a cosmos as that of Eudoxus proved inadequate. The data base kept improving—with the conquest of Babylon by Alexander the Great in 330 B.C., the Greeks gained access to such Babylonian astronomical records as had previously eluded them, while continuing to make at least intermittent observations of their own—and Eudoxus' model failed to explain the subtleties revealed by this more ample and refined information. Thus began the phoenixlike cycle of the *science* of cosmology, where theories, however grand, are held hostage to empirical data that has the power to ruin them.

The next round fell, for better or worse, to Aristotle. Routinely described in the textbooks as an empiricist alternative to Plato, Aristotle was, indeed, relatively devoted to observation; he is said, for instance, to have spent his honeymoon collecting specimens of marine life. But he was also addicted to explanation and intolerant of ambiguity, qualities not salutary in science. A physician's son, he inherited a doctor's bedside habit of having a confident and reassuring answer to every anxious question. When pressed, this cast of mind made him credulous (women, he asserted, have fewer teeth than men) and propelled him to the extremities of empty categorizing, as when he observed that "animals are to be divided into three parts, one that by which food is taken in, one that by which excrement is discharged, and the third the region intermediate between them."[7] Aristotle wrote and lectured on logic, rhetoric, poetry, ethics, economics, politics, physics, metaphysics, natural history, anatomy, physiology, and the weather, and his thinking on many of these subjects was subtle as dewfall, but he was not a man to whose lips sprang readily the phrase, "I do not know." His mind was a killing jar; everything that he touched he both illuminated and anesthetized.

Nobody really likes a man who knows everything, and Aristotle became the first known victim of the world's first academic politics. Though he was an alumnus of the Academy and its most celebrated teacher, and clearly the man best qualified to succeed Plato as its director, he was twice passed over for the post. He then took the only satisfactory course open to a man of his stature, and stalked off to teach at another institution. As there *was* no other academic institution, he was obliged to found one; such was the origin of the Lyceum.

When it came time for Aristotle to declaim on the structure of the universe, he based his model on the heavenly spheres of Eudoxus, whom he had esteemed at the Academy for his moderate character as well as for his peerless accomplishments in astronomy. As his research assistant on the cosmology project Aristotle chose the astronomer Callippus, a native of Eudoxus' adopted home of Cyzicus. Together Aristotle and Callippus produced a model—consistent, symmetrical, expansive, and graceful to contemplate —that ranks among the most stirring of history's many errant cosmologies. Enshrined in Aristotle's book *De Caelo* (*On the Heavens*), it was to beguile and mislead the world for centuries to come.

Its details need not detain us; they consisted principally of adding spheres and adjusting their parameters, with the result that the universe now sported fully fifty-five glistening, translucent spheres. Beyond its outermost sphere, Aristotle argued on exquisite epistemological grounds, nothing could exist, not even space. At its center sat an immobile Earth, the model's shining diadem and its fatal flaw.

Confronted with an inevitable disparity between theory and observation, cosmologists who worked from the geocentric hypothesis had little choice but to keep making their models ever more complicated. And so cosmology was led into a maze of epicycles and eccentrics in which it would remain trapped for over a thousand years. The virtuoso of this exploration was Claudius Ptolemy.

He was born in the second century A.D. in Ptolemais on the Nile, and funding for his astronomical studies came from the Ptolemaic dynasty via the museum of Alexandria. Whatever his shortcomings—and many have been exposed, including evidence that he laundered some of his data—he was a hardworking astronomer and no armchair theorist. He charted the stars from an observatory at Canopus, a city named for a star, situated fifteen miles east of Alexandria, and was acquainted with atmospheric refraction and extinction and many of the other tribulations that bedevil the careful observer. He titled his principal cosmological work *Mathematical Syntaxis*, meaning "the mathematical composition," but it

Aristotle's universe consisted of spheres nested within spheres, their axes and directions of rotation adjusted to approximate the observed motions of the sun, moon, and stars across the sky. (Not to scale.)

has come down to us *Almagest*, Arabic for "the greatest." What it did so splendidly was predict the motions of the sun, moon, and stars more accurately than had its predecessors.

The epicycles and eccentrics by which Ptolemy sought to reconcile theory and observation had been introduced by the geometer Apollonius of Perga and refined by the astronomer Hipparchus. Epicycles were little circular orbits imposed upon the orbits of the planets: If a planet for Aristotle circled the earth like an elephant on a tether, the same planet for Ptolemy described the path of a stone whirled on a string by the elephant's rider. Eccentrics further improved the fit between the inky page and the night sky, by moving the presumptive center of the various heavenly spheres to one side of the center of the universe. To these motions Ptolemy added another, circular motion pursued by the center of the planetary spheres: The elephant's tether pole itself now orbited the center of the universe, hauling the whole system of spheres and epicycles back and forth so that planets could first approach the earth and then recede from it.

The system was ungainly—it had lost nearly all the symmetry that had commended celestial spheres to the aesthetics of Aristotle—but it worked, more or less. Wheeling and whirring in Rube Goldberg fashion, the Ptolemaic universe could be tuned to predict almost any observed planetary motion—and when it failed, Ptolemy fudged the data to make it fit. In its elaboration, and in the greater elaborations that later astronomers were obliged to add, it made predictions accurate enough to maintain its reputation as "the greatest" guide to heavenly motion from Ptolemy's day down to the Renaissance.

The price Ptolemy's followers paid for such precision as his model acquired was to forsake the claim that it represented physical reality. The Ptolemaic system came to be regarded, not as a mechanical model of the universe, but as a useful mathematical fiction. All those wheels within wheels were not actually out there in space—any more than, say, the geometrical boundary lines recorded in the Alexandrian land office represented real lines drawn across the silted farmland along the Nile. As the fifth-century Neoplatonist Proclus noted, "These circles exist only in thought. . . . They account for natural movements by means of things which have no existence in nature."[8] Ptolemy himself took the position that the complexities of the model simply reflected those found in the sky; if the solution was inelegant, he noted, so was the problem:

> So long as we attend to these models which we have put together, we find the composition and succession of the various motions awkward. To set them up in such a way that each motion can freely be accomplished hardly seems feasible. But when we study what happens in the sky, we are not at all disturbed by such a mixture of motions.[9]

The aim of the theory, then, was not to depict the actual machinery of the universe, but merely to "save the appearances." Much fun has been made of this outlook, and much of it at Ptolemy's expense, but science today has frequent recourse to intangible abstractions of its own. The "space-time continuum" depicted by the general theory of relativity is such a concept, and so is the quantum number called "isospin," yet both have been highly successful in predicting and accounting for events in the observed world. It should be said in Ptolemy's defense that he at least had the courage to admit to the limitations of his theory.

The phrase to "save the appearances" is Plato's, and its ascension via the Ptolemaic universe marked a victory for Platonic idealism and a defeat for empirical induction. Plato shared with his teacher Socrates a deep skepticism about the ability of the human mind to comprehend nature by studying objects and events. As Socrates told his friend Phaedrus while they strolled along the Ilissus, "I can't as yet 'know myself,' as the inscription at Delphi enjoins, and so long as that ignorance remains it seems to me ridiculous to inquire into extraneous matters."[10] Among these "extraneous matters" was the question of the structure of the universe.

Aristotle loved Plato, who seems not entirely to have returned his devotion; their differences went beyond philosophy, and sounded to the depths of style. Plato dressed plainly, while Aristotle wore tailored robes and gold rings and expensive haircuts. Aristotle cherished books; Plato was wary of men who were too bookish.* With a touch of irony that has survived the centuries, Plato called Aristotle "the brain."

Aristotle, for all his empirical leanings, never lost his attachment to the beauty of Plato's immortal geometrical forms. His universe of lucid spheres was a kind of heaven on earth, where his spirit and Plato's might live together in peace. Neither science nor philosophy has yet succeeded where Aristotle failed. Consequently his ghost and Plato's continue to contend, on the pages of the philosophical and scientific journals and in a thousand laboratories and schoolrooms. When philosophers of science today wrestle with

---

*In Plato's *Phaedrus*, Socrates recounts an old story of how the legendary King Thamus of Egypt had declined the god Theuth's offer to teach his subjects how to write. "What you have discovered is a recipe not for memory, but for reminder," says King Thamus. "And it is no true wisdom that you offer your disciples, but only its semblance, for by telling them of many things without teaching them you will make them seem to know much, while for the most part they know nothing, and as men filled, not with wisdom, but with the conceit of wisdom, they will be a burden to their fellows." This remains one of the most prophetic denunciations of the perils of literacy ever enunciated—although, of course, it is thanks to the written word that we know of it.

such questions as whether subatomic particles behave determinist-
ically, or whether ten-dimensional space-time represents the genuine architecture of the early universe or is instead but an interpretive
device, they are in a sense still trying to make peace between old
broadshoulders and his bright brash student, "the brain."

2

RAISING (AND LOWERING) THE ROOF

Aristarchus of Samos supposed that the heavens remained immobile and that the earth moved through an oblique circle, at the same time turning about its own axis.

—Plutarch

Now see that mind that searched and made
All Nature's hidden secrets clear
Lie prostrate prisoner of night.

—Boethius

The earth-centered universes of Eudoxus, Aristotle, Callippus, and Ptolemy were small by today's standards. Ptolemy's appears to have been the most generous. Certainly he thought it grand, and he liked to remark, with an astronomer's fondness for wielding big numbers, that in his universe the earth was but "a point" relative to the heavens. And, indeed, it was enormous by the standards of a day when celestial objects were assumed to be small and to lie close at hand; Heraclitus and Lucretius thought the sun was about the size of a shield, and Anaxagoras the atomist was banished for impiety when he suggested that the sun might be larger than the Peloponnesus. Nevertheless, the Ptolemaic universe is estimated to have measured only some fifty million miles in radius, meaning that it could easily fit inside what we now know to be the dimensions the earth's orbit around the sun.

The diminutive scale of these early models of the cosmos resulted from the assumption that the earth sits, immobile, in the center of the universe. If the earth does not move, then the stars do: The starry sphere must rotate on its axis once a day in order to bring the stars trooping overhead on schedule, and the larger the sphere, the faster it must rotate. Were such a cosmos very large, the speed mandated for the celestial sphere would become unreasonably high. The stars of Ptolemy's universe already were obliged to hustle along at better than ten million miles per hour, and were the celestial sphere imagined to be a hundred times larger it would have to be turning faster than the velocity of light. One did not have to be an Einstein, or even to know the velocity of light, to intuit that *that* was too fast—a point that had begun to worry cosmologists by the sixteenth century. All geocentric, immobile-earth cosmologies tended to inhibit appreciation of the true dimensions of space.

To set the earth in motion would be to expand the universe, a step that seemed both radical and counterintuitive. The earth does not *feel* as if it is spinning, nor does the observational evidence suggest any such thing: Were the earth turning on its axis, Athens and all its citizens would be hurtling eastward at a thousand miles per hour. If so, the Greeks reasoned, gale-force easterlies ought constantly to sweep the world, and broad jumpers in the Olympics would land in the stands well to the west of their jumping-off points. As no such effects are observed, most of the Greeks concluded that the earth does not move.

The problem was that the Greeks had only half the concept of inertia. They understood that objects at rest tend to remain at rest—a context we retain today when we speak of an "inert object" and mean that it is immobile—but they did not realize that objects in motion, including broad jumpers and the earth's atmosphere, tend to remain in motion. This more complete conception of inertia would not be achieved until the days of Galileo and Newton. (Even with amendments by Einstein and intimations of others by the developing superunified theories, plenty of mystery remains in the idea of inertia today.) Its absence was a liability for the ancient Greeks, but it was not the same thing as the religious prejudice to which many schoolbooks still ascribe the motives of rational and irrational geocentrists alike.

If one goes further and imagines that the earth not only spins on its axis but orbits the sun, then one's estimation of the dimensions of the cosmos must be enlarged even more. The reason for this is that if the earth orbits the sun, then it must alternately approach and withdraw from one side of the sphere of stars—just as, say, a child riding a merry-go-round first approaches and then recedes from the gold ring. If the stellar sphere were small, the differing distance would show up as an annual change in the apparent brightness of stars along the zodiac; in summer, for instance, when the earth is on the side of its orbit closer to the star Spica, its proximity would make Spica look brighter than it does in winter, when the earth is on the far side of its orbit. As no such phenomenon is observed, the stars must be *very* far away, if indeed the earth orbits the sun.

The astonishing thing, then, given their limited understanding of physics and astronomy, is not that the Greeks thought of the universe in geocentric terms, but that they did not *all* think of it that way. The great exception was Aristarchus, whose heliocentric cosmology predated that of Copernicus by some seventeen hundred years.

Aristarchus came from Samos, a wooded island near the coast of Asia Minor where Pythagoras, three centuries earlier, had first proclaimed that all is number. A student of Strato of Lampsacus, the head of the Peripatetic school founded by Aristotle, Aristarchus was a skilled geometer who had a taste for the third dimension, and he drew, in his mind's eye, vast geometrical figures that stretched not only across the sky but out into the depths of space as well. While still a young man he published a book suggesting that the sun was nineteen times the size and distance of the moon; his conclusions were quantitatively erroneous (the sun actually is four hundred times larger and farther away than the moon) but his methods were sound.

It may have been this work that first led Aristarchus to contemplate a sun-centered cosmos: Having concluded that the sun was larger than the earth, he would have found that for a giant sun to orbit a smaller earth was intuitively as absurd as to imagine that

In a small, heliocentric universe, the earth would be much closer to a summer star like Spica in summer than in winter, making Spica's brightness vary annually. As there is no observable annual variation in the brightness of such stars, Aristarchus concluded that the stars are extremely distant from the earth.

a hammer thrower could swing a hammer a hundred times his own weight. The evolution of Aristarchus' theory cannot be verified, however, for his book proposing the heliocentric theory has been lost. We know of it from a paper written in about 212 B.C. by Archimedes the geometer.

Archimedes' paper was titled "The Sand Reckoner," and its purpose was to demonstrate that a system of mathematical notation he had developed was effective in dealing with large numbers. To make the demonstration vivid, Archimedes wanted to show that he could calculate even such a huge figure as the number of grains of sand it would take to fill the universe. The paper, addressed to his friend and kinsman King Gelon II of Syracuse, was intended as but a royal entertainment or a piece of popular science writing. What makes it vitally important today is that Archimedes, wanting to make the numbers as large as possible, based his calculations on the dimensions of the most colossal universe he had ever heard of —the universe according to the novel theory of Aristarchus of Samos.

Archimedes, a man of strong opinions, had a distaste for loose talk of "infinity," and he begins "The Sand Reckoner" by assuring King Gelon that the number of grains of sand on the beaches of the world, though very large, is not infinite, but can, instead, be both estimated and expressed:

> I will try to show you, by means of geometrical proofs, which you will be able to follow, that, of the numbers named by me . . . some exceed not only the number of the mass of sand equal in magnitude to the earth filled up in the way described, but also that of a mass equal in magnitude to the universe.[1]

Continuing in this vein, Archimedes adds that he will calculate how many grains of sand would be required to fill, not the relatively cramped universe envisioned in the traditional cosmologies, but the much larger universe depicted in the new theory of Aristarchus:

> Aristarchus of Samos brought out a book consisting of certain hypotheses, in which it appears, as a consequence of the assumptions made, that the universe is many times greater [in size] than that now so called. His hypotheses are that the fixed stars and the sun remain unmoved, that the earth revolves about the sun in the circumference of a circle, the sun lying in the middle of the orbit, and that the sphere of the fixed stars, situated about the same center as the sun, is so great that the circle in which he supposes the earth to revolve bears such a proportion to the distance of the fixed stars as the center of the sphere bears to its surface.[2]

Here Archimedes has a problem, for Aristarchus is being hyperbolic when he says that the size of the universe is as much larger than the orbit of the sun as is the circumference of a sphere to its center. "It is easy to see," Archimedes notes, "that this is impossible;

for, since the center of the sphere has no magnitude, we cannot conceive it to bear any ratio whatever to the surface of the sphere."[3] To plug hard numbers into Aristarchus' model, Archimedes therefore takes Aristarchus to mean that the ratio of the size of the earth to the size of the universe is comparable to that of the orbit of the earth compared to the sphere of stars. Now he can calculate. Incorporating contemporary estimates of astronomical distances, Archimedes derives a distance to the sphere of stars of, in modern terminology, about six trillion miles, or one light-year.*

This was a stupendous result for its day—a heliocentric universe with a radius more than a hundred thousand times larger than that of the Ptolemaic model, proposed four centuries before Ptolemy was born! Although we know today that one light-year is but a quarter of the distance to the nearest star, and less than one ten-billionth of the radius of the observable universe, Aristarchus' model nonetheless represented a tremendous increase in the scale that the human mind had yet assigned to the cosmos. Had the world listened, we today would speak of an Aristarchian rather than a Copernican revolution in science, and cosmology might have been spared a millennium of delusion. Instead, the work of Aristarchus was all but forgotten; Seleucus the Babylonian championed the Aristarchian system a century later, but appears to have been lonely in his enthusiasm for it. Then came the paper triumph of Ptolemy's shrunken, geocentric universe, and the world stood still.

Writing "The Sand Reckoner" was one of the last acts of Archimedes' life. He was living in his native Syracuse on the southeast coast of Sicily, a center of Greek civilization, and the city was besieged by the Roman general Marcus Claudius Marcellus. Though his last name means "martial" and he was nicknamed the Sword of Rome, Marcellus for all his mettle was getting nowhere in Syracuse. Credit for holding his army at bay went to the frightening machines of war that Archimedes had designed. Roman ships approaching

---

*Archimedes concluded that it would take 10^{63} grains of sand to fill the Aristarchian universe. The American cosmologist Edward Harrison points out that 10^{63} grains of sand equals 10^{80} atomic nuclei, which is "Eddington's number"—the mass of the universe as calculated in the 1930s by the English astrophysicist Arthur Stanley Eddington. So Archimedes, in underestimating the size of the universe but imagining it to have a matter density much higher than it does, arrived at a total amount of cosmic matter that wasn't far from Eddington's twentieth-century estimate.

the city walls were seized in the jaws of giant Archimedean cranes, raised high into the air while the terrified marines aboard clung to the rails, then dashed on the rocks below. Troops attacking on foot were crushed by boulders rained down on them by Archimedean catapults. As Plutarch recounts the siege, the Romans soon were so chagrined that "if they did but see a little rope or a piece of wood from the wall, instantly crying out, that there it was again, Archimedes was about to let fly some engine at them, they turned their backs and fled."[4]

"Who," Marcellus asked in his frustration as the siege wore on, "is this Archimedes?"

A good question. The world remembers him as the man who ran naked through the city streets shouting "Eureka" after having realized, while lowering himself into a bath, that he could measure the specific gravity of a gold crown (a gift to King Hieron, one that he suspected of being adulterated) by submerging it and weighing the amount of water it displaced. Remembered, too, is his invention of the Archimedes' screw, still widely used to pump water today, and his fascination with levers and pulleys. "Give me a place to stand," he is said to have boasted to King Hieron, "and I shall move the earth."[5] The king requested a demonstration on a smaller scale. Archimedes commandeered a ship loaded with freight and passengers—one that normally would have required a gang of strong men to warp from the dock—and pulled the ship unassisted, employing a multiple pulley of his own design. The king, impressed, commissioned Archimedes to build the engines of war that were to hold off the Romans.

Plutarch writes that although he was famous for his technological skills, Archimedes disdained "as sordid and ignoble the whole trade of engineering, and every sort of art that lends itself to mere use and profit," preferring to concentrate upon pure mathematics. His passion for geometry, Plutarch adds,

made him forget his food and neglect his person, to that degree that when he was occasionally carried by absolute violence to bathe, or have his body anointed, he used to trace geometrical figures in the ashes of the fire, and diagrams in the oil on his body, being in a state of entire preoccupation, and, in the truest sense, divine possession with his love and delight in science.[6]

Archimedes determined the value of pi to three decimal places, proved that the area of the surface of a sphere equals four times that of a circle of the same size (the rule of $4\pi r^2$), and discovered that if a sphere is circumscribed within a cylinder, the ratio of their volumes and surfaces is 3:2. (He was so proud of this last feat that he asked friends to have a sphere within a cylinder inscribed on his tombstone. Cicero, quaestor of Sicily in 75 B.C., located and restored the tomb; it has since vanished.)

Marcellus' invasion came while the Syracusans were celebrating the feast of Diana, traditionally an excuse for heavy drinking. Marcellus had ordered that no free citizens be injured, but his men had seen many of their compatriots killed by Archimedes' war machines, and they were not in a conciliatory mood. As the story is told, Archimedes was absorbed in calculations when a Roman soldier approached and addressed him in an imperative tone. Archimedes was seventy-five years old and no fighter, but he was also one of the freest men who ever lived, and unaccustomed to taking orders. Drawing geometrical diagrams in the sand, Archimedes waved the soldier aside, or told him to go away, or otherwise dismissed him, and the angry man cut him down. Marcellus damned the soldier as a murderer, writes Plutarch, adding that "nothing afflicted Marcellus so much as the death of Archimedes."[7]

Greek science was mortal, too. By the time of Archimedes' death the world center of intellectual life already had shifted from Athens to Alexandria, the city Alexander the Great had established a century earlier with the charter—inspired, I suppose, by his boyhood tutor Aristotle—that it be a capital of learning modeled on the Greek ideal. Here Ptolemy I, the Macedonian general and biographer of Alexander, established with the wealth of empire a vast library and a museum where scientists and scholars could carry on their studies, their salaries paid by the state. It was in Alexandria that Euclid composed his *Elements* of geometry, that Ptolemy constructed his eccentric universe, and that Eratosthenes measured the circumference of the earth and the distance of the sun to within a few percent of the correct values. Archimedes himself had studied at Alexandria, and had often ordered books from the library there to be sent to Syracuse. But the tree of science grew poorly in Alexandrian soil, and within a century or two had hardened into the deadwood of pedantry. Scholars continued to study and annotate the great books of the past, and roomfuls of copiers laboriously duplicated them, and historians owe a great debt to the anonymous clerks of the library of Alexandria, but they were the pallbearers of science and not its torchbearers.

The Romans completed their conquest of the known world on the day in 30 B.C. that Cleopatra, last of the Ptolemies, bared her breast to the asp. Theirs was a nonscientific culture. Rome revered authority; science heeds no authority but that of nature. Rome excelled in the practice of law; science values novelty over precedent. Rome was practical, and respected technology, but science at the cutting edge is as impractical as painting and poetry, and is exemplified more by Archimedes' theorems than by his catapults. Roman surveyors did not need to know the size of the sun in order to tell time by consulting a sundial; nor did the pilots of Roman galleys concern themselves overmuch with the distance of the moon, so long as it lit their way across the benighted Mediterranean. Ceramic stars ornamented the ceilings of the elegant dining rooms of Rome; to ask what the real stars were made of would have been as indelicate as asking one's host how the roast pig on the table had been slaughtered. When a student Euclid was tutoring wondered aloud what might be the use of geometry, Euclid told his slave, "Give him a coin, since he must gain from what he learns."[8] This story was not popular in Rome.

Roman rule engendered among those it oppressed a growing scorn for material wealth, a heightened regard for ethical values, and a willingness to imagine that their earthly sufferings were but a preparation for a better life to come. The conflict between this essentially spiritual, otherworldly outlook and the stolid practicality of Rome crystallized in the interrogation that Pontius Pilate, a prefect known for his ruthlessness and legal acumen, conducted of the obscure Jewish prophet Jesus of Nazareth.

The world knows the story. Pilate asked Jesus, "Are you the king of the Jews?"

"My kingdom is not of this world," Jesus replied.

"Are you a king, then?"

"You say I am a king," Jesus replied. "To this end was I born, and for this cause I came into the world, that I should bear witness to the truth. Everyone that is of the truth hears my voice."

"What is truth?" asked Pilate.[9]

Jesus said nothing, and was led off to execution, and his few followers dropped from sight. Yet within two centuries his eloquent

silence had swallowed up the words of the law, and Christianity
had become the state religion of Rome.

Science, however, fared no better in Christian than in pagan
Rome. Christianity, in its emphasis upon asceticism, spirituality,
and contemplation of the afterlife, was inherently uninterested in
the study of material things. What difference did it make whether
the world was round or flat, if the world was corrupt and doomed?
As Saint Ambrose put it in the fourth century, "To discuss the
nature and position of the earth does not help us in our hope of
the life to come." Wrote Tertullian the Christian convert, "For us,
curiosity is no longer necessary."

To the Christians, the fall of Rome illustrated the futility of
putting one's trust in the here and now. "Time was when the world
held us fast to it by its delight," declaimed Pope Gregory the Great,
seated on a marble chair amid the flickering candles of the chapel
of the Catacomb of St. Domitilla in Rome at the close of the sixth
century (by which time the city had been sacked five times). "Now
'tis full of such monstrous blows for us, that of itself it sends us
home to God at last. The fall of the show points out to us that it
was but a *passing* show," he said, advising the somber celebrants to
"let your heart's affections wing their way to eternity, that so de-
spising the attainments of this earth's high places, you may come
unto the goal of glory which ye shall hold by faith through Jesus
Christ, our Lord."[10]

Christian zealots are alleged to have burned the pagan books
in the library of Alexandria, and Muslims to have burned the Christian books, but the historical record of this great crime is subject
to dispute on both counts; in any event, the books went up in
smoke. The old institutions of learning and philosophy, most of
them already in decline, collapsed under the rising winds of change.
Plato's Academy was closed by Justinian in A.D. 529; the Sarapeum
of Alexandria, a center of learning, was razed to the ground by
Christian activists in A.D. 391; and in 415 the geometer Hypatia,
daughter of the last known associate of the museum of Alexandria,
was murdered by a Christian mob. ("They stripped her stark naked," an eyewitness reported. "They raze[d] the skin and ren[t] the
flesh of her body with sharp shell, until the breath departed out of
her body; they quarter[ed] her body; they [brought] her quarters
unto a place called Cinaron and burn[ed] them to ashes."[11])

Scholars fled from Alexandria and Rome and headed for

Byzantium—followed closely by the Roman emperor himself, after whom the city was renamed Constantinople—and the pursuit of science devolved to the province of Islam. Encouraged by the Koran to practice *taffakur*, the study of nature, and *taskheer*, the mastery of nature through technology, Islamic scholars studied and elaborated upon classics of Greek science and philosophy forgotten in the West. Evidence of their astronomical research is written in the names of stars—names like Aldebaran, from *Al Dabaran*, "the follower"; Rigel, from *Rijl Jauzah al Yusra*, "the left leg of the Jauzah"; and Deneb, from *Al Dhanab al Dajajah*, "the hen's tail."

But the Arabs were enchanted by Ptolemy, and envisioned no grander cosmos. Aristarchus' treatise on astronomical distances was translated in the early tenth century by a Syrian-Greek scholar named Questa ibn Luqa, and an Arabic secret society known as the Brethren of Purity published an Aristarchian table of wildly inaccurate but robustly expansive planetary distances, but otherwise little attention was paid to the concept of a vast universe. The generally accepted authority on the scale of what we today call the solar system was al-Farghani, a ninth-century astronomer who, by assuming that the Ptolemaic epicycles fit as tightly as ball bearings between the planetary spheres—"there is no void between the heavens," he asserted—estimated that Saturn, the outermost known planet, was eighty million miles away.[12] Its true distance is more than ten times that.

The Islamic devotees of Ptolemy, however, inadvertently undermined the very cosmology they cherished, by transmuting Ptolemaic abstractions into real, concrete celestial spheres and epicycles. So complex and unnatural a system, palatable if regarded as purely symbolic, became hard to swallow when represented as a genuine mechanism that was actually out there moving the planets around. The thirteenth-century monarch King Alfonso ("the Learned") of Castile is said to have remarked, upon being briefed on the Ptolemaic model, that if this was really how God had built the universe, he might have given Him some better advice.

But that was many long, dark centuries later. The last classical scholar in the West was Ancius Boethius, who enjoyed power and prestige in the court of the Gothic emperor Theodoric at Ravenna until he backed the losing side in a power struggle and was jailed. In prison he wrote *The Consolation of Philosophy*, a portrait of the life of the mind illuminated by the fading rays of a setting sun. There,

Boethius contrasts the constancy of the stars with the unpredictability of human fortune:

> Creator of the starry heavens,
> Lord on thy everlasting throne,
> Thy power turns the moving sky
> And makes the stars obey fixed laws
>
> . . . .
>
> All things thou holdest in strict bounds,—
> To human acts alone denied
> Thy fit control as Lord of all.
> Why else does slippery Fortune change
> So much, and punishment more fit
> For crime oppress the innocent?[13]

In words the Greek Stoics would have appreciated, the muse of philosophy upbraids Boethius for his self-pity. "You are wrong if you think Fortune has changed towards you," she tells him. "Change is her normal behavior, her true nature. In the very act of changing she has preserved her own particular kind of constancy towards you."[14]

In Boethius, the universe of Ptolemy is reduced to a symbol of resignation to the vicissitudes of fate:

> Consider how thin such fame is and how unimportant. It is
> well known, and you have seen it demonstrated by astrono-
> mers, that beside the extent of the heavens, the circumference
> of the earth has the size of a point; that is to say, compared
> with the magnitude of the celestial sphere, it may be thought
> of as having no extent at all. The surface of the world, then,
> is small enough, and of it, as you have learnt from the geog-
> rapher Ptolemy, approximately one quarter is inhabited by
> living beings known to us. If from this quarter you subtract
> in your mind all that is covered by sea and marshes and the
> vast area of desert by lack of moisture, then scarcely the small-
> est of regions is left for men to live in. This is the tiny point
> within a point, shut in and hedged about, in which you think
> of spreading your fame and extending your renown.[15]

Boethius was executed in 524, and with the extinguishing of that last guttering lamp the darkness closed in. The climate during the Dark Ages grew literally colder, as if the sun itself had lost interest in the mundane. The few Western scholars who retained any interest in mathematics wrote haltingly to one another, trying to recall such elementary facts of geometry as the definition of an interior angle of a triangle. The stars came down: Conservative churchmen modeled the universe after the tabernacle of Moses; as the tabernacle was a tent, the sky was demoted from a glorious sphere to its prior status as a low tent roof. The planets, they said, were pushed around by angels; this obviated any need to predict celestial motions by means of geometrical or mechanical models. The proud round earth was hammered flat; likewise the shimmering sun. Behind the sky reposed eternal Heaven, accessible only through death.

THE
DISCOVERY
OF THE
EARTH

There will come a time in the later years when
Ocean shall loosen the bonds by which we
have been confined, when an immense land
shall be revealed . . . and Thule will no longer
be the most remote of countries.

—Seneca

The sea was like a river.

—Christopher Columbus

The reawakening of informed inquiry into the na-
ture of cosmological space that we associate with the Renaissance
had its roots in an age of terrestrial exploration that began at about
the time of Marco Polo's adventures in China in the thirteenth
century and culminated two hundred years later with Columbus's
discovery of America. Astronomy and the exploration of the earth
had of course long been related. Navigators had been steering by
the stars for millennia, as evidenced by the Chinese practice of
calling their blue-water junks "starry rafts" and by the legend that
Jason the Argonaut was the first man to employ constellations as

an aid to memorizing the night sky. When Magellan crossed the Pacific, his fleet following an artificial star formed by a blazing torch set on the stern of his ship, he was navigating waters that had been traversed thousands of years earlier by the colonizers of Micronesia, Australia, and New Guinea—adventurers in dugout canoes who, like Jason, carried their star maps in their heads. Virgil emphasized the importance of sighting the stars in his account of Aeneas' founding of Rome:

> And not yet Night, whirled onward by the hours,
> Was reaching her mid-course; from his couch
> The ever-watchful Palinurus rose,
> Examines every wind, and on his ear
> Catches the breeze; within the silent heavens
> Marks all stars that swim the sky,
> Arcturus, and the rainy Hyades,
> And the twin bears, and Orion armed in gold.
> And when all he sees is calm within the cloudless sky,
> From off the stern he gives the signal clear.
> We strike the camp, essay our course again
> And spread our sail-wings.
>   And already Dawn
> Was reddening as the stars were put to flight,
> When far off we beheld the shadowy hills,
> And Italy low-lying. "Italy!"[1]

Explorers of dry land found the stars useful, too; American Indians lost in the woods took comfort in the presence of Father Sky, his hands the great rift that divides the Cygnus-Sagittarius zone of the Milky Way, and escaped slaves making their way north through the scrub pines of Georgia and Mississippi were admonished to "follow the drinking gourd," meaning the Big Dipper. Ptolemy employed his considerable knowledge of geography to aid his studies of astronomy; his assertion that the earth is but a point compared to the celestial sphere was based in part upon the testimony of travelers who ventured south into central Africa or north toward Thule and reported seeing no evidence that their wanderings had brought them any closer to the stars in those quarters of the sky.

Thus, though the principal motive for the new wave of European exploration was economic—European adventurers stood to make a fortune if they could "orient" themselves, by navigating an ocean route to the East—it is not surprising to learn that one of its instigators was an astronomer. He was a Florentine named Paolo dal Pozzo Toscanelli, and he emphasized that knowledge as well as wealth was to be found in the East. Asia, Toscanelli wrote enticingly to Christopher Columbus,

> is worthy to be sought by the Latins not only because immense wealth can be had in the form of gold, silver, gems of every kind, and spices which are never brought to us; but also because of the learned men, wise philosophers and astrologers by whose genius and arts those mighty and magnificent provinces are governed.[2]

Much of the romance that colored the Western image of the East had come from Marco Polo's extraordinary book recounting his equally extraordinary travels in China. Marco came from Venice, itself no backwater, but nothing had prepared him for the likes of Hangchow, which he visited in 1276 and from which he never quite recovered. "The greatest city in the world," he called it, "where so many pleasures may be found that one fancies himself to be in Paradise." Hangchow stood on a lake amid jumbled, misty mountains, the literal depiction of which by Sung landscape painters still strikes Western eyes as almost too good to be true. "In the middle of the lake," Marco reported,

> there are two islands, on each of which stands a palatial edifice with an incredibly large number of rooms and separate pavilions. And when anyone desired to hold a marriage feast, or to give a big banquet, it used to be done at one of these palaces. And everything would be found there ready to order, such as dishes, napkins and tablecloths and whatever else was needed. These furnishings were acquired and maintained at common expense by the citizens in these palaces constructed by them for this purpose.[3]

Ornately carved wooden boats were available for hire, the largest of them capable of serving multiple-course banquets to scores of diners at a sitting. Skiffs maneuvered alongside the larger boats, carrying little orchestras and "sing-song girls" in bright silk dresses and boatmen selling chestnuts, melon seeds, lotus roots, sweetmeats, roast chicken, and fresh seafood. Other boats carried live shellfish and turtles, which in accordance with Buddhist custom one purchased and then threw back into the water alive. The lake was clear, thanks to strict antipollution ordinances, and its banks were given over to public parks—this a legacy of Hangchow's revered prefect Su Tung-p'o, a gifted poet who was often in trouble with the authorities. Wrote Su:

> Drunk, I race up Yellow Grass Hill,
> Slope strewn with boulders like flocks of sheep.
> At the top collapse on a bed of stone,
> Staring at white clouds in a bottomless sky.
> My song wings to the valley on long autumn winds.
> Passers-by look up, gaze southeast,
> Clap their hands and laugh: "The governor's gone mad!"[4]

All of which was a long way from the cold stone walls and plainsongs of northern Europe, and even from the commercial bustle and guile of Venice.

Bolstering the travelers' tales was tangible evidence of Asian glory, in the form of silks and lacquer boxes and spices and drugs that had reached Europe overland. The Silk Road by which these treasures arrived, however, had long been a costly bucket-brigade of middlemen and brigands, and was now being constricted by the Black Death and the retreat of the Mongol khanates before an expanding Islamic empire. By the fifteenth century the European powers were ready to try reaching the East on their own, by sea.

The epicenter of this venturesome new spirit was Sagres, a spit of land at the southwesternmost tip of Europe that juts out into the ocean like a Renaissance Cape Canaveral. There, in 1419, a spaceport of sorts was established by Prince Henry the Navigator. A devout, monomaniacal Christian in a hair shirt, his eyes baggy with the fatigue of overwork and the vexation of debt, Henry was the first to explore the coast of Africa and to exploit its riches in gold, sugar, and slaves, and the first to navigate a seaway around Africa to Asia.

His library at Sagres contained an edition of Marco Polo (translated by his wandering brother Pedro) and a number of other books that encouraged Henry's belief that Africa could be circumnavigated, opening up a seaway to the East. The evidence, though fragmentary, was tantalizing. Herodotus in the fifth century B.C.

recounted (though he did not believe it) a story that Phoenician expeditionaries had rounded Africa from the east, eventually finding that while sailing west they had the sun on their *right*—which Henry understood, as Herodotus did not, to mean that they were south of the equator. Two centuries later, Eudoxus of Cyzicus (no relation to the astronomer) was reported in a book by Strabo the geographer to have found, in Ethiopia, the sculptured prow of a wrecked ship that the natives said had come from the west; Eudoxus took the prow home with him to Egypt and was told by the local sailors and traders that it belonged to a vessel that had sailed out through the Columns of Hercules, never to be seen again. In the *Periplus of the Erythraean Sea*, an anonymous geography dating from the first century A.D., Henry could read that "beyond the town of Rhapta"—i.e., opposite Zanzibar—"the unexplored coast curves away to the west and mingles with the Western Ocean."[5]

Emboldened by these and similar accounts, Henry installed on Sagres's windswept promontory an astronomical observatory and navigational institute, staffed by German mathematicians, Italian cartographers, and Jewish and Muslim scholars who were put to work determining the circumference of the earth and drawing improved maps. He drew on the stars for spiritual as well as for navigational guidance; his horoscope had predicted that he was fated to direct the conquest of unknown lands. He did not sail himself, but rather dispatched his expeditions, more than a dozen of them, down the coast of Africa.

His captains proceeded with understandable trepidation. Many believed, on the authority of the ancient geographers, that the Torrid Zone to the south was too hot to endure and that it was guarded by a Green Sea of Darkness that was perpetually enshrouded in fog. Nor did the realities prove to be much less unpleasant than the fables. The sea off Cape Non opposite the Canaries did indeed turn blood red (from ruddy sands blown off the deserts near the coast) and farther south the waters turned green, and there was, to be sure, plenty of fog. At Cape Bojador, called by the ancients "the end of the world," the coast rose up in a seemingly interminable wall of harborless cliffs. Fifty-foot waves threatened to smash the explorers' caravels against the rocks of Cape Juby. One landing party stumbled on elephant chips nearly the size of a man. Another was attacked by natives shooting poisoned arrows; only five of the twenty-five man crew survived. Several of the captains turned back, only to be chastened and threatened by Henry and refitted and sent south again.

In 1455 a Venetian in Henry's service, Alvise da Cadamosto, watched anxiously as the pole star, theretofore the guiding light of all European navigators, sank from sight beneath the northern horizon. But he was cheered when, as if by way of compensation, the "six large and wonderfully bright stars" of the Southern Cross hove up into view. In 1488, twenty-eight years after Prince Henry's death, Bartholomeu Diaz finally rounded the Cape of Good Hope, and ten years later Vasco da Gama reached India, after a stormy, ninety-five-hundred-mile voyage that consumed ten months and twelve days. Asked what he was seeking, Da Gama answered, "Christians and spices."[6]

The investment paid off, and by the end of the century the Portuguese annually were importing seven hundred kilograms of gold and ten thousand slaves from Africa. They traded wheat for the gold; the slaves generally could be obtained for free. Recalled one of Henry's men who took part in a raiding party:

> Our men, crying out, "Sant' Iago! San Jorge! Portugal!" fell upon them, killing or capturing all they could. There you might have seen mothers catch up their children, husbands their wives, each one escaping as best he could. Some plunged into the sea; others thought to hide themselves in the corners of their hovels; others hid their children under the shrubs . . . where our men found them. And at last our Lord God, Who gives to all a due reward, gave to our men that day a victory over their enemies; and in recompense for all their toil in His service they took 165 men, women, and children, not counting the slain.[7]

In all, over one million slaves were captured and brought to Europe by the Portuguese.

Unknown to the Europeans, the Chinese, rulers of the greatest land in the fabled East, were trading along Africa's east coast while the Portuguese were exploring its west coast. Theirs was a more venerable and less violent campaign. They mounted expeditions of thousands of men in fleets of junks each five times or more the size of the Portuguese caravels, conducted peaceful trade backed by this show of force, and are recorded to have resorted to violence on only three occasions in a century of exploration. But the Chinese furled

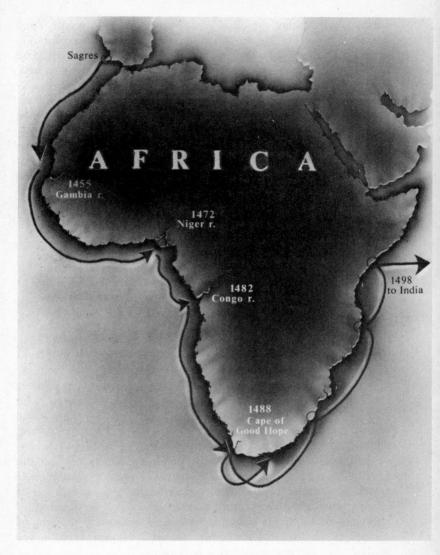

Henry the Navigator's reconnaissance of Africa, A.D. 1455–1498.

their sails following the death of the adventurous emperor Yung Lo. By the time Da Gama reached India the Chinese antiexploration faction had made it a crime to build an oceangoing junk and had burned the ships' logbooks—some of which are thought to have contained accounts of voyages extending across the Pacific as far as to the Americas—on grounds that they contained "deceitful exaggerations of bizarre things."[8] (Which, by the way, was just what Western critics said of Marco Polo's account of China.)

The Portuguese, in contrast, were smaller in number but fierce with the torch and the sword. The first colonist in Portugal's first colony, Joad Goncalves of Madeira, set the island afire. Da Gama and his successor Pedro Cabral "tortured helpless fishermen," writes R. S. Whiteway in his *The Rise of Portuguese Power in India, 1497–1550*. Whiteway adds that

> Almeida tore out the eyes of a Nair who had come in with a safe-conduct because he suspected a design on his own life; Albuquerque cut off the noses of the women and the hands [of the men] who fell into his power on the Arabian coast. To follow the example of Almeida and sail into an Indian harbor with the corpses of unfortunates, often not fighting-men, dangling from the yards, was to proclaim oneself a determined fellow.[9]

Columbus was a fighting man, shaped, as we might expect, more in the Portuguese than in the Chinese mold. His destiny, he felt, had been sealed on August 13, 1476, when he floated to shore just up the coast from Prince Henry's institute at Sagres, clutching an oar and leaving behind the burning wreck of the ship in which he had been fighting in the battle of Cape St. Vincent (on the Portuguese side, against his native Genoa). To be wringing the salt water out of his shirt on the beach near Sagres was just the sort of thing Columbus expected from a life he believed to be directed by the hand of God. He took his first name seriously, thought of himself as *Christophoros*, the "Christ carrier," whose mission it was to discover "a new heaven and a new earth."

He was already something of an anachronism—a dead-reckoning navigator in an epoch of ever improving charts and navigational instruments, a sometime pirate in an age when violence at sea was busily being turned into a state monopoly, an amateur scholar in an era of growing professionalism. "Neither reason nor mathematics nor maps were any use to me," he wrote of his discovery of America, which he died believing was Asia. "Fully accomplished were the words of Isaiah."[10] He had in mind Isaiah 11:11: "And it shall come to pass in that day, that the Lord shall set his hand again the second time to recover the remnant of his people, which shall be left, from Assyr'-i-a, and from E'-gypt, and from Path'-ros, and from Cush, and from E'-lam, and from Shi'-nar, and from Ha'-math"—and here came the part that spoke most vividly to Columbus—"and from the islands of the sea." The "islands of the sea" were the Indies. To "recover the remnant of his people" was what the Portuguese slavers had been doing in Africa, reclaiming lost souls for Christ. Cruel work in the short term, it was thought to be worth it in the end. The chronicler Gomez Eannes de Azurara observed that when Prince Henry, "mounted upon a powerful steed," picked out 46 slaves for himself from a cargo of 223 men, women, and children huddled wretchedly in a field in Lagos, Portugal, an act that required that he "part fathers from sons, husbands from wives, brothers from brothers," he "reflected with great pleasure upon the salvation of those souls that before were lost. And certainly his expectation was not in vain, since . . . as soon as they understood our language, they turned Christians with very little ado."[11]

Columbus was to carry on a similar crusade in the New World. He longed to reach the East for the usual reasons: Out there was a rich continent, the conquest of which could bring a man wealth and glory and (if Toscanelli could be believed) even wisdom. The brave and irresponsible argument by which he persuaded Queen Isabella of Spain to finance his expedition was not that the world was round—every educated person knew that—but that it was small.* "I have made it my business to read all that has been written on geography, history, philosophy, and other sciences,"[12] Columbus said, but the lamp of his learning cast its narrow beam only on those maps and old geographies that most severely underestimated the dimensions of the terrestrial globe. By marshaling a total of eight different geographical arguments, all tending to make the globe smaller and Asia larger than they really are, Columbus arrived

---

*The myth that Columbus was out to prove the world round was invented 130 years after the fact, and subsequently was popularized by Washington Irving.

at the extraordinary conclusion that the distance from the Canary
Islands to the Indies was only 3,550 nautical miles—less than one
third the actual figure. "Thus Our Lord revealed to me that it was
feasible to sail from here to the Indies, and placed in me a burning
desire to carry out this plan," Columbus wrote.[13] His position was
simple: God was right and the professional geographers were wrong.

Columbus's plan appeared foolhardy to anyone who possessed
a realistic sense of the dimensions of the earth. To sail westward
to Asia, as the geographers of the court at Castile took pains to
inform Columbus, would require a voyage lasting approximately
three years, by which time he and his men would surely be dead
from starvation or scurvy.* The voyage had been attempted twice
before, by Moorish explorers out of Lisbon and by the Vivaldi
brothers of Genoa in the thirteenth century; none had been heard
from since. Columbus endured ten years of rejection on such grounds
by the geographers of the leading courts of Europe. "All who knew
of my enterprise rejected it with laughter and mockery," he recalled,
but the pilot light of his destiny shone on undimmed. He replied
to the scorn of the experts with his collection of shrunken-earth
maps, Aristotle's assertion "that there is continuity between the
parts about the pillars of Hercules and the parts about India,"[15]
and Seneca's prophecy that "an immense land" lay beyond Ultima
Thule. All this Columbus delivered up with thundering certitude;
one searches his writings in vain for any trace of the skeptical,
empirical temper of the scientist. He would be admiral of the ocean
sea, the man who opened, to the west, a shorter route to the wealth
of Asia than the Portuguese had managed to eke out by sailing south
and east.†

*The circumnavigation of the earth by Ferdinand Magellan would prove the
geographers right. In the course of that grueling, three-year voyage Magellan was
killed, most of his men died, and his collaborator, the cosmographer Rui Faleiro,
went insane. Wrote Magellan's shipmate Antonio Pigafetta of the privations suffered during their crossing of the Pacific, "I believe that nevermore will any man
undertake to make such a voyage."[14]

†As Columbus was a practical, unbookish man and not (yet) insane, presumably
he had some reason other than the old geographies to think his voyage would
succeed. We do not know what this was, but can speculate that he heard sailors'
tales of sighting the coast of South America when driven west by winds while
trying to round the Cape of Good Hope, or knew that the Gulf Stream, which
flows east, carries fresh horsebeans and other signs of a reasonably proximate land-
mass. The explorer Thor Heyerdahl even proposes that Columbus heard of Leif
Erikson's discovery of America, either from Vatican sources or during a visit to
Iceland that Columbus is said by his son to have made at the age of twenty-six.[16]

The queen decided to give him a shot at it, and Columbus sailed in 1492, a pillar of unblinking zeal. He set his hourglass (inaccurately) by observing transits of the sun and noting the position of the Little Dipper. He navigated (accurately) by watching the compass. He corrected for variations in magnetic north by sighting the north star at both its easternmost and westernmost excursions—this a precaution that Columbus himself had developed, and one more important in 1492, when Polaris stood 3.3 degrees from the pole, than today, when the precession of the earth's axis has brought it to within 1 degree of true north.

Once embarked on the path of his destiny, Columbus was unshakable in his resolve to persevere. When his crewmen threatened to mutiny after a month at sea, he told them, as his son Ferdinand recorded his words, "that it was useless to complain, he had come [to go] to the Indies, and so had to continue until he found them, with the help of Our Lord."[17] Had America not intervened, he would certainly have led them to their deaths. Instead, at 2:00 A.M. on the night of October 12, 1492, Rodrigo de Triana, lookout aboard the *Pinta*, squinting westward toward where the bright star Deneb was setting, saw in the moonlight a distant spit of land, cried out, "*Tierra! Tierra!*," and claimed his reward as the first to sight India. The natives who beheld Columbus's three ships by the first light of dawn ran from hut to hut, shouting, "Come see the people from the sky!"

"They bear no arms, nor know thereof," Columbus noted, "for I showed them swords and they grasped them by the blade and cut themselves through ignorance."[18] He insisted that the natives be treated "lovingly," but business was business, and soon many were on their way to the Old World in chains.

Columbus on his subsequent voyages wandered from paradise to hell, laying eyes on some of the most beautiful islands on Earth but also suffering from thirst, starvation, and attacks by the "Indians." As the years passed and evidence for the true dimensions of the earth mounted, he took refuge in the unique hypothesis that the earth was small toward the north, where he had rounded it, and large elsewhere: Perhaps, he wrote, the world "is not round as it is described, but is shaped like a pear, which is round everywhere except near the stalk where it projects strongly; or it is like a very round ball with something like a woman's nipple in one place, and this projecting part is highest and the one nearest heaven"—the breast being where other navigators measured the circumference of the globe, and the "nipple . . . nearest heaven" being where Columbus sailed.[19]

Toward the end Columbus roamed the coasts of the New World in a state of gathering madness. He kept a gibbet mounted on the taffrail of his ship from which to hang mutineers, and made use of it so frequently that at one point he had to be recalled to Cadiz in chains. Crewmen on his final voyage watched warily as their captain hobbled around the deck, his body twisted by arthritis, his wild eyes peering out from under an aurora of tangled hair, searching endless coastlines for the mouth of the River Ganges. He threatened to hang anyone who denied they were in India. He sent back shiploads of slaves, which alarmed his queen, and cargos of gold, which delighted them both. "O, most excellent gold!" Columbus wrote. "Who has gold has a treasure with which he gets what he wants, imposes his will in the world, and even helps souls to paradise."[20] He died poor.

Gold outweighed the stars in the balance sheets of the exploratory enterprise. Montezuma II, emperor of the Aztecs, sent Cortez a gold disk the size of a cartwheel representing the sun, and another of silver representing the moon; soon he was Cortez's prisoner, and soon thereafter dead. Atahualpa of Peru sued for his freedom by filling his cell with gold higher than a man could reach, but Pizarro had him strangled nevertheless; he would have burned him had Atahualpa not agreed to accept baptism.

The New World's loss was the Old World's gain. As the traders and explorers had hoped, Portugal and Spain—and, through Spain, Holland and Britain—prospered at the expense of Africa and America. The greatest profits, however, came not in coin but in knowledge, tools, and dreams; Toscanelli, in a skewed way, had been right. Blue-water sailing called for improved navigational instruments and better charts of the earth, sea, and sky, all of which promoted the development of geography and astronomy. Schools of navigation were established in Portugal, Spain, England, Holland, and France, and their graduates joined a growing professional class adept at applied mathematics and steering by the stars. The independent, self-reliant spirit of the explorers touched those on land as well, eroding medieval confidence in ancient authority; wrote one of Prince Henry's captains, "With all due respect to the renowned Ptolemy, we found everything the opposite of what he said."[21]

More importantly though less distinctly, the great explorations opened up the human imagination, encouraging Western thinkers to regard not only the continents and seas but the entire planet from a more generous perspective. The dimensions of the known world had doubled by the year 1600, prompting a corresponding expansion in the cosmos of the mind. Heartened by the decline of the old authorities and by the adventuresome spirit of Columbus and the other explorers, the scholars of what would become the Renaissance began to imagine themselves traveling not only across the surface of the earth but also up into space. As Leon Frobenius was to write in a later century, "Our view is confined no longer to a spot of space on the surface of this earth. It surveys the whole of the planet. . . . This lack of horizon is something new." Nicholas of Cusa pointed out that "up" and "down" are relative terms, postulated that each star might be its own center of gravity, and suggested that if we lived on another planet we might assume that we occupied the center of the universe. Leonardo da Vinci was forty years old when Columbus reached the continent to which Leonardo's friend Amerigo Vespucci would lend his name, and he was a friend as well to Paolo Toscanelli, the astronomer who urged Columbus on his way. Imbued with an explorer's vision, Leonardo cast his mind's eye out into space and imagined that the earth from a distance would look like the moon:

> If you were where the moon is, it would appear to you that the sun was reflected over as much of the sea as it illumines in its daily course, and the land would appear amid this water like the dark spots that are upon the moon, which when looked at from the earth presents to mankind the same appearance that our earth would present to men dwelling in the moon.[22]

Copernicus was a student at the University of Cracow when Columbus landed in the Indies. He was forty-nine years old when Magellan's ship completed its circumnavigation of the globe. He sent *his* mind's eye journeying to the sun, and what he saw turned the earth into a ship under sail, assaying oceanic reaches of space undreamed of since the days of Aristarchus of Samos.

4

THE SUN WORSHIPERS

There is no new thing under the sun.
                        —Ecclesiastes

Amazed, and as if astonished and stupefied,
I stood still, gazing for a certain length of time
with my eyes fixed intently upon it. . . . When
I had satisfied myself that no star of that kind
had ever shone forth before, I was led into
such perplexity by the unbelievability of the
thing that I began to doubt the faith of my
own eyes.
        —Tycho, on the supernova of 1572

Mikolai Kopernik, though rightly esteemed as a great astronomer, was never much of a stargazer. He did some observing in his student days, assisting his astronomy professor at Bologna, Dominico Maria de Novara, in watching an occultation of the star Aldebaran by the moon, and he later took numerous sightings of the sun, using an instrument of his own devising that reflected the solar disk onto a series of graph lines etched into a wall outside his study. But these excursions served mainly to confirm what Kopernik and everybody else already knew, that the Ptolemaic system was inaccurate, making predictions that often proved to be wrong by hours or even days.

Kopernik drew inspiration less from stars than from books. In this he was very much a man of his time. The printing press—invented just thirty years before he was born—had touched off a communications revolution comparable in its impact to the changes wrought in the latter half of the twentieth century by the electronic computer. To be sure, Greek and Roman classics had been making their way from the Islamic world to Europe for centuries, and with enlightening effect—the first universities had been founded principally to house the books and study their contents—but the books themselves, each laboriously copied out by hand, were rare and expensive, and frequently were marred by transcription errors. All this changed with the advent of cheap, high-quality paper (a gift of Chinese technology) and the press. Now a single competent edition of Plato or Aristotle or Archimedes or Ptolemy could be reproduced in considerable quantities; every library could have one, and so could many individual scholars and more than a few farmers and housewives and tradespeople. As books spread so did literacy, and as the number of literate people increased, so did the market for books. By the time Kopernik was thirty years old (and printing itself but sixty years old), some six to nine million printed copies of more than thirty-five thousand titles had been published, and the print shops were working overtime trying to satisfy the demand for more.

Kopernik was as voracious a reader as any, at home in law, literature, and medicine as well as natural philosophy. Born in 1473 in northern Poland, he had come under the sponsorship of his powerful and calculating uncle Lucas Waczenrode, later bishop of Warmia, who gave him books and sent him to the best schools. He attended the University of Cracow, then ventured south into the Renaissance heartland to study at the universities of Bologna and Padua. He read Aristotle, Plato, Plutarch, Ovid, Virgil, Euclid, Archimedes, and Cicero, the restorer of Archimedes' grave. Steeped in the literature and science of the ancients, he returned home with a Latinized name, as Nicolaus Copernicus.

Like Aristotle, Copernicus collected books; unlike Aristotle, he did not have to be wealthy to do so. Thanks to the printing press, a scholar who was only moderately well off could afford to read widely, at home, without having to beg admission to distant institutions of learning where the books were kept chained to the reading desks. Copernicus was one of the first scholars to study printed books in his own library, and he studied none more closely than Ptolemy's *Almagest*. Great was his admiration for Ptolemy, whom he admired as a thoroughly professional astronomer, mathematically sophisticated and dedicated to fitting his cosmological model to the observed phenomena. Indeed, Copernicus's *De Revolutionibus* (*On the Revolutions*), the book that would set the earth into motion around the sun and bring about Ptolemy's downfall, otherwise reads like nothing so much as a sustained imitation of Ptolemy's *Almagest*.

It is widely assumed that Copernicus proposed his heliocentric theory in order to repair the inaccuracies of the Ptolemaic model. Certainly it must have become evident to him, in his adulthood if not in his student days, that the Ptolemaic system did not work very well: "The mathematicians are so unsure of the movements of the sun and moon," notes the preface to *De Revolutionibus*, "that they cannot even explain or observe the constant length of the seasonal year."[1] Prior to the advent of the printing press, the failings of Ptolemy's *Almagest* could be attributed to errors in transcription or translation, but once reasonably accurate printed editions of the book had been published, this excuse began to evaporate. Copernicus owned at least two editions of *Almagest*, and had read others in libraries, and the more clearly he came to understand Ptolemy's model, the more readily he could see that its deficiencies were inherent, not incidental, to the theory. So considerations of accuracy may indeed have helped convince him that a new approach was required.

But by "new," Copernicus the Renaissance man most often meant the rediscovery of something old. *Renaissance*, after all, means "*re*-birth," and Renaissance art and science in general sprang more from classical tradition than from innovation. The young Michelangelo's first accomplished piece of sculpture—executed in the classical style—was made marketable by rubbing dirt into it and palming it off, in Paris, as a Greek relic. Petrarch, called the founder of the Renaissance, dreamed not of the future but of the day when "our grandsons will be able to walk *back* into the pure radiance *of the past*"[2] (emphasis added); when Petrarch was found dead, at the age of seventy, slumped at his desk after an all-night study session, his head was resting not on a contemporary volume but on a Latin edition of his favorite poet, Virgil, who had lived fourteen centuries earlier. Copernicus similarly worked in awe of the ancients, and his efforts, like so much of natural philosophy then and since, can be read as a continuation of the academic dialogues of Plato and Aristotle.

Aristotle, the first of the Greeks to have been rediscovered in the West, was so widely revered that he was routinely referred to as "the philosopher," much as lovers of Shakespeare were to call him "the poet." Much of his philosophy had been incorporated into the world view of the Roman Catholic Church. (Most notably by Thomas Aquinas—at least until the morning of December 6, 1273, when, while saying mass in Naples, Thomas became enlightened and declared that "I can do no more; such things have been revealed to me that all I have written seems as straw, and I now await the end of my life.") From Aristotle, Copernicus acquired an enthusiasm for the universe of crystalline spheres—although, like Aristotle, Copernicus never could decide whether the spheres actually existed or were but a useful abstraction.

Copernicus also read Plato, as well as many of the Neoplatonic philosophers whose work ornaments and obfuscates medieval thought, and from them absorbed the Platonic conviction that there must be a simple underlying structure to the universe. It was just this unitary beauty that the Ptolemaic cosmology lacked. "A system of this sort seemed neither sufficiently absolute nor sufficiently pleasing to the mind," Copernicus wrote.[3] He was after a grasp of the more central truth. He called it "the principal thing—namely the shape of the universe and the unchangeable symmetry of its parts."[4]

Rather early on, perhaps during his student days in sunny Italy, Copernicus decided that the "principal thing" was to place the sun at the center of the universe. He may have drawn encouragement from reading, in Plutarch's *Morals*, that Aristarchus of Samos "supposed that the heavens remained immobile and that the earth moved through an oblique circle, at the same time turning about its own axis."[5] (He mentions Aristarchus in *De Revolutionibus*, though not in this context.) Possibly he encountered more recent speculations about the motion of the earth, as in Nicole Oresme, the fourteenth-century Parisian scholar who pointed out that

if a man in the heavens, moved and carried along by their daily motion, could see the earth distinctly and its mountains, valleys, rivers, cities, and castles, it would appear to him that the earth was moving in daily motion, just as to us on earth it seems as though the heavens are moving. . . . One could then believe that the earth moves and not the heavens.[6]

Copernicus was influenced by Neoplatonic sun worship as well. This was a popular view at the time—even Christ was being modeled by Renaissance painters on busts of Apollo the sun god —and decades later, back in the rainy north, Copernicus remained effusive on the subject of the sun.* In *De Revolutionibus* he invokes the authority of none other than Hermes Trismegistus, "the thrice-great Hermes," a fantastical figure in astrology and alchemy who had become the patron saint of the new sun-worshipers: "Trismegistus calls [the sun] a 'visible god,' Sophocles' Electra, 'that which gazes upon all things.' "[7] He quotes the Neoplatonist mystic Marsilio Ficino's declaration that "the sun can signify God himself to you, and who shall dare to say the sun is false?"[8] Finally, Copernicus tries his hand at a solar paean of his own:

In this most beautiful temple, who would place this lamp in another or better position than that from which it can light up everything at the same time? For the sun is not inappropriately called by some people the lantern of the universe, its mind by others, and its ruler by still others.[9]

Trouble arose not in the incentive for the Copernican cosmology, but in its execution. (The devil, like God, is in the details.) When Copernicus, after considerable toil, managed to complete a fully realized model of the universe based upon the heliocentric hypothesis—the model set forth, eventually, in *De Revolutionibus*— he found that it worked little better than the Ptolemaic model. One difficulty was that Copernicus, like Aristotle and Eudoxus before him, was enthralled by the Platonic beauty of the sphere—"The sphere," he wrote, echoing Plato, "is the most perfect . . . the most capacious of figures . . . wherein neither beginning nor end can be found"[10]—and he assumed, accordingly, that the planets move in

---

*One could write a plausible intellectual history in which the decline of sun worship, the religion abandoned by the Roman emperor Constantine when he converted to Christianity, was said to have produced the Dark Ages, while its subsequent resurrection gave rise to the Renaissance.

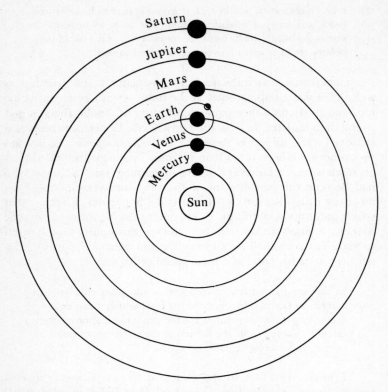

Copernicus's model of the solar system is generally portrayed in simplified form, as in this illustration based upon one in his *De Revolutionibus*. In its details, however, it was as complex as Ptolemy's geocentric model.

circular orbits at constant velocities. Actually, as Kepler would establish, the orbits of the planets are elliptical, and planets move more rapidly when close to the sun than when distant from it. Nor was the Copernican universe less intricate than Ptolemy's: Copernicus found it necessary to introduce Ptolemaic epicycles into his model and to move the center of the universe to a point a little away from the sun. Nor did it make consistently more accurate predictions, even in its wretchedly compromised form; for many applications it was less useful.

This, in retrospect, was the tragedy of Copernicus's career—

that while the beauty of the heliocentric hypothesis convinced him that the planets ought to move in perfect circles around the sun, the sky was to declare it false. Settled within the stone walls of Frauenburg Cathedral, in a three-story tower that afforded him a view of Frisches Haff and the Gulf of Danzig below and the wide (though frequently cloudy) sky above—"the most remote corner of the earth,"[11] he called it—Copernicus carried out his sporadic astronomical observations, and tried, in vain, to perfect the heliocentric theory he had outlined while still a young man. For decades he turned it over in his thoughts, a flawed jewel, luminous and obdurate. It would not yield.

As Darwin would do three centuries later, Copernicus wrote and privately circulated a longhand sketch of his theory. He called it the "ballet of the planets." It aroused interest among scholars, but Copernicus published none of it. He was an old man before he finally released the manuscript of *De Revolutionibus* to the printer, and was on his death bed by the time the final page proofs arrived.

One reason for his reluctance to publish was that Copernicus, like Darwin, had reason to fear censure by the religious authorities. The threat of papal disapproval was real enough that the Lutheran theologian Andreas Osiander thought it prudent to oil the waters by writing an unsigned preface to Copernicus's book, as if composed by the dying Copernicus himself, reassuring its readers that divine revelation was the sole source of truth and that astronomical treatises like this one were intended merely to "save the phenomena." Nor were the Protestants any more apt to kiss the heliocentric hem. "Who will venture to place the authority of Copernicus above that of the Holy Spirit?" thundered Calvin,[12] and Martin Luther complained, in his voluble way, that "this fool wishes to reverse the entire science of astronomy; but sacred Scripture tells us that Joshua commanded the sun to stand still, and not the earth."[13]*

---

*Modern myth to the contrary, little of the ecclesiastical opposition to Copernicanism appears to have derived from fear that the theory would "dethrone" humanity from a privileged position at the center of the universe. The center of the universe in Christian cosmology was hell, and few mortals would have felt disaccommodated at being informed that they did not live there. Heaven was the place of distinction, for Christian and pagan thinkers alike. As Aristotle put it, "The superior glory of . . . nature is proportional to its distance from this world of ours." When Leonardo da Vinci suggested that the earth "is not in the center of the universe," he intended no slander of Earth, but was suggesting that our planet is due the same dignity—*noblesse*—as are the stars.

The book survived, however, and changed the world, for much the same reason that Darwin's *Origin of Species* did—because it was too technically competent for the professionals to ignore it. In addition to presenting astronomers with a comprehensive, original, and quantitatively defensible alternative to Ptolemy, *De Revolutionibus* was full of observational data, much of it fresh and some of it reliable. Consequently it was consulted regularly by astronomers—even by non-Copernicans like Erasmus Reinhold, who employed it in compiling the widely consulted *Prutenic Tables*—and thus remained in circulation for generations.

To those who gave it the benefit of the doubt, Copernicanism offered both a taste of the immensity of space and a way to begin measuring it. The minimum radius of the Copernican sphere of stars (given the unchanging brightnesses of the zodiacal stars) was estimated in the sixteenth century to be more than 1.5 million times the radius of the earth. This represented an increase in the volume of the universe of at least 400,000 times over al-Farghani's Ptolemaic cosmos. The maximum possible size of the Copernican universe was indefinite, and might, Copernicus allowed, be infinite: The stars, he wrote, "are at an immense height away," and he expressed wonderment at "how exceedingly vast is the godlike work of the Best and Greatest Artist!"[14]

Interplanetary distances in Ptolemy were arbitrary; scholars who ventured to quantify them did so by assuming that the various orbits and epicycles fit snugly together, like nested Chinese boxes. The Copernican theory, however, precisely stipulated the relative dimensions of the planetary orbits: The maximum apparent separation of the inferior planets Mercury and Venus from the sun yields the relative diameters of their orbits, once we accept that both orbit the sun and not the earth. Since the relative sizes of all the orbits were known, if the actual distance of any one planet could be measured, the distances of all the others would follow. As we will see, this advantage, though purely theoretical in Copernicus's day, was to be put to splendid use in the eighteenth century, when astronomical technology reached the degree of sophistication required to measure directly the distances of nearby planets.

The immediate survival of Copernicanism was due less to any compelling evidence in its favor than to the waning fortunes of the Ptolemaic, Aristotelian model. And that, as it happened, was prompted in large measure by changes in the sky—by the apparition of comets, and, most of all, by the fortuitous appearance of two brilliant *novae*, or "new stars," during the lifetimes of Tycho, Kepler, and Galileo.

Integral to Aristotle's physics was the hypothesis that the stars never change. Aristotle saw the earth as composed of four elements—earth, water, fire, and air—each of which naturally moves in a vertical direction: The tendency of earth and water is to fall, while that of fire and air is to rise. The stars and planets, however, move neither up nor down, but instead wheel across the sky. Aristotle concluded that since objects in the sky do not partake of the vertical motion characteristic of the four terrestrial elements, they must be made of another element altogether. He called this fifth element "aether," from the Greek word for "eternal," and invested it with all his considerable reverence for the heavens. Aether, he argued, never ages or changes: "In the whole range of time past," he writes, in his treatise *On the Heavens*, "so far as our inherited records reach, no change appears to have taken place either in the whole scheme of the outermost heaven or in any of its proper parts."[15]

Aristotle's segregation of the universe into two realms—a mutable world below the moon and an eternal, unchanging world above—found a warm welcome among Christian theologians predisposed by Scriptures to think of heaven as incorruptible and the earth as decaying and doomed. The stars, however, having heard neither of Aristotle nor of the Church, persisted in changing, and the more they changed, the worse the cosmology of Aristotle and Ptolemy looked.

Comets were an old problem for the Aristotelians, since no one could anticipate when they would appear or where they would go once they showed up.* (It was owing to their unpredictability that comets acquired a reputation as heralds of disaster—from the

---

*Comets are chunks of ice and dirt that fall in from the outer solar system, sprouting long, glowing "tails" of vapor and dust blown off by the sun's heat and by solar wind. The appearance of new comets cannot be predicted even today; they appear to originate in a cloud that lies near the outer reaches of the solar system, about which little is understood. Their orbits, altered by encounters with the planets and by the kick of their own vapor jets, remain difficult to predict as well.

Latin *dis-astra*, "against the stars.")* Aristotle swept comets under
the rug—or under the moon—by dismissing them as atmospheric
phenomena. (He did the same with meteors, which is why the
study of the weather is known as "meteorology.")

But when Tycho Brahe, the greatest observational astronomer
of the sixteenth century, studied the bright comet of 1577, he found
evidence that Aristotle's explanation was wrong. He triangulated
the comet, by charting its position from night to night and comparing his data with those recorded by astronomers elsewhere in
Europe on the same dates. The shift in perspective produced by
the differing locations of the observers would have been more than
sufficient to show up as a difference in the comet's position against
the background stars, were the comet nearby. Tycho found no such
difference. This meant that the comet was well beyond the moon.
Yet Aristotle had held that nothing superlunar could change.

The other great empirical challenge to Aristotle's cosmological
hegemony came with the opportune appearance, in the late sixteenth and early seventeenth centuries, of two violently exploding
stars—what we today call *supernovae*. A star that undergoes such a
catastrophic detonation can increase a hundred million times in
brightness in a matter of days. Since only a tiny fraction of the
stars in the sky are visible without a telescope, supernovae almost
always seem to have appeared out of nowhere, in a region of the
sky where no star had previously been charted; hence the name
nova, for "new." Supernovae bright enough to be seen without a
telescope are rare; the next one after the seventeenth century did
not come until 1987, when a blue giant star exploded in the Large
Magellanic Cloud, a neighboring galaxy to the Milky Way, to the
delight of astronomers in Australia and the Chilean Andes. The
two supernovae that graced the Renaissance caused quite a stir,
inciting not only new sights but new ideas.

Tycho spotted the supernova of 1572 on the evening of No-

---

*The cometary stigma persisted into the early twentieth century, when
millions bought patent medicines to protect themselves from the evil effects
of comet Halley during its 1910 visitation. Several fatalities were reported,
among them a man who died of pneumonia after jumping into a frozen creek
to escape the ethereal vapors. A deputation of sheriffs intervened to prevent
the sacrifice of a virgin, in Oklahoma, by a sect called the Sacred Followers
who were out to appease the comet god.

vember 11, while out taking a walk before dinner, and it literally stopped him in his tracks. As he recalled the moment:

> Amazed, and as if astonished and stupefied, I stood still, gazing for a certain length of time with my eyes fixed intently upon it and noticing that same star placed close to the stars which antiquity attributed to Cassiopeia. When I had satisfied myself that no star of that kind had ever shone forth before, I was led into such perplexity by the unbelievability of the thing that I began to doubt the faith of my own eyes.[16]

The next supernova came only thirty-two years later, in 1604. Kepler observed it for nearly a year before it faded from view, and Galileo lectured on it to packed halls in Padua.

Scrutinized week by week through the pinholes and lensless sighting-tubes of the sixteenth- and seventeenth-century astronomers, the two supernovae stayed riveted in the same spot in the sky, and none revealed any shift in perspective when triangulated by observers at widely separated locations. Clearly the novae, too, belonged to the starry realm that Aristotle had depicted as inalterable. Wrote Tycho of the 1572 supernova:

> That it is neither in the orbit of Saturn . . . nor in that of Jupiter, nor in that of Mars, nor in that of any one of the other planets, is hence evident, since after the lapse of several months it has not advanced by its own motion a single minute from that place in which I first saw it; which it must have done if it were in some planetary orbit. . . . Hence this new star is located neither . . . below the Moon, nor in the orbits of the seven wandering stars but in the eighth sphere, among the other fixed stars.[17]

The shock dealt to the Aristotelian world view could not have been greater had the stars bent down and whispered in the astronomers' ears. Clearly there was something new, not only under the sun but beyond it.*

---

*Twentieth-century radio astronomers using Renaissance star charts have located the wreckage of both Tycho's supernova, now designated 3C 10 in the Cambridge catalog of radio sources, and of Kepler's, known as 3C 358. Also located is the remnant of the Vela supernova, which blazed forth in the southern skies some six

Tycho was no Copernican. It was through Ptolemy that his
passion for astronomy had crystallized, when, on August 21, 1560,
at the age of thirteen, he watched a partial eclipse of the sun and
was amazed that it had been possible for scholars, consulting the
Ptolemaic tables, accurately to predict the day (though not the hour)
of its occurrence. It struck him, he recalled, as "something divine
that men could know the motions of the stars so accurately that
they could long before foretell their places and relative positions."[18]

But when Tycho began making observations of his own, he
soon became impressed by the *in*accuracy of Ptolemy's predictions.
He watched a spectacular conjunction of Saturn and Jupiter on
August 24, 1563, and found that the time of closest approach—
which in this case was so close that the two bright planets appeared
almost to merge—was days away from the predictions of the Ptolemaic tables. He emerged from the experience with a lifelong passion for accuracy and exactitude and a devotion to the verdict of
the sky.

To compile more accurate records of the positions of the stars
and planets required state-of-the-art equipment, and that cost money.
Fortunately, Tycho *had* money. His foster father had saved King
Frederick II from drowning, dying of pneumonia as a result, and
the grateful king responded with a hefty grant to the young astronomer. With it, Tycho built Uraniburg, a fabulous observatory
on an island in the Sund between Elsinor Castle (Hamlet's haunt)
and Copenhagen. He ransacked Europe in search of the finest astronomical instruments, complemented them with improved quadrants and armillaries of his own design, and deployed them atop
the turrets of a magnificent castle that he equipped with a chemical
laboratory, a printing plant supplied by its own paper mill, an
intercom system, flush toilets, quarters for visiting researchers, and
a private jail. The grounds sported private game preserves, sixty
artificial fishponds, extensive gardens and herbariums, and an ar-
boretum with three hundred species of trees. The centerpiece of

---

to eight *thousand* years ago, casting long shadows across the plains of Eden. (The
word *Eden* is Sumerian for "flatland," and is thought to refer to the fertile, rock-
free plains of the Tigris-Euphrates.) The Sumerians identified that supernova with
the god Ea (in Egypt, Seshat), whom they credited with the invention of writing
and agriculture. The Ea myth thus suggests that the creation of agriculture and
the written word were attributed by the ancients to the incentive provided by the
sight of an exploding star.

the observatory was a gleaming brass celestial globe, five feet in diameter, on which a thousand stars were inscribed, one by one, as Tycho and his colleagues remapped the visible sky.

No dilettante, Tycho drove himself and his assistants in a ceaseless pursuit of the most accurate possible observations, charting the positions of the stars and the courses of the planets night after night for over twenty years. The resulting data were more than twice as accurate as those of the preceding astronomers—precise enough, at last, to unlock the secrets of the solar system.

Tycho, however, was an observer and not a theorist. His chief contribution to theoretical cosmology—a compromise geocentric model in which the planets orbit the sun, which in turn orbits the earth—created as many problems as it solved. Needed was someone with the ingenuity and perservance to compose Tycho's tables into a single, accurate and simple theory.

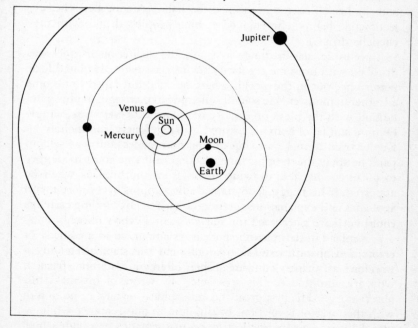

Tycho proposed a compromise between the Copernican and Ptolemaic models in which the sun orbited the earth, and was in turn orbited by the other planets. (Not to scale.)

Amazingly, just such a man turned up. He was Johannes Kepler, and on February 4, 1600, he arrived at Benatek Castle near Prague, where Tycho had moved his observatory and retinue after his benefactor King Frederick drank himself to death. Tycho and Kepler made for unlikely collaborators, with each other or anybody else. Tycho was an expansive, despotic giant of a man, who sported a belly of Jovian proportions and a gleaming, metal-alloy nose (the bridge of his original nose having been cut off in a youthful duel). Heroically passionate and wildly eccentric, he dressed like a prince and ruled his domain like a king, tossing scraps to a dwarf named Jepp who huddled beneath the dinner table. Kepler, for his part, was a prototypical outsider. Myopic, sickly, and "doglike" in appearance (his words) he came from the antipodes of nobility. His father was a mercenary soldier and a dipsomaniac wife-beater. His mother had been raised by an aunt who was burned alive as a witch, and she herself narrowly escaped the stake. (Among her other objectionable habits, she enjoyed spiking people's drinks with psychedelic drugs.)

Neurotic, self-loathing, arrogant, and vociferous, Kepler was drubbed with tiresome regularity by his classmates. He fared little better once out in the world, where he tried but failed to become a Lutheran minister. He sought solicitude in marriage, but his wife, he said with the bleak objectivity of a born observer, was "simple of mind and fat of body . . . stupid, sulking, lonely, melancholy."[19] Kepler tried to make a living casting horoscopes, but was seldom paid; he spent much of his time trekking from one court to another to plead for his fee, drawing titters from the flunkies when he appeared, in his baggy, food-stained suit, tripping over himself with apologies and explanations, getting nowhere. His lifetime earnings could not have purchased the star-globe in Tycho's library.

Kepler's initial scientific endeavors amounted to a comedy of errors and absurdities. He tried to sight the stars using only a wooden staff suspended from a rope: "Hold your laughter, friends, who are admitted to this spectacle," he wrote of his makeshift observatory.[20] His first great theoretical idea—which came to him with the force of revelation, halting him in mid-sentence while he was delivering a soporific lecture on mathematics in a high school in Graz, Austria—was that the intervals between the orbits of the planets describe a nest of concentric Platonic solids. They do not.

Yet this was the man who would discern the architecture of

the solar system and discover the phenomenological laws that govern the motions of the planets, thus curing the Copernican cosmology of its pathologies and flinging open the door to the depths
of cosmic space. An extraordinarily perspicacious theorist—no less
exacting a critic than Immanuel Kant called him "the most acute
thinker ever born"[21]—Kepler was blessed with an ecstatic conviction that the world that had treated him so harshly was, nonetheless,
fundamentally beautiful. He never lost either this faith or the clearheaded empiricism with which it was tempered, and the combination eventually rewarded him with some of the most splendid
insights into the workings of the universe ever granted a mortal
mind.

Kepler's chief source of inspiration was the Pythagorean doctrine of celestial harmony, which he had encountered in Plato. "As
our eyes are framed for astronomy, so our ears are framed for the
movements of harmony," Plato wrote, "and these two sciences are
sisters, as the Pythagoreans say and we agree."[22] In the final book
of the *Republic*, Plato portrays with great beauty a voyage into space,
where the motion of each planet is attended to by a Siren singing

> one sound, one note, and from all the eight there was a concord
> of a single harmony. And there were three others who sat
> round about at equal intervals, each one on her throne, the
> Fates, daughters of Necessity, clad in white vestments with
> garlands on their heads, Lachesis, and Clotho, and Atropos,
> who sang in unison with the music of the Sirens, Lachesis
> singing the things that were, Clotho the things that are, and
> Atropos the things that are to be.[23]

Aristotle found all this a bit much. "The theory that the movement
of the stars produces a harmony, i.e., that the sounds they make
are concordant, in spite of the grace and originality with which it
has been stated, is nevertheless untrue," he wrote.[24] Kepler sided
with Plato. The muddy tumult of the world, he felt, was built
upon harmonious and symmetrical law; if the motions of the planets
seem discordant, that is because we have not yet learned how to
hear their song. Kepler wanted to hear it before he died. At this
he succeeded, and the sunlight of his success banished the gloom
of his many failures.

The doctrine of celestial harmony was, literally, in the air, in the new music and poetry of Kepler's generation and those that immediately followed it. Milton, who was always ransacking science for promising themes, celebrated it in verses like this one:

> Ring out ye Crystall sphears,
> Once bless our human ears,
>   (If ye have power to touch our senses so)
> And let your silver chime
> Move in melodious time;
>   And let the Base of Heav'ns deep Organ blow,
> And with your ninefold harmony
> Make up full consort to th' Angelike symphony.[25]

Even Shakespeare, who was rather unsympathetic toward astronomy, found room in the *Merchant of Venice* for a nod to Pythagoras:

> Sit, Jessica. Look how the floor of heaven
> Is thick inlaid with patens of bright gold.
> There's not the smallest orb which thou behold'st
> But in his motion like an angel sings,
> Still quiring to the young-eyed cherubims;
> Such harmony is in immortal souls,
> But whilest this muddy vesture of decay
> Doth grossly close it in, we cannot hear it.[26]

The churches of the day rang with approximations of the music of the spheres. The plainsongs and chants of the medieval cathedrals were being supplanted by polyphony, the music of many voices that would reach an epiphany in the fugues—the word *fugue* means "flight"—of Johann Sebastian Bach. For Kepler, polyphony in music was a model for the voices sung by the planets as they spun out their Pythagorean harmonies: "The ratio of plainsong or monody . . . to polyphony," he wrote,

> is the same as the ratio of the consonances which the single planets designate to the consonances of the planets taken together. . . .
>   . . . The movements of the heavens are nothing except a certain ever-lasting polyphony (intelligible, not audible). . . .
> Hence it is no longer a surprise that man, the ape of his Creator, should finally have discovered the art of singing polyphoni-

cally, which was unknown to the ancients, namely in order
that he might play the everlastingness of all created time in
some short part of an hour by means of an artistic concord of
many voices and that he might to some extent taste the satisfaction of God the Workman with His own works, in that very
sweet sense of delight elicited from this music which imitates
God.[27]

Kepler's interest in astronomy, like Tycho's, dated from his
boyhood, when his mother took him out in the evening to see the
great comet of 1577 and, three years later, to behold the sanguine
face of the eclipsed moon. He was introduced to heliocentric cosmology at the University of Tübingen, by Michael Mastlin, one of
the few Copernican academics of his day. Attracted to it partly out
of mystical, Neoplatonic motives like those that had inspired Copernicus himself, Kepler wrote of sunlight in terms that would have
brought a smile to the countenance of Marsilio Ficino:

> Light in itself is something akin to the soul. . . . And so it is
> consonant that the solar body, wherein the light is present as
> in its source, is endowed with a soul which is the originator,
> the preserver, and the continuator. And the function of the
> sun in the world seems to persuade us of nothing else except
> that just as it has to illuminate all things, so it is possessed of
> light in its body; and as it has to make all things warm, it is
> possessed of heat; as it has to make all things live, of a bodily
> life; and as it has to move all things, it itself is the beginning
> of the movement; and so it has a soul.[28]

But Kepler's penchant for Platonic ecstasy was wedded to an
acid skepticism about the validity of all theories, his own included.
He mocked no thinker more than himself, tested no ideas more
rigorously than his own. If, as he avowed in 1608, he was to "in-
terweave Copernicus into the revised astronomy and physics, so
that either both will perish or both be kept alive," he would need
more accurate observational data than were available to Ptolemy or
to Copernicus. Tycho had those data. "Tycho possesses the best
observations," Kepler mused. ". . . He only lacks the architect who
would put all this to use according to his own design."[29] Tycho
was "superlatively rich, but he knows not how to make proper use
of it as is the case with most rich people. Therefore, one must try

to wrest his riches from him."[30] Suiting action to intention, Kepler wrote adoring letters to Tycho, who in reply praised his theories as "ingenious" if rather too a priori, and invited him to come and join the staff at Benatek Castle.

There the two quarreled constantly. Tycho, justly fearful that the younger and more incisive Kepler would eclipse him, played his cards close to his chest. "Tycho did not give me the chance to share his practical knowledge," Kepler recalled, "except in conversation during meals, today something about the apogee, tomorrow something about the nodes of another planet."[31] Kepler threw fits and threatened to leave; at one point he had packed his bags and boarded a stage before Tycho finally summoned him back.

Realizing that he would have to give his young colleague something of substance to work on if he wanted to keep him on staff, Tycho devised a scheme redolent with the enmity that Kepler seemed to attract like lightning to a summit pine. "When he saw that I possess a daring mind," Kepler wrote, "he thought the best way to deal with me would be to give me my head, to let me choose the observations of one single planet, Mars."[32] Mars, as Tycho knew and Kepler did not, presented an almost impossible challenge. As Mars lies near the earth, its position in the sky had been ascertained with great exactitude; for no planet were the inadequacies of both the Ptolemaic and Copernican models rendered more starkly. Kepler, who did not at first appreciate the difficulties involved, brashly prophesied that he would solve the problem of determining the orbit of Mars in eight days. Tycho must have been cheerful at dinner that night. Let the Platonist take on Mars. Kepler was still working on the problem eight years later.

Tycho, though, was out of time. He died on October 24, 1601, as the result of a burst bladder suffered while drinking too much beer at a royal dinner party from which he felt constrained by protocol from excusing himself. "Let me not seem to have died in vain," he cried repeatedly that night.[33]

Kepler was to grant his dying wish. Named Tycho's successor as imperial mathematician (albeit, as befitting his lesser status, at a much reduced stipend), he pressed on in his search for a single, straightforward theory to account for the motion of Mars. If every great achievement calls for the sacrifice of something one loves, Kepler's sacrifice was the perfect circle. "My first mistake was in having assumed that the orbit on which planets move is a circle," he recalled. "This mistake showed itself to be all the more baneful in that it had been supported by the authority of all the philosophers, and especially as it was quite acceptable metaphysically."[34] In all, Kepler tested *seventy* circular orbits against Tycho's Mars data, all to no avail. At one point, performing a leap of the imagination like Leonardo's to the moon, he imagined himself on Mars, and sought to reconstruct the path the *earth's* motion would trace out across the skies of a Martian observatory; this effort consumed nine hundred pages of calculations, but still failed to solve the major problem. He tried imagining what the motion of Mars would look like from the sun. At last, his calculations yielded up their result: "I have the answer," Kepler wrote to his friend the astronomer David Fabricius. ". . . The orbit of the planet is a perfect ellipse."

Now everything worked. Kepler had arrived at a fully realized Copernican system, focused on the sun and unencumbered by epicycles or crystalline spheres. (In retrospect one could see that Ptolemy's eccentrics had been but attempts to make circles behave like ellipses.)

Fabricius replied that he found Kepler's theory "absurd," in that it abandoned the circles whose symmetry alone seemed worthy of the heavens. Kepler was unperturbed; he had found a still deeper and subtler symmetry, in the *motions* of the planets. "I discovered among the celestial movements the full nature of harmony," he exclaimed, in his book *The Harmonies of the World*, published eighteen years after Tycho's death.

> I am free to give myself up to the sacred madness, I am free to taunt mortals with the frank confession that I am stealing the golden vessels of the Egyptians, in order to build of them a temple for my God, far from the territory of Egypt. If you pardon me, I shall rejoice; if you are enraged, I shall bear up. The die is cast.[35]

And so on. The cause of his celebration was his discovery of what are known today as Kepler's laws. The first contained the news he had communicated to Fabricius—that each planet orbits the sun in an ellipse with the sun at one of its two foci. The second law revealed something even more astonishing, a Bach fugue in the sky. Kepler found that while a planet's velocity changes during its year, so that it moves more rapidly when close to the sun and more

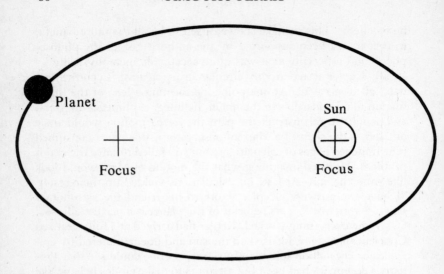

Kepler's first law: The orbit of each planet describes an ellipse, with the sun at one of its foci.

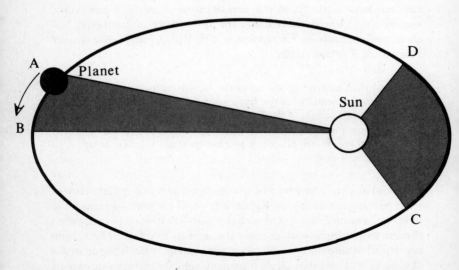

Kepler's second law: If time *AB* = time *CD*, area *ABSun* = area *CDSun*.

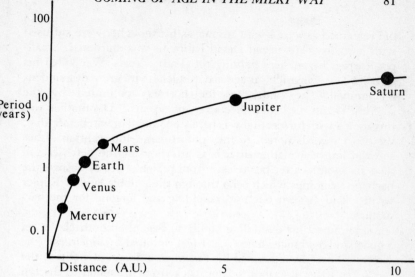

Kepler's third law: The cube of the distance of each planet from the sun is proportional to the square of its orbital period.

slowly when distant from the sun, its motion obeys a simple mathematical rule: Each planet sweeps out equal areas in equal times. The third law came ten years later. It stated that the cube of the mean distance of each planet from the sun is proportional to the square of the time it takes to complete one orbit. Archimedes would have liked that one. Newton was to employ it in formulating his law of universal gravitation.

Here at last was "the principal thing" of which Copernicus had dreamed, the naked kinematics of the sun and its planets. "I contemplate its beauty with incredible and ravishing delight," Kepler wrote.[36] Scientists have been contemplating it ever since, and Kepler's laws today are utilized in studying everything from binary star systems to the orbits of galaxies across clusters of galaxies. The intricate etchings of Saturn's rings, photographed by the Twin Voyager spacecraft in 1980 and 1981, offer a gaudy display of Keplerian harmonies, and the Voyager phonograph record, carried aboard the spacecraft as an artifact of human civilization, includes a set of computer-generated tones representing the relative velocities of the planets—the music of the spheres made audible at last.

But the sun of learning is paired with a dark star, and Kepler's life remained as vexed with tumult as his thoughts were suffused with harmony. His friend David Fabricius was murdered. Smallpox carried by soldiers fighting the Thirty Years' War killed his favorite son, Friedrich, at age six. Kepler's wife grew despondent —"numbed," he said, "by the horrors committed by the soldiers"—and died soon thereafter, of typhus.[37] His mother was threatened with torture and was barely acquitted of witchcraft (due, the court records noted, to the "unfortunate" intervention of her son the imperial mathematician as attorney for the defense) and died six months after her release from prison. "Let us despise the barbaric neighings which echo through these noble lands," Kepler wrote, "and awaken our understanding and longing for the harmonies."[38]

He moved his dwindling family to Sagan, an outback. "I am a guest and a stranger here. . . . I feel confined by loneliness," he wrote.[39] There he annotated his *Somnium*, a dream of a trip to the moon. In it he describes looking back from the moon to discern the continent of Africa, which, he thought, resembled a severed head, and Europe, which looked like a girl bending down to kiss that head. The moon itself was divided between bright days and cold dark nights, like Earth a world half darkness and half light.

Dismissed from his last official post, as astrologer to Duke Albrecht von Wallenstein, Kepler left Sagan, alone, on horseback, searching for funds to feed his children. The roads were full of wandering prophets declaring that the end of the world was at hand. Kepler arrived in Ratisbon, hoping to collect some fraction of the twelve thousand florins owed him by the emperor. There he fell ill with a fever and died, on November 15, 1630, at the age of forty-eight. On his deathbed, it was reported, he "did not talk, but pointed his index finger now at his head, now at the sky above him."[40] His epitaph was of his own composition:

> Mensus eram coelos, nunc terrae metior umbras
> Mens coelestis erat, corporis umbra iacet.
>
> I measured the skies, now I measure the shadows
> Skybound was the mind, the body rests in the earth.

The grave has vanished, trampled under in the war.

5

THE WORLD
IN
RETROGRADE

Pure logical thinking cannot yield us any
knowledge of the empirical world; all knowledge of reality starts from experience and ends
in it. . . . Because Galileo saw this, and particularly because he drummed it into the scientific world, he is the father of modern
physics—indeed, of modern science altogether.

—Einstein

What if the Sun
Be Center to the World, and . . .
The Planet Earth, so stedfast though she seem,
Insensibly three different Motions move?

—Milton, *Paradise Lost*

History plays on the great the trick of calcifying
them into symbols; their legend becomes like the big house on the
hill, whose owner is much talked about but seldom seen. For no
scientist has this been more true than for Galileo Galilei. Galileo
dropping a cannonball and a musket ball from atop the Leaning

Tower of Pisa, thus demonstrating that objects of unequal weight fall at the same rate of acceleration, has come to symbolize the growing importance of observation and experiment in the Renaissance. Galileo fashioning the first telescope symbolizes the importance of technology in opening human eyes to nature on the large scale. Galileo on his knees before the Inquisition symbolizes the conflict between science and religion.

Such mental snapshots, though useful as cultural mnemonic devices, extract their price in accuracy. The story of Galileo at the Leaning Tower is almost certainly apocryphal. It appears in a romantic biography written by his student Vincenzio Viciani, but Galileo himself makes no mention of it, and in any event the experiment would not have worked: Owing to air resistance, the heavier object would have hit the ground first. Nor did Galileo invent the telescope, though he improved it, and applied it to astronomy. And, while Galileo was indeed persecuted by the Roman Catholic Church, and on trumped-up charges at that, he did as much as anyone outside of a few hard-core Vatican extremists to lay his body across the tracks of martyrdom.

Still, these distortions in the popular conception of Galileo work to his favor, and that would have pleased him. A devoted careerist with a genius for public relations, he was ahead of his time in more ways than one. His mission, as he put it, was "to win some fame."[1]

Galileo was born in Pisa, on February 15, 1564, twenty years after the publication of Copernicus's *On the Revolutions*. From his father, Vincenzo Galilei, a professional lute player and amateur mathematician, Galileo inherited a biting wit, a penchant for the dialogue form of argument, and a vehement distrust of authority. Vincenzo had written a book, *Dialogue of Ancient and Modern Music*, that encouraged Kepler in his search for Pythagorean harmonies. One of the characters in it utters a declaration that could have been the motto of the younger Galileo:

> It appears to me that they who in proof of any assertion rely simply on the weight of authority, without adducing any argument in support of it, act very absurdly. I, on the contrary, wish to be allowed freely to question and freely to answer you without any sort of adulation, as well becomes those who are in search of truth.[2]

Galileo prospered so long as he remained true to that independent creed. Disaster beset him once he neglected it and began demanding that questions be decided on the pronouncements of his *own* authority.

As a young man, however, Galileo waged glorious campaigns against those who, as he was to write, "think that our intellect should be enslaved to that of some other man."[3] An incandescent speaker and pamphleteer, he was known during his student days at the University of Pisa as "the wrangler" for the sarcastic aplomb with which he skewered the Scholastic professors.

At his parents' behest Galileo studied medicine, but he found little there to gratify his appetite for empirical knowledge. Medical lecturers typically taught from a volume of Galen, who had been dead for fifteen hundred years, and their laboratory sessions were hindered by a Church prohibition against dissection of human bodies. Galileo soon dropped out. He then spent four irresponsible, productive years lazing about at home, reading Virgil and Ovid, building little machines, and studying mathematics with a tutor, Ostilio Ricci, with whom he shared a devotion to the works of Archimedes.

Galileo was twenty-five years old when a scientifically inclined nobleman, Francesco Cardinal del Monte, took an interest in his abilities and got him appointed professor of mathematics at Pisa. There he lectured on astronomy, poetry, and mathematics and resumed his hectoring of the Aristotelians, at one point circulating a satirical poem poking fun at the Scholastics' habit of coming to school in togas, like little wax Aristotles. The students were delighted but the Scholastics were in the majority on the faculty, and when Galileo's contract expired he was let go.

He then managed to gain an appointment to the chair of mathematics at the University of Padua, in the free Republic of Venice.*

---

*Ruled not by a feudal aristocracy but by a thriving merchant class, Venice was relatively liberal, innovative, and inquisitive, an excellent place for a freethinker like Galileo. The difference was evident in the way the anatomy classes were conducted: The proscription against dissection, generally obeyed in Pisa, was circumvented at Padua by means of a laboratory table that could be lowered to an underground river, where corpses brought to the university by boat in the dark of night were raised into the hall for dissection in the advanced anatomy class. Proctors kept a lookout, and if the authorities approached the body was lowered away, its place was taken by the usual volume of Hippocrates or Galen, and the lecturer resumed teaching in the conventional fashion.

(Another applicant for the post was Giordano Bruno, but he was in chains by the time Galileo arrived at the university in September 1592 and was burned alive eight years later for refusing to abjure many heresies, among them his insistence that the stars are suns.) Galileo remained at Padua for eighteen years, writing, lecturing, conducting experiments, and inventing scientific instruments, among them the thermometer.

During this time his financial troubles, always onerous, became insupportable. His father had died in 1591, leaving Galileo to pay his two sisters' dowries, each of which equaled several years' worth of his university salary. In addition he was obliged to send money to his brother Michelangelo, a wandering musician who demonstrated his contempt for cash by squandering it as rapidly as he could get his hands on it. By the age of forty-five, Galileo was a respected scientist and teacher with a couple of books to his credit, but his contract was coming up for renewal, his debts were mounting, and he needed something to elevate his career from the creditable to the extraordinary. It came to him in 1609. It was the telescope.

During one of his frequent visits to nearby Venice, Galileo learned that telescopes were being constructed in Holland. Quick to grasp the principles involved, he returned home to Padua and built a telescope for himself. "Placing my eye near the concave lens," he recalled, "I perceived objects satisfactorily large and near, for they appeared three times closer and nine times larger than when seen with the naked eye alone. Next I constructed another one, more accurate, which represented objects as enlarged more than sixty times."[4]

Galileo did not need to be told that the telescope would have great practical value. Venice was an unwalled city, and its citizens depended for their defense upon their ability to spot approaching enemy ships in time to dispatch a fleet to engage them while they were still at sea; the telescope would greatly improve this early-warning system. The Venetians, furthermore, made their living from sea trade, and frequently kept an anxious watch, from the lookout towers (*campanili*) that dotted the city, for galleys returning with their holds full of cornmeal from the Levant, spices from Constantinople, and silver from Spain; an investor might be ruined if his ship were lost, or double his money once "his ship came in."

A lookout using a telescope could spot the flag flying from an incoming trading ship much sooner than with the unaided eye.

Galileo accordingly arranged a demonstration for the authorities. On August 25, 1609, he led a procession of Venetian senators across the Piazza San Marco and up the Campanile for their first look through his first telescope. As he recalled the scene:

> Very many were the patricians and senators who, although aged, have more than once climbed the stairs of the highest campanili of Venice, to detect sails and vessels on the sea, so far away that coming under full sail toward the harbor, two hours or more passed before they could be seen without my eyeglass; because in fact the effect of this instrument is to represent an object that is, for example, fifty miles off, as large and near as if it were only five miles away.[5]

The senators, suitably impressed, doubled Galileo's salary and granted him a lifelong appointment at Padua; as we would say today, Galileo got tenure. But his triumph was darkened by a cloud of deception. He permitted the senators to assume that he had *invented* the telescope. This was not strictly true, and his silence as to the stimulus of his greatest invention became embarrassing once telescopes produced by Dutch and Italian spectacle-makers began turning up in the marketplaces of Venice. In Bertolt Brecht's play *Galileo*, Priuli the Venetian curator upbraids Galileo for his guile:

CURATOR: There it is—your "miraculous optical tube." Do you know that this invention he so picturesquely termed "the fruit of seventeen years' research" will be on sale tomorrow for two scudi apiece at every street corner in Venice? A shipload of them has just arrived from Holland.

SAGREDO: Oh, dear! *Galileo turns his back and adjusts the telescope.*

CURATOR: When I think of the poor gentlemen of the Senate who believed they were getting an invention they could monopolize for their own profit. . . . Why, when they took their first look through the glass, it was only by the merest chance that they didn't see a peddler, seven times enlarged, selling tubes exactly like it at the corner of the street.[6]

But while the senators trained their telescopes on the horizon, Galileo trained his on the night skies. He was the first scientist to do so (or one of the first; Thomas Harriot in England observed the moon through a telescope that same summer) and what he saw spelled the beginning of the end of the closed, geocentric cosmos, and the opening up of the depths of space.

As beginning observers have done ever since, Galileo looked first at the moon, and the sight of its mountains and craters immediately impressed him with the fact that it was not a wafer composed of heavenly aether, but a rocky, dusty, sovereign world. Aristotle to the contrary, the moon is "not robed in a smooth and polished surface," wrote Galileo, but is ". . . rough and uneven, covered everywhere, just like the earth's surface, with huge prominences, deep valleys, and chasms."[7]

Turning his telescope to Jupiter, Galileo discovered four moons orbiting that giant planet, their positions changing perceptibly in the course of just a few hours' observation. Jupiter, he was to conclude, constituted a Copernican solar system in miniature, and proof as well that the earth is not unique in having a moon. Galileo called it

> a fine and elegant argument for quieting the doubts of those who, while accepting with tranquil mind the revolutions of the planets about the sun in the Copernican system, are mightily disturbed to have the moon alone revolve about the earth and accompany it in an annual rotation about the sun. Some have believed that this structure of the universe should be rejected as impossible. But now we have not just one planet rotating about another while both run through a great orbit around the sun; our own eyes show us the four stars [i.e., *satellites*, a term coined by Kepler] which wander around Jupiter as does the moon around the earth, while all together trace out a grand revolution about the sun in the space of twelve years.[8]

When Galileo observed the bright white planet Venus, he found that it exhibits phases like those of the moon, and that it appears much larger when in the crescent phase than when almost full. The obvious explanation was that Venus orbits the sun and not the earth, exhibiting a crescent face when it stands nearer to the earth than does the sun and a gibbous face when it is on the far side of the sun. "These things leave no room for doubt about the orbit of Venus," Galileo wrote. "With absolute necessity we shall conclude, in agreement with the theories of the Pythagoreans and of Copernicus, that Venus revolves about the sun just as do all the other planets."[9]

The greatest surprise was the stars. The telescope suggested, as the unaided eye could not, that the sky has *depth*, that the stars are not studded along the inner surface of an Aristotelian sphere, but range out deep into space. "You will behold through the telescope a host of other stars, which escape the unassisted sight, so numerous as to be almost beyond belief," Galileo reported. Moreover, the stars were organized into definite structures, of which the most imposing was the Milky Way:

> I have observed the nature and the material of the Milky Way. . . . The galaxy is, in fact, nothing but a congeries of innumerable stars grouped together in clusters. Upon whatever part of it the telescope is directed, a vast crowd of stars is immediately presented to view. Many of them are rather large and quite bright, while the number of smaller ones is quite beyond calculation.[10]

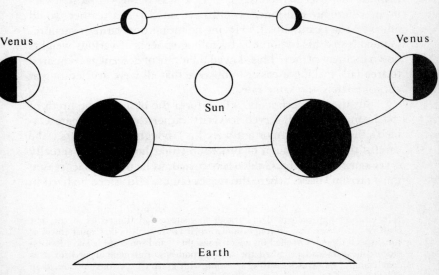

The phases of Venus, observed by Galileo through his telescope, proved that Venus lies closer to the sun than does the earth.

Galileo's account of his visions through the telescope were first published in March 1610, in his *Sidereus Nuncius*, or *Starry Messenger*. The book was an instant success, and soon readers as far away as China were reading its reports of the rocky reality of the moon, the satellites of Jupiter, and the multitude of previously unseen stars in the sky. Here was observational evidence that we live in a Copernican solar system in a gigantic universe.

Galileo, who was principally a physicist and had been a Copernican before he ever looked through a telescope, understood that the task now facing science was to bring physics into accord with the reality of a moving Earth. The old anti-Copernican arguments had been turned inside out: Given that the earth really *does* rotate on its axis, why *don't* arrows shot into the air fly off to the west, or east winds constantly blow across the land? Why, in short, does a moving Earth act *as if* it were at rest? Finding the answers to these questions would require a greatly improved understanding of the concepts of gravitation and inertia. Galileo struggled with both.

In Aristotelian physics, heavy objects were said to fall faster than light ones. Early on, probably while still at Pisa, Galileo had realized that this commonsensical view was wrong—that in a vacuum, where air resistance would have no effect, a feather would fall as fast as a cannonball.* Having no means of creating a vacuum, Galileo tested his hypothesis by rolling spheres of various weights down inclined planes. This slowed their rate of descent as compared to free fall, making it easier to observe that all were accelerating at approximately the same rate.

But these experiments, which form the basis for the myth of the Leaning Tower, served to verify rather than to instigate Galileo's thesis. More important were his "thought experiments," the careful thinking through of procedures that he could not actually carry out. To be sure, Galileo recognized, as he put it, that "reason must step in" only "where the senses fail us." But since he lived in

---

*He was not unprecedented in making this suggestion. Lucretius in the first century B.C. wrote that "through undisturbed vacuum all bodies must travel at equal speed though impelled by unequal weights," and some of Galileo's Renaissance colleagues had proposed the same hypothesis. But none argued for it as convincingly, or investigated the question with greater experimental care, than did Galileo. And, in any event, there is more to science than precedence. As Whitehead remarked, "Everything of importance has been said before by somebody who did not discover it."[11]

a time when the senses were aided by none but the most rudimentary experimental apparatus—he had, for instance, no timepiece more accurate than his pulse—Galileo found that reason had to step in rather often. In the words of Albert Einstein, the greatest all-time master of the thought experiment, "The experimental methods of Galileo's disposal were so imperfect that only the boldest speculation could possibly bridge the gaps between empirical data."[12] Consequently it was more by thinking than by experimentation that Galileo arrived at new insights into the law of falling bodies.

His reasoning went something like this: Suppose that a cannonball takes a given time—say, two pulse beats—to fall from the top of a tower to the ground. Now saw the cannonball in half, and let the two resulting demiballs fall. If Aristotle is right, each demiball, since it weighs only half as much as the full cannonball, should fall more slowly than did the original, full-size cannonball. If, therefore, we drop the two demiballs side by side, they should descend at an identical, relatively slow velocity. Now tie the demiballs together, with a bit of string or a strand of hair. Will this object, or "system," in Galileo's words, fall fast, as if it knew it were a reconstituted cannonball, or slowly, as if it still thought of itself as consisting of two half cannonballs?

Galileo phrased his *reductio ad absurdum* this way, in his *Dialogues Concerning Two New Sciences:*

> [Were Aristotle right that] a large stone moves with a speed of, say, eight while a smaller moves with a speed of four, then when they are united, the system will move with a speed of less than eight; but the two stones when tied together make a stone larger than that which before moved with a speed of eight. Hence the heavier body moves with less speed than the lighter; an effect which is contrary to [Aristotle's] supposition. Thus you see how, from your assumption that the heavier body moves more rapidly than the lighter one, I infer that the heavier body moves more slowly.[13]

This line of reasoning pointed directly to the second major question facing post-Copernican physics, that of inertia. If a cannonball and a feather fall at the same rate in a vacuum, then what is the difference between them? There must be *some* difference: The cannonball, after all, weighs more than the feather, will make more of an impression if dropped on one's head from atop the Leaning

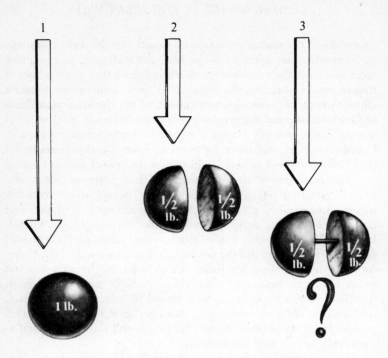

Galileo's thought experiment: According to Aristotle, if a one-pound can-nonball falls a given distance in a given time (1), then if the ball is cut in half, each half-pound ball should fall less far in the same interval (2). But, reasoned Galileo, what happens if the two half-balls are attached, by a thread or a stick (3)? Thus was Aristotle's physics of falling bodies reduced to absurdity.

Tower, and is harder to kick along the ground. We would say today that the feather and the cannonball have differing *mass*, and that the amount of their mass determines their *inertia*—their tendency to resist changes in their state of motion. It is precisely because the heavier object possesses greater inertia that it takes longer for gravity to get it going, which is why it falls no faster than the lighter object. But these are Newtonian conceptions, unknown to Galileo, who had to make his way on his own.

Aristotle had defined half of the concept of inertia, that bodies at rest tend to remain at rest. This was sufficient for dealing with an immobile Earth, but was of no use in explicating the physics of an earth in motion in a Copernican universe. Galileo groped his way toward the other half of the concept—that bodies in motion tend to remain in motion, i.e., that the cannonball's inertial mass makes it just as difficult to stop as to start. Sometimes he came close, as in his charming comparison of the residents of planet Earth with voyagers aboard a ship:

> Shut yourself up with some friend in the main cabin below decks on some large ship, and have with you there some flies, butterflies, and other small flying animals. Have a large bowl of water with some fish in it; hang up a bottle that empties drop by drop into a wide vessel beneath it. With the ship standing still, observe carefully how the little animals fly with equal speed to all sides of the cabin. The fish swim indifferently in all directions; the drops fall into the vessel beneath; and, in throwing something to your friend, you need throw it no more strongly in one direction than another, the distances being equal; jumping with your feet together, you pass equal spaces in every direction. When you have observed all these things carefully . . . have the ship proceed with any speed you like, so long as the motion is uniform and not fluctuating this way and that. You will discover not the least change in all the effects named, nor could you tell from any of them whether the ship was moving or standing still.[14]

But here Galileo bogged down. He was still a captive of Aristotle's erroneous supposition that the behavior of objects results from an internal tendency, or "desire," rather than simply from their inertial mass and the application of force:

> I seem to have observed that physical bodies have physical inclination to some motion (as heavy bodies downward), which motion is exercised by them through an intrinsic property and without need of a particular external mover, whenever they are not impeded by some obstacle. And to some other motion they have a repugnance (as the same heavy bodies to motion upward), and therefore they never move in that manner unless thrown violently by an external mover. Finally, to some movements they are indifferent, as are these same heavy bodies to horizontal motion, to which they have neither inclination . . . or repugnance. . . . And therefore, all external impediments removed, a heavy body on a spherical surface concentric with

the earth will be indifferent to rest and to movements toward
any part of the horizon. And it will maintain itself in that state
in which it has once been placed; that is, if placed in a state
of rest, it will conserve that; and if placed in movement toward
the west (for example) it will maintain itself in that movement.[15]

Some of these words anticipate Newton's explanation of inertia;
bodies "placed in movement" tend to remain in motion, those "at
rest" to remain at rest. Others remain ensnared in Aristotle's dusty
web, as when Galileo asserts that objects have an inherent "inclination" or "repugnance" for certain sorts of motion. Galileo never
really freed himself of confusion on this point, and his "law" of
falling bodies, stated in 1604 and often called the first law of classical
physics, was fraught with error.

    Galileo might have made more progress in understanding in-
ertia and gravitation had he collaborated with Kepler. Kepler, too,
had only part of the answer; he, like Galileo, thought of inertia
chiefly as a tendency of objects to remain at rest, and, consequently,
he conceived of gravity as having not only to hold planets in thrall
to the sun but also to tug them along in their orbits. But he was
ahead of Galileo in some ways, as when he proposed that the
gravitational attraction of the moon is responsible for the tides.
Galileo dismissed Kepler's theories of gravity as mere mysticism.
"I am . . . astonished at Kepler," he wrote. ". . . Despite his open
and acute mind, and though he has at his fingertips the motions
attributed to the earth, he has nevertheless lent his ear and his
assent to the moon's dominion over the waters, to occult properties,
and to such puerilities."[16]

    The differences between the two men were pronounced.
Galileo was an urbane gentleman who loved wine (which he described as "light held together by moisture"), women (he had three
children by his mistress, Marina Gamba), and song (he was an
accomplished musician). Kepler sneezed when he drank wine, had
little luck with women, and heard his music in the stars. The deep
organ-tones of religiosity and mysticism that resounded through
Kepler's works struck Galileo as anachronistic and more than a bit
embarrassing. Kepler suspected as much, and pled with Galileo to
please "not hold against me my rambling and my free way of
speaking about nature." Galileo never answered his letter. Einstein
remarked near the end of his life that "it has always hurt me to

think that Galilei did not acknowledge the work of Kepler. . . . That, alas, is vanity," Einstein added. "You find it in so many scientists."[17]

Nowhere is Galileo's disdain for Kepler more painful to recount than in the matter of the telescope. Kepler was by this time recognized as the most accomplished astronomer in the world, and his enthusiastic endorsement of Galileo's *Starry Messenger* had helped stave off criticism by those who dismissed the telescope as a kaleidoscopelike toy that produced not magnification but illusion. (This was not an entirely unreasonable suspicion; Galileo's early telescopes produced spurious colors, and they presented such a dim image, in so narrow a field of view, that it was not immediately obvious that they magnified at all.) But astronomy hereafter would require telescopes, and Kepler, though he understood the optical principles involved much better than Galileo did, could not obtain lenses of quality in Prague. With his customary earnestness and lack of restraint, Kepler wrote to Galileo in 1610, asking him for a telescope or at least a decent lens, "so that at last I too can enjoy, like yourself, the spectacle of the skies."

> O telescope, instrument of much knowledge, more precious than any scepter! . . . How the subtle mind of Galileo, in my opinion the first philosopher of the day, uses this telescope of ours like a sort of ladder, scales the furthest and loftiest walls of the visible world, surveys all things with his own eyes, and, from the position he has gained, darts the glances of his most acute intellect upon these petty abodes of ours—the planetary spheres I mean—and compares with keenest reasoning the distant with the near, the lofty with the deep.[18]

Galileo ignored Kepler's entreaties. Possibly he feared that his observations might be eclipsed by what an astronomer of Kepler's abilities could accomplish if he, too, had a telescope at hand. In any event, he had other fish to fry. He was busy parlaying his rapidly growing celebrity into a position at Cosimo de' Medici's court in Tuscany. He passed the request along to Cosimo's ambassador, who advised him to, by all means, send the estimable Kepler a spyglass. Galileo instead told Kepler that he had no telescopes to spare, and that to make a new one would require too much time. Meanwhile, he was making presents of telescopes to royal patrons whose favor might advance his career. One of the beneficiaries of Galileo's gifts, the elector of Cologne, summered in Prague that year and loaned Kepler his telescope. For one month, Kepler could gaze with delight at the craters of the moon and the stars of the Milky Way. Then the elector left town, taking the telescope with him.

Just when Galileo might have done the most to help bring physics to a Copernican maturity, he instead diverted his efforts to a quixotic campaign aimed at converting the Roman Catholic Church to the Copernican cosmology. Politics did not suit him, and soon he was demanding, like any blustering campaigner, that Copernicanism be accepted for little better reason than that he said it was correct. The old anti-Aristotelian was asking to be regarded as the new Aristotle, urging that it was now acceptable to ignore the planets in favor of the decree of a book, so long as the book was his own.

His situation grew more precarious when he abandoned the Venetian Republic for the glittering court at Tuscany, where he was named chief mathematician and philosopher to the grand duke. His friend Giovanni Sagredo warned him that he was making a mistake. "Who knows what the infinite and incomprehensible events of the world may cause if aided by the impostures of evil and envious men," he wrote Galileo in a letter from the Levant, where he was serving as the Venetian consul. ". . . I am very much worried by your being in a place where the authority of the friends of the Jesuits counts heavily."[19] But Galileo could resist neither the glory nor the wealth of the Medician court, nor the prospect of being relieved of his teaching duties at Padua: "I deem it my greatest glory to be able to teach princes," he wrote. "I prefer not to teach others."[20]

The initial reaction against Galileo's campaign came less from priests than from pedants. The reactionaries whom the world remembers for their obstinate refusal to look through his telescope —"pigeons" and "blockheads" as Galileo called them—were not clerics but professors, and they were worried less about impiety than about threats to their academic authority. The Church, initially, was more tolerant. The Vatican praised Galileo's research with the telescope and honored him with a day of ceremonies at the Jesuit Roman College, and when a Dominican monk named Thommaso Caccini preached a sermon against Galileo in Florence, he was promptly rebuked by the preacher general of the Dominican Order, Father Luigi Maraffi, who apologized to Galileo for the fact that he was sometimes obliged "to answer for all the idiocies that thirty or forty thousand brothers may or do actually commit."[21]

But Galileo's appetites had evolved from praise to power. Carried away by zeal for his cause, he began insisting that the Copernican cosmology was sufficiently well established scientifically that Scriptures must be conformed to it. Cardinal Robert Bellarmine, Master of Controversial Questions at the Roman College and the greatest theologian of the day, had reservations on this score. He agreed, he wrote in a letter dated April 4, 1615, "that, if there were a real proof that the sun is the center of the universe, that the earth is in the third sphere, and that the sun does not go round the earth but the earth round the sun, then we should have to proceed with great circumspection in explaining passages of Scripture which appear to teach the contrary. . . . But," he added, "I do not think there is any such proof since none has been shown to me."[22] In the absence of such a demonstration, Bellarmine cautioned Galileo, to teach Copernicanism as bald fact would be "a very dangerous attitude and one calculated not only to arouse all scholastic philosophers and theologians but also to injure our holy faith by contradicting the Scriptures."[23]

Galileo replied that he could prove that Copernicus was right, "but how can I do this, and not be merely wasting my time, when those Peripatetics who must be convinced show themselves incapable of following even the simplest and easiest of arguments?"[24]

This was pure sophistry. Galileo did not, in fact, have definitive proof of the Copernican theory. What he proffered instead were a series of analogies (the planets go around the sun as Jupiter's moons go around Jupiter, each is a world just as the moon evidently is a world, etc.) and the phases of Venus, which could be explained as readily by the geocentric model of Tycho as by the heliocentric model of Copernicus.

When in Rome, Galileo ridiculed the anti-Copernicans at every opportunity, and promised that he would finally reveal his irrefutable proof of the Copernican theory. This turned out to be his erroneous account of the tides—Kepler's more nearly correct theory having, as usual, been ignored by Galileo. His friends, ecclesiastical and secular alike, warned him not to press the point too far. "This is no place to come to argue about the moon," the Florentine ambassador cautioned him. Galileo persisted, regardless. "I cannot and must not neglect that assistance which is afforded to me by my conscience as a zealous Christian and Catholic," he wrote.[25]

The result of his efforts was that the pope referred the matter to the Holy Office, which declared Copernicanism contrary to Scriptures and put Copernicus's *De Revolutionibus* on the Index of forbidden books. Kepler, for once, lost patience. "Some," he fumed, "through their imprudent behavior, have brought things to such a point that the reading of the work of Copernicus, which remained absolutely free for eighty years, is now prohibited."[26]

Enjoined by the Church against espousing Copernicanism, Galileo returned to Florence and there wrote *Il Saggiatore* (*The Assayer*) a sarcastic attack on the Jesuit thinker Horatio Grassi. In doing so he added to his growing list of enemies many Jesuits who had been among his allies. (Cardinal Bellarmine, the most powerful of the Jesuits sympathetic to Galileo, had by this time died.)

In 1623, in what seemed a stroke of good fortune, Galileo's friend and admirer Maffeo Barberini was elected pope. Intelligent, vital, learned, and vain, Barberini had much in common with Galileo. As Galileo's biographer Arthur Koestler writes, the pope's "famous statement that he 'knew better than all the Cardinals put together' was only equalled by Galileo's that he alone had discovered everything new in the sky. They both considered themselves supermen and started on a basis of mutual adulation—a type of relationship which, as a rule, comes to a bitter end."[27] Galileo enjoyed six audiences with the new pope, Urban VIII, and was rewarded with lavish gifts and a declaration of "fatherly love" for "this great man, whose fame shines in the heavens."[28] Warmed by the newly risen papal sun, Galileo spent the next four years writing an exposition of the Copernican manifesto, his *Dialogue . . . Concerning the Two Chief World Systems, Ptolemaic and Copernican*. Cleared by Church censors, chief among whom now was Galileo's former pupil Father Niccoló Riccardi, it was published in 1632.

The dialogue form was a device, transparent as Aristotle's crystalline spheres, through which Galileo could argue for Copernicanism without violating the letter of the papal edict. Two of the conversants, Salviati and Sagredo, are learned gentlemen who sympathize with the Copernican scheme; they serve to speed the argument along on wheels of mutual agreement. Simplicio, the third participant, represents the Scholastics, and is presented as little better than a fool. In a typical passage, Simplicio maintains that "if the terrestrial globe must move in a year around the circumference of a circle—that is, around the zodiac—it is impossible for it at the same time to be in the center of the zodiac. But the earth is at that center, as is proved in many ways by Aristotle, Ptolemy, and others." To which Salviati, dripping sarcasm, replies: "Very well argued. There can be no doubt that anyone who wants to have the earth move along the circumference of a circle must first prove that it is not at the center of that circle."[29]

Galileo's enemies were quick to point out to the pope that the official cosmology of the Roman Catholic Church had been put into the mouth of the Simplicio the simpleton. It is Simplicio, for instance, who gives voice to a (scientifically accurate, by the way) statement that the pope had ordered inserted into the manuscript, to the effect that Galileo's theory of the tides does not prove that the earth revolves on its axis. The pope, angered, ordered an investigation, and in August 1632, the Inquisition banned further sales of the *Dialogue* and ordered all extant copies confiscated.

Galileo responded with the political naïveté that was fast becoming his hallmark. He prevailed upon his protector, the grand duke of Tuscany, to send the pope a strongly worded objection to the ban. The pope, who had been elected with the support of Francophile cardinals, was under attack from pro-Spanish factions in the Vatican—a controversy sufficiently heated that he feared assassination—and Galileo's duke supported Spain. The letter presented the pope with an irresistible opportunity to demonstrate his resolve by quashing an ally of the Francs. The only cost would be his friendship with Galileo, a brilliant but increasingly troublesome old man.

Thus was the clutch released from the wheels of persecution.* Galileo was ordered to appear before the Inquisition in Rome, either voluntarily or to be brought "to the prisons of this supreme tribunal in chains." He confidently awaited intervention by his friend the pope; it never came. He took refuge for a time in the thought that

*Pietro Redoni argues, in his book *Galileo: Heretic*, that Vatican objections to Galileo may have had less to do with Copernicanism than with his advocacy of atomism and a corpuscular theory of light. Certainly the motives behind Galileo's persecution were complicated, and are likely to be debated among historians for some time yet to come.

"everyone will understand that I have been moved to become involved in this task only by zeal for the Holy Church, and to give to its ministers that information which my long studies have brought to me." The ambassador, whose predecessor had warned him that Rome was "no place to argue about the moon," quietly acquainted Galileo with the facts of life. There would be no debate concerning the scientific merits of the Copernican system. The issue was obedience. Too late, Galileo realized his position. "He is much afflicted about it," the ambassador reported back to Florence. "I myself have seen him from yesterday to the present time so dejected that I have feared for his very life."[30]

Galileo, now seventy years old, was interrogated at length and threatened with torture. The case against him was sealed by forged "minutes" of his 1616 meeting with Cardinal Bellarmine, reporting that he had been enjoined from holding, teaching, or defending Copernicanism in any way, even as a hypothesis. This was stronger than the warning that had in truth been given him at the time. Left defenseless, Galileo took the only reasonable option available to him, and on June 22, 1633, he recited the prescribed abjuration, from his knees, in the great hall of the Dominican convent of Santa Maria Sopre Minera:

> Wishing to remove from the minds of your Eminences and of every true Christian this vehement suspicion justly cast upon me, with sincere heart and unfeigned faith I do abjure, damn, and detest the said errors and heresies, and generally each and every other error, heresy, and sect contrary to the Holy Church; and I do swear for the future that I shall never again speak or assert, orally or in writing, such things as might bring me under similar suspicion. . . .[31]*

Galileo spent the remaining eight years of his life under house arrest in his villa outside Florence. There he wrote his finest book, the *Dialogues Concerning Two New Sciences*, a study of motion and

---

*Three centuries later, in 1980, Pope John Paul II ordered a reexamination of the case of Galileo. Speaking at a ceremony honoring the centenary of Einstein's birth, the Pope declared that Galileo had "suffered at the hands of men and institutions of the Church," adding that "research performed in a truly scientific manner can never be in contrast with faith because both profane and religious realities have their origin in the same God."[32]

inertia. His daughter Sister Marie Celeste, whom he had sent to a convent against her wishes twenty-three years earlier, stayed with him and said the seven daily psalms of penitence ordered by the Holy Office as part of his sentence. He observed the moon and planets through his telescope up until only a few months before he lost his sight, in 1637. "This universe that I have extended a thousand times . . . has now shrunk to the narrow confines of my own body," he wrote.[33]

Milton visited Galileo, and may have gained from him something of the sense of vast spaces that permeates *Paradise Lost*. Milton's universe, however, remained earth-centered, and his poem contains a warning against cosmological presumption. In it, a Miltonic angel advises Adam:

> Sollicit not thy thoughts with matters hid,
> Leave them to God above, him serve and feare;
> Of other Creatures, as him pleases best,
> Wherever plac't, let him dispose: joy thou
> In what he gives to thee, this Paradise
> And thy fair Eve: Heav'n is for thee too high
> To know what passes there; be lowlie wise:
> Think onely what concernes thee and thy being;
> Dream not of other Worlds.[34]

But *that* paradise had indeed been lost. Humankind was awakening from a dream of immobility to find itself in a waking fall, its planet plummeting through boundless space. The weight of authority that brought Galileo to his knees succeeded only in halting the growth of science in the Mediterranean. Thereafter, the great advances came in the north countries. The physics of the Copernican universe was to be elucidated by Isaac Newton, born in Woolsthorpe, Lincolnshire, on Christmas Day, 1642, the year of Galileo's death.

6

NEWTON'S
REACH

Watch the stars, and from them learn.
To the Master's honor all must turn,
each in its track, without a sound,
forever tracing Newton's ground.*
                                            —Einstein

Nearer the gods no mortal may approach.
                                    —Edmond Halley,
                                    on Newton's *Principia*

Newton created a mathematically quantified ac-
count of gravitation that embraced terrestrial and celestial phenomena alike. In doing so he demolished the Aristotelian bifurcation of
the universe into two realms, one above and one below the moon,
and established a physical basis for the Copernican universe. The
thoroughness and assurance with which he accomplished this task
were such that his theory came to be regarded, for more than two
centuries thereafter, as something close to the received word of
God. Even today, when Newtonian dynamics is viewed as but a
part of the broader canvas painted by Einstein's relativity, most of
us continue to think in Newtonian terms, and Newton's laws still

*Translation by Dave Fredrick.

work well enough to guide spacecraft to the moon and planets. ("I think Isaac Newton is doing most of the driving now," said astronaut Bill Anders, when asked by his son who was "driving" the Apollo 8 spacecraft carrying him to the moon.)

Yet the man whose explication of the cosmos lives on in a billion minds was himself one of the strangest and most remotely inaccessible individuals who ever lived. When John Maynard Keynes purchased a trunk full of Newton's papers at auction, he was startled to find that it was full of notes on alchemy, biblical prophecy, and the reconstruction from Hebraic texts of the floor plan of the temple of Jerusalem, which Newton took to be "an emblem of the system of the world." "Newton was not the first of the age of reason," a shaken Keynes told a gathering at the Royal Society. "He was the last of the magicians, the last of the Babylonians and Sumerians."[1] Newton was isolated, too, by the singular power of his intellect. Richard Westfall spent twenty years writing a highly perceptive scholarly biography of Newton, yet confessed, in the first paragraph of its preface, that

> The more I have studied him, the more Newton has receded from me. It has been my privilege at various times to know a number of brilliant men, men whom I acknowledge without hesitation to be my intellectual superiors. I have never, however, met one against whom I was unwilling to measure myself, so that it seemed reasonable to say that I was half as able as the person in question, or a third or a fourth, but in every case a finite fraction. The end result of my study of Newton has served to convince me that with him there is no measure. He has become for me wholly other, one of the tiny handful of supreme geniuses who have shaped the categories of the human intellect, a man not finally reducible to the criteria by which we comprehend our fellow beings.[2]

Newton was an only child, the posthumous son of an illiterate yeoman. Born prematurely—so small, his mother used to say, that he could have fit in a quart bottle—he was not expected to survive. His mother, a widow with a farm to manage, soon remarried, and her new husband, the Reverend Barnabus Smith, sent the child off to be raised by his maternal grandmother; there he grew up, only a mile and a half away, within sight of the house where dwelt his loving mother and usurping stepfather. The product of all this—

a fatherless birth on Christmas Day, survival against the odds, separation from his mother, and possession of a mind so powerful that he was as much its vassal as its master—was a brooding, simmering boy, sullen and bright and quick to anger. At age twenty Newton compiled a list of his youthful sins; among them were "threatening my father and mother Smith to burne them and the house over them," "peevishness with my mother," "striking many," and "wishing death and hoping it to some."[3]

The young Newton was as sensitive to the rhythms of nature as he was indifferent to those of men. As a child he built clocks and sundials and was known for his ability to tell time by the sun, but he habitually forgot to show up for meals, a trait that persisted throughout his life, and he was far too fey to help out reliably on the farm. Sent to gather in livestock, he was found an hour later standing on the bridge leading to the pasture, gazing fixedly into a flowing stream. On another occasion he came home trailing a leader and bridle, not having noticed that the horse he had been leading had slipped away. A sometime practical joker, he alarmed the Lincolnshire populace one summer night by launching a hot-air flying saucer that he constructed by attaching candles to a wooden frame beneath a wax paper canopy.* He seldom studied and customarily fell behind at grammar school, but applied himself at the end of each term and surpassed his classmates on final examinations, a habit that did little to enhance his popularity. A contemporary of Newton's reported that when the boy left for Cambridge, the servants at Woolsthorpe Manor "rejoiced at parting with him, declaring, he was fit for nothing but the 'Versity."[4]

At college he filled his lonely life with books. "*Amicus Plato amicus Aristoteles magis amica veritas,*" he wrote in his student notebook—"Plato is my friend, Aristotle is my friend, but my greatest friend is truth."[5] He seems to have made the acquaintance of only one of his fellow students, John Wickins, who found him walking in the gardens "solitary and dejected" and took pity on him. Newton's studies, like those of many a clever undergraduate, were eclectic—he looked into everything from universal languages to perpetual motion machines—but he pursued them with a unique intensity. Nothing, least of all his personal comfort, could deter him when he was on to a question of interest: To investigate the

*I have tried this one myself and can testify that, like many of Newton's inventions, it works very well indeed.

anatomy of the eye he stuck a bodkin "betwixt my eye and the bone as near to the backside of my eye as I could," and he once stared at the sun for so long that it took days of recuperation in a dark room before his vision returned to normal.

For a time he drew inspiration from the books of René Descartes, a kindred spirit. Descartes like Newton had been a frail child, brought up by his grandmother, and both men were seized by an all-embracing vision while in their early twenties: Newton's epiphany was universal gravitation; Descartes's involved nothing less than a science of all human knowledge. Descartes died in 1650, more than a decade before Newton arrived at Cambridge, but his works were very much alive among the "brisk part" of the faculty—those whose intellectual horizons were not bounded by Aristotle's.*

But if Newton learned a great deal from Descartes's *Principia Philosophiae*—which included, among many other things, an assertion that inertia involves resistance to changes in motion and not just to motion itself—he was always happiest in contention, and Descartes's philosophy promoted in him an equal and opposite reaction. Descartes's disapproval of atomism helped turn Newton into a confirmed atomist. Descartes's vortex theory of the solar system became the foil for Newton's demonstration that vortices could not account for Kepler's laws of planetary motion. Descartes's emphasis on depicting motion algebraically encouraged Newton to develop a dynamics written in terms of algebra's alternative, geometry; as this was not yet mathematically feasible, Newton found it necessary to invent a new branch of mathematics, the calculus. Infinitesimal calculus set geometry in motion: The parabolas and hyperbolas Newton drew on the page could be analyzed as the product of a moving point, like the tip of the stick with which Archimedes drew figures in the sand. As Newton put it, "Lines are described, and thereby generated not by the opposition of parts, but by the continued motion of points." Here the unbending Newton danced.

---

*A devotee of warmth who had experienced his transcendent moment in an overheated room he called "the oven," Descartes succumbed at age fifty-two to the impetuous attentions of the twenty-three-year-old Queen Christina of Sweden, who insisted that he brave the Nordic chill to tutor her in science and philosophy each morning at five. The less accommodating Newton declined most invitations, never traveled abroad, and lived to be eighty-five.

Newton had completed this work by the time he received his bachelor's degree, in April 1665. It would have established him as the greatest mathematician in Europe (and as the most accomplished undergraduate in the history of education) but he published none of it. Publication, he feared, might bring fame, and fame abridge his privacy. As he remarked in a letter written in 1670, "I see not what there is desirable in public esteem, were I able to acquire and maintain it. It would perhaps increase my acquaintance, the thing which I chiefly study to decline."[6]

Soon after his graduation the university was closed owing to an epidemic of the plague, and Newton went home. There he had ample time to think. One day (and it seems quite plausibly to have dawned on him all at once) he hit upon the grand theory that had eluded Kepler and Galileo—a single, comprehensive account of how the force of gravitation dictates the motion of the moon and planets. As he recounted it:

> In those days I was in the prime of my age for invention & minded Mathematics & Philosophy more than at any time since. . . . I began to think of gravity extending to the orb of the Moon & . . . from Kepler's rule of the periodical times of the Planets being in sesquialterate proportion of their distances from the center of their Orbs, I deduced that the forces which keep the planets in their Orbs must [be] reciprocally as the squares of their distances from the centers about which they revolve: & thereby compared the force requisite to keep the Moon in her Orb with the force of gravity at the surface of the earth, & found them answer pretty nearly.[7]

Newton is said to have recalled, near the end of his life, that this inspiration came to him when he saw an apple fall from the tree in front of his mother's house. The story may be true—Newton's desk in his bedroom looked out on an apple orchard, and even a Newton must occasionally have interrupted his work to gaze out the window—and it serves, in any event, to trace how he arrived at a quantitative description of gravitation that drew together the physics of the heavens and the earth.

Suppose, as Newton did that day, that the same gravitational force responsible for the apple's fall extends "to the orb of the Moon," and that its force decreases by the square of the distance over which it propagates.* The radius of the earth is 4,000 miles, meaning that Newton and his apple tree were located 4,000 miles from a point at the center of the earth from which (and this was one of Newton's key insights) the gravitational force of the earth emanates. The moon's distance from the center of the earth is 240,000 miles—60 times farther than that of the apple tree. If the inverse-square law holds, the falling apple should therefore experience a gravitational force 60^2, or 3,600, times stronger than does the moon. Newton assumed, from the principle of inertia, that the moon would fly away in a straight line, were it not constantly tugged from that path by the force of the earth's gravity. He calculated how far the moon "falls" toward the earth—i.e., departs from a straight line in order to trace out its orbit—every second. The answer was 0.0044 feet per second. Multiplying 0.0044 by 3,600 to match the proposed strength of gravitation at the earth's surface yielded 15.84 feet per second, or "pretty nearly" the 16 feet per second that an apple, or anything else, falls on Earth. This agreement confirmed Newton's hypothesis that the same gravitational force that pulls the apple down pulls at the moon, too.

Having done the calculation, Newton silently set it aside. Various explanations can be offered for his quietude: The calculations fit "pretty nearly" but not perfectly, owing to inaccuracies in the estimated distance to the moon; Newton was interested in other matters, among them the binomial series and the nature of color; and, in any event, he seldom felt any impulse to call attention to himself: He didn't publish the calculus, either, for twenty-seven years, and then anonymously.

---

*This, the "inverse square" law, can be arrived at intuitively if we imagine the force of gravity as being spread out across the surface of a sphere. Consider two planets orbiting a star in such a way that the distance of planet B from the star is twice that of planet A. Let each planet rest on the surface of an imaginary sphere centered on the star. Since the radius of the sphere encompassing the orbit of planet B is twice that for planet A, its surface area is equal to the square of the surface area of planet B's sphere. (The area of the surface of a sphere equals 4 pi r^2, where r is the radius of the sphere.) This means that the total amount of gravitational force emanating from the star must be spread out over sphere B with a surface equal to that of sphere A squared. The gravitational force experienced by planet B will, therefore, be the inverse square of that experienced by planet A. Newton derived this much from Kepler's third law, but Kepler himself had failed to obtain it, evidently because he thought of gravitation as being propagated in only two dimensions, not three.

The young Newton's realization on universal gravitation went as follows: If the moon is 60 times as far from the center of the earth as is the apple (4,000 miles for the apple, 240,000 miles for the moon), and gravitation diminishes by the square of the distance, then the apple is subject to a gravitational force 60^2, or 3,600, times that experienced by the moon. The moon, therefore, should "fall" along the curve of its orbit $\frac{1}{3,600}$th as far each second as does the apple. And so it does (time AB = time CD).

A few academic colleagues did become acquainted with elements of Newton's research, however, and two years after returning to Trinity College, Cambridge, he was named Lucasian Professor of Mathematics. (The position had been vacated by his favorite teacher, the blustery and witty mathematician Isaac Barrow, who left to take up divinity studies and died seven years later of an opium overdose.) But Newton the teacher had little more in common with his colleagues than had Newton the student. Numerous among the professors were the so-called "wet epicures," their lives spent, wrote the satirist Nicholas Amherst, "in a supine and regular course of eating, drinking, sleeping, and cheating the juniors."[8]

Others were known as much for their eccentricities as for their scholarship; the master of Trinity, for one, was an effeminate shut-in who kept enormous house spiders in his rooms as pets. Not that Newton had any difficulty holding his own when it came to idio-syncrasies. Gaunt and disheveled, his wig askew, he dressed in run-down shoes and soiled linen, seldom stopped working, and frequently forgot to sleep. Once, puzzling over why he seemed to be losing his mental agility while working on a problem, he reflected on the matter, realized that he had not slept for days, and reluctantly went to bed. He forgot to eat as well, often rising from his desk at dawn to breakfast on the congealed remains of the dinner that had been brought to him and left untouched the night before. His rare efforts at conviviality fared poorly; one night while entertaining a few acquaintances he went to his room to fetch a bottle of wine, failed to return, and was found at his desk, hunched over his papers, wine and guests forgotten.

As the years passed, Newton elaborated the calculus, advanced the art of analytical geometry, did pioneering work in optics, and conducted innumerable experiments in alchemy (possibly poisoning himself in the process; some of the symptoms of a mental breakdown he suffered in 1693 are consistent with those of acute mercury toxemia). All this he did in silence. Occasionally he reported on his research in his lectures, but few of the professors and fewer among the students could follow his train of thought, and so few came. Sometimes nobody at all showed up, whereupon Newton, confronted with the empty hall, would trudge back to his rooms, evidently unperturbed.

The outer world eventually intruded nonetheless. In the case of Newton, who shunned notoriety, as in that of Galileo, who welcomed it, the agency responsible was the telescope.

Newton was handy, and liked to build experimental devices. (A good thing, said a colleague, for he took no exercise and had no hobbies and would otherwise have killed himself with overwork.) He wanted a telescope with which to observe comets and the planets. The only type of telescope in use at the time was the refractor— the sort that Galileo built, with a large lens at the front end to gather light. Newton disliked refractors; his extensive studies of optics had acquainted him with their tendency to introduce spurious colors. To overcome this defect he invented a new kind of telescope, one that employed a mirror rather than a lens to collect light.

Efficient, effective, and cheap, the "Newtonian reflector" was to become the most popular telescope in the world. It brought Newton's name to the attention of the Royal Society of London, which elected him to membership and prevailed upon him to publish a short paper he had written on colors. This decision he soon regretted; the paper drew twelve letters, prompting Newton to complain to Henry Oldenburg, the society's secretary, that he had "sacrificed my peace, a matter of real substance."[9]

The Royal Society was the most influential of the several scientific societies that had sprung up in the seventeenth century, each devoted to the empirical study of nature without interference by Church or State. The first of these, the Italian Academy of the Lynx, was founded in 1603 and had formed a platform from which Galileo, its most famous member, conducted his polemics. Founded under the amateur physicist King Charles II, the Royal Society

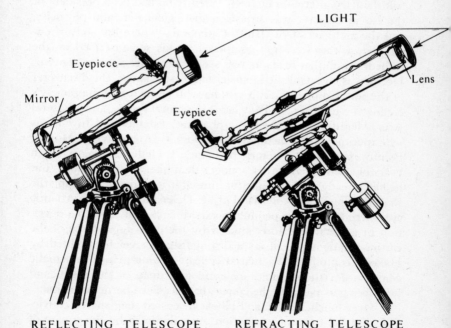

REFLECTING TELESCOPE   REFRACTING TELESCOPE

Reflecting telescopes gather light by means of a curved mirror, refracting telescopes by a curved lens.

was too poor to afford a laboratory or even an adequate headquarters, but was fiercely independent and proudly unfettered by tradition or superstition. Its temper had been expressed by Oldenburg in a letter to the philosopher Benedict Spinoza:

> We feel certain that the forms and qualities of things can best be explained by the principles of mechanics, and that all effects of Nature are produced by motion, figure, texture, and the varying combinations of these and that there is no need to have recourse to inexplicable forms and occult qualities, as to a refuge from ignorance.[10]

This clear new cast of mind was personified by the three members of the Royal Society—Edmond Halley, Christopher Wren, and Robert Hooke—who lunched together in a London tavern one cold January afternoon in 1684. Wren, who had been president of the Royal Society, was an astronomer, geometer, and physicist, and the architect of St. Paul's Cathedral—where his body is entombed, with an epitaph composed by his son inscribed on the cathedral wall that reads, IF YOU SEEK A MONUMENT, LOOK AROUND. Hooke was an established physicist and astronomer, the discoverer of the rotation of Jupiter; it was he who had worded the society's credo: "To improve the knowledge of natural things, and all useful Arts, Manufactures, Mechanic practices, Engines and Inventions by Experiments (not meddling with Divinity, Metaphysics, Morals, Politics, Grammar, Rhetoric or Logic)."[11] Halley at twenty-seven years old was a generation younger than his two companions, but he had already made a name for himself in astronomy, charting the southern skies from the island of St. Helena in the South Atlantic and there conducting pendulum experiments that showed a deviation in gravitational force caused by the centrifugal force of the earth's rotation. Ahead lay a distinguished career highlighted by Halley's compiling of actuarial tables, drawing maps of magnetic compass deviations and a meteorological map of the earth, and identifying as periodic the comet that has since borne his name.

Over lunch, Halley and Hooke discussed their shared conviction that the force of gravitation must diminish by the square of the distance across which it is propagated. They felt certain that the inverse-square law could explain Kepler's discovery that the planets move in elliptical orbits, each sweeping out an equal area within its orbit in an equal time. The trouble was, as Halley noted, that he could not demonstrate the connection mathematically. (Part of the problem was that nobody, except the silent Newton, had realized that the earth's gravitational force could be treated as if it were concentrated at a point at the center of the earth.) Hooke brashly asserted that he had found the proof, but preferred to keep it a secret so that others might try and fail and thus appreciate how hard it had been to arrive at it. Perhaps he meant to echo Descartes's *Geometry*, which ends with the infuriating declaration that the author has "intentionally omitted" elements of his proofs "so as to leave to others the pleasure of discovery."[12] In any event, Wren had his doubts about Hooke's mathematical ability if not Descartes', and he offered as a prize to Hooke or Halley a book worth up to forty shillings—an expensive book—if either could produce such a demonstration within two months. Hooke immediately agreed, but the two months passed and he failed to come up with the proof. Halley tried, and failed, but kept thinking about the matter.

The man who might be able to answer it, he realized, was Newton. Newton was forbidding, to be sure; his amanuensis, Humphrey Newton (no relation), said he had seen his master laugh only once in five years, this when Newton inquired of an acquaintance what he thought of a copy of Euclid he had loaned him, and the man asked what use or benefit its study might be in his life, "upon which Sir Isaac was very merry."[13] But when the two men had met a couple of years earlier, Newton pumping Halley for data on the great comet of 1680, they had got along reasonably well. So, in August, while visiting Cambridge, Halley stopped in to see Newton again.

What, Halley asked Newton, would be the shape of the orbits of the planets if the gravitational force holding them in proximity to the sun decreased by the square of their distance from the sun?

An ellipse, Newton answered without hesitation.

Halley, in a state of "joy and amazement" as Newton recalled the moment, asked Newton how he knew this answer to be true.

Newton replied that he had calculated it.

Halley asked if he might see the calculation.

Newton searched through some of the stacks of papers that littered his quarters. There were thousands of them. Some bore the spiderweb tracings of his diagrams in optics. Others, adorned with medieval symbols and ornate diagrams of the philosophers' stone, recorded his explorations of alchemy. A paper crammed with columns of notes compared twenty different versions of the Book of Revelations, part of the theological research Newton had conducted in substantiating his opposition to the doctrine of the Trinity—this a deep secret for the Lucasian Professor of Mathematics at Trinity College. Other pages were devoted to Newton's attempts to show that the Old Testament prophets had known that the universe is centered on the sun, and that the geocentric cosmology upheld by the Roman Catholic Church was therefore a corruption. But, Newton said, he could not find his calculations connecting the inverse-square law to Kepler's orbits. He told Halley he would write them out anew and send them to him.

Newton had calculated elliptical orbits five years earlier, upon his return from a stay of nearly six months at his mother's farm in Woolsthorpe, where he had gone when he learned that she had fallen mortally ill with a fever. His behavior there displayed a tenderness we do not normally associate with this frosty man: "He attended her with a true filial piety, sat up whole nights with her, gave her all her Physic himself, dressed all her blisters with his own hands, and made use of that manual dexterity for which he was so remarkable to lessen the pain which always attends the dressing," reported John Conduitt, who wrote notes for a memoir on Newton's life.[14] The semiliterate Hannah Newton Smith could not have understood much of what her firstborn son did and thought, but her devotion to him was unwavering. A letter she wrote him shortly before his graduation from Cambridge survives; one edge has been burned away (perhaps by Newton, who destroyed many of his papers) and a few words are missing, but what remains contains the word "love" three times in two lines:

> *Isack*
>
> *received your leter and I perceive you*
> *letter from mee with your cloth but*
> *none to you your sisters present thai*
> *love to you with my motherly lov*
> *you and prayers to god for you I*
> *your loving mother*
>
> *hanah*[15]

She was buried on June 4, 1679. Conduitt described her as a woman of "extraordinary . . . understanding and virtue."

When Newton returned to Cambridge after his mother's death, he returned as well to the study of universal gravitation. He had paid little attention to the problem since the time, years before, when he had watched the apple fall outside the window of his room in his mother's farmhouse. But now he was blessed with an antagonist—none other than Hooke himself, the tight-lipped claimant to the inverse-square law, who had written him with questions concerning the trajectory described by an object falling straight toward a gravitationally attractive body. Newton, aloof as usual, replied by declining Hooke's invitation to correspond further, but took the trouble to answer Hooke's questions, and in so doing made a mistake. Hooke seized upon the error, pointing it out in a letter of reply. Furious at himself, Newton concentrated on the matter for a time, and in the process verified to his own satisfaction that gravity obeying an inverse-square law could be shown to account for the orbits of the planets. Then he put his calculations aside. These were the papers he referred to when Halley came calling.

But they, too, turned out to contain an error—which may explain why the cautious Newton said he was "unable" to find them in the first place—and so Newton was obligated to resume work on the problem in order to satisfy his promise to Halley. This he did, and three months later, in November, he sent Halley a paper that successfully derived all three of Kepler's laws from the precept of universal gravitation obeying an inverse-square law. Halley, immediately recognizing the tremendous importance of Newton's accomplishment, hastened to Cambridge and urged him to write a book on gravitation and the dynamics of the solar system. Thus was born Sir Isaac Newton's *Mathematical Principles of Natural Philosophy and His System of the World*—the *Principia*.

Work on the book took over Newton's life. "Now I am upon this subject," he wrote the astronomer John Flamsteed, in a letter soliciting data on the orbits of Saturn's satellites, "I would gladly know the bottom of it before I publish my papers."[16] The effort only intensified his air of preoccupation. His amanuensis Humphrey Newton observed that

> he ate very sparingly, nay, ofttimes he has forget [*sic*] to eat at all, so that going into his Chamber, I have found his Mess

> untouched of which when I have reminded him, [he] would
> reply, Have I; and then making to the Table, would eat a bit
> or two standing. . . . At some seldom Times when he design'd
> to dine the Hall, would turn to the left hand, & go out into
> the street, where making a stop, when he found his Mistake,
> would hastily turn back, & then sometimes instead of going
> into the Hall, would return to his Chamber again.[17]

Newton still wandered alone in the gardens, as he had since his
undergraduate days, and when fresh gravel was laid in the walks
he drew geometric diagrams in it with a stick (his colleagues carefully stepping around the diagrams so as not to disturb them). But
now his walks were more often interrupted by bolts of insight that
sent him running back to his desk in such haste, Humphrey Newton
noted, that he would "fall to write on his Desk standing, without
giving himself the Leisure to draw a Chair to sit down in."[18]

Newton's surviving drafts of the *Principia* support Thomas
Edison's dictum that genius is one percent inspiration and ninety-
nine percent perspiration. Like Beethoven's drafts of the opening
bars of the Fifth Symphony, they are characterized less by sudden
flashes of insight than by a constant, indefatigable hammering away
at immediate, specific problems; when Newton was asked years
later how he had discovered his laws of celestial dynamics, he
replied, "By thinking of them without ceasing."[19] Toil was trans-
muted into both substance and veneer, and the finished manuscript,
delivered to Halley in April 1686, had the grace and easy assurance
of a work of art. For the modern reader the *Principia* shares with
a few other masterworks of science—Euclid's *Elements* among them,
and Darwin's *Origin of Species*—a kind of inevitability, as if its conclusions were self-evident. But the more we put ourselves into the
mind-set of a seventeenth-century reader, the more it takes on the
force of revelation. Never before in the history of empirical thought
had so wide a range of natural phenomena been accounted for so
precisely, and with such economy.

Gone forever was Aristotle's misconception that the dynamics
of objects depended upon their elemental composition, so that water,
say, had a different law of motion from fire. In the Newtonian
universe every object is described by a single quantity, its *mass*—
Newton invented this concept—and mass possesses *inertia*, the
tendency to resist any change in its state of motion. This is New-

ton's first law—that "every body perseveres in its state of rest, or of uniform motion in a right [i.e., straight] line, unless it is compelled to change that state . . ."[20]

Whenever an immobile object is set into motion, or a moving object changes its velocity or direction of motion, Newton infers that a *force* is responsible. Such a change may be expressed as *acceleration*, the rate of change of velocity with time. This is Newton's second law—that force equals mass times acceleration:

$$F = ma$$

The price paid for the application of force is that the action it produces must also result in an equal and opposite reaction. Thus, Newton's third law—that "to every action there is always opposed an equal reaction."[21]

Applied to the motions of the planets, these concepts explicated the entire known dynamics of the solar system. The moon circles the earth; the law of inertia tells us that it would move in a straight line unless acted upon by an outside force; as it does *not* move in a straight line, we can infer that a force—gravity—is responsible for bending its trajectory into the shape of its orbit. Newton demonstrates that gravitational force diminishes by the square of the distance, and establishes that this generates Kepler's laws of planetary motion. It is because gravitation obeys the inverse-square law that Halley's Comet or the planet Mars moves rapidly when near the sun and moves more slowly when far from the sun, sweeping out equal areas along its orbital plane in equal times. The amount of gravitational force exerted by each body is directly proportional to its mass. (From these considerations Newton was able to account for the tides as being due to the gravitational tug of both the sun and the moon, thus clearing up Galileo's confusion on that score.)

From Newton's third law (for every action an equal and opposite reaction) we can deduce that gravitational force is *mutual*. The earth not only exerts a gravitational force on the moon, but is subjected to a gravitational force *from* the moon. The mutuality of gravitational attraction introduces complexities into the motions of the planets. Jupiter, for instance, harbors 90 percent of the mass of all the planets, and so perturbs the orbits of the nearby planet Saturn to a degree "so sensible," Newton comments dryly, "that astronomers are puzzled with it." With the publication of the *Principia*, their puzzlement was at an end. Newton had provided the key to deciphering all observed motion, whether cosmic or mundane.

Halley had to exert himself to get the *Principia* published in financially thirsty times. The Royal Society had taken a loss the year before by publishing John Ray's *History of Fishes*, a handsome book that nevertheless had not exactly flown from the booksellers' shelves. Unsold copies lay stacked in the society storeroom, and at one point, Halley's salary was being paid in copies of the *History of Fishes*. Further complications arose when Hooke proposed, groundlessly, that Newton had stolen the theory of universal gravitation from him, and Newton responded by threatening to leave the *Principia* unfinished by omitting Part Three, a more popularized section that Halley hoped would "much advance the sale" of the book.*

But Halley persisted, paying the printing costs out of his own pocket, and the *Principia* appeared in 1687, in an edition of some three or four hundred copies. The book was (and is) difficult to read, owing in part to Newton's having, as he told his friend William Derham, "designedly made his *Principia* abstruse . . . to avoid being baited by little Smatterers in Mathematicks."[22] But Halley promoted it tirelessly, sending copies to leading philosophers and scientists throughout Europe, presenting King James II with a gloss of it, and going so far as to review it himself, for the *Philosophical Transactions of the Royal Society*. Thanks in large measure to his efforts, the *Principia* had a resounding impact. Voltaire wrote a popular account of it, and John Locke, having verified with Christian Huygens that Newton's mathematics could be trusted, mastered its contents by approaching it as an exercise in logic. Even those who could not understand the book were awed by what it accomplished; the Marquis de l'Hopital, upon being presented with a copy by Dr. John Arbuthnot, "asked the Doctor every particular thing about Sir Isaac," recalled a witness to their exchange, "even

---

*"He was of an active, restless, indefatigable Genius even almost to the last, and always slept little to his death, seldom going to Sleep till two three, or four a Clock in the Morning, and seldomer to Bed, often continuing his Studies all Night, and taking a short Nap in the Day. His Temper was Melancholy. . . ." Sound familiar? That's Hooke, not Newton, as described by a contemporary. Inevitably, we tend to quarrel most bitterly with those who most nearly resemble ourselves.

to the color of his hair, said does he eat & drink & sleep. Is he like other men?"[23]

The answer, of course, was no. Newton was a force of nature, brilliant and unapproachable as a star. "As a man he was a failure," wrote Aldous Huxley, "as a monster he was superb." We remember the monster more than the man, and the specter of a glacial Newton portraying the universe as a machine has furthered the impression that science itself is inherently mechanical and inhuman. Certainly Newton's personality did little to alleviate this misconception. Indifferent to the interdependence of science and the humanities, Newton turned a deaf ear to music, dismissed great works of sculpture as "stone dolls," and viewed poetry as "a kind of ingenious nonsense."[24]

He spent his last forty years in the warming and stupefying embrace of fame, his once lean face growing pudgy, the dark luminous eyes becoming puffy, the wide mouth hardening from severity to petulance. His penetrating gaze and unyielding scowl became the terror of the London counterfeiters he enjoyed interrogating as warden of the mint, sending many to the gallows. He denied requests for interviews submitted by the likes of Benjamin Franklin and Voltaire. He was friendlier with Locke, with whom he studied the Epistles of Saint Paul, and with the diarist Samuel Pepys, who had been president of the Royal Society, but alarmed them when in 1693 he succumbed to full-scale insomnia and suffered a mental breakdown, writing them strange, paranoid letters in a spidery scrawl in which he implied that Pepys was a papist and told Locke that "being of opinion that you endeavoured to embroil me with woemen & by other means I was so much affected with it as that when one told me you were sickly & would not live I answered twere better if you were dead."[25] Newton was confined to bed by friends who, unable otherwise to assess the health of an intellect so far above the timberline, judged him well when at last he regained the ability to make sense of his own *Principia*. Elected to Parliament, he is said during the 1689–1690 session to have spoken but once, when, feeling a draft, he asked an usher to close the window. He died a virgin.

Newton cast a long shadow, and is said to have retarded the progress of science by seeming to settle matters that might otherwise have been further investigated. But he himself was acutely aware that the *Principia* left many questions unanswered, and he was forthright in confronting them.

Of these, none was more puzzling than the mystery of gravitation itself. If nature operated according to cause and effect, its paradigm the cue ball that scatters the billiard balls, then how did the force of gravitation manage to make itself felt across gulfs of empty space, without benefit of any medium of contact between the planets involved? This absence of a causal explanation for gravity in Newton's theory prompted sharp criticism: Leibniz branded Newton's conception of gravity "occult," and Huygens called it "absurd."

Newton agreed, calling the idea of gravity acting at a distance "so great an absurdity, that I believe no man who has in philosophical matters a competent faculty of thinking, can ever fall into it,"[26] and conceding that he had no solution to the riddle: "The Cause of Gravity is what I do not pretend to know," he said.[27] In the *Principia* appears his famous phrase *Hypotheses non fingo*—"I have not been able to discover the cause of those properties of gravity from phenomena, and [so] I frame no hypothesis."[28] He would have approved of the quatrain that adorned one of his portraits:

> See the great Newton, he who first surveyed
> The plan by which the universe was made;
> Saw Nature's simple yet stupendous laws,
> And proved the effects, though not explained the cause.

One might say, then, that evidence of Newton's genius survives in his questions as well as in his answers. Human understanding of gravitation has been greatly improved by Einstein's conception of gravity as a manifestation of the curvature of space, but the road to full comprehension still stretches on ahead; its next, dimly perceived way station is thought to be a hyperdimensional unified theory or a quantum account of general relativity. Until that goal is achieved, and perhaps even thereafter, Newton's prudent tone will remain the byword of gravitational physics.

Newton was equally straightforward in pointing out that he could not hope to calculate all the minute variations in the orbits of the planets produced by their mutual gravitational interactions. As he put it in the *Principia*:

The orbit of any one planet depends on the combined motion
of all the planets, not to mention the action of all these on each
other. But to consider simultaneously all these causes of motion
and to define these motions by exact laws allowing of convenient calculation exceeds, unless I am mistaken, the force of the
entire human intellect.[29]

Today this is known as the "many body problem," and it remains
unsolved, just as Newton foresaw. Calculation of the precise interactions of all the planets in the solar system—much less that of
all the stars in the Milky Way—may as Newton prophesied forever
elude "the force of the entire human intellect," or it may one day
yield, if not to the mind, then to the inhuman power of giant
electronic computers. No one knows. For now, let Einstein pronounce Newton's eulogy: *"Genug davon. Newton verzeih' mir,"* Einstein wrote, in his "Autobiographical Notes," after discussing
weaknesses in Newton's assumptions:

> Enough of this. Newton, forgive me; you found the only
> way which, in your age, was just about possible for a man of
> highest thought and creative power. The concepts, which you
> created, are even today still guiding our thinking physics, al-
> though we now know that they will have to be replaced by
> others farther removed from the sphere of immediate experi-
> ence, if we aim at a profounder understanding.[30]

In any case, the ultimate unsolved questions were for Newton
not scientific but theological. His career had been one long quest
for God; his research had spun out of this quest, as if by centrifugal
force, but he had no doubt that his science like his theology would
redound to the greater glory of the Creator. "When I wrote my
treatise upon our System I had an eye upon such Principles as
might work with considering men for the belief of a Deity & nothing
can rejoice me more than to find it useful for that purpose," he
replied to a query from a young chaplain, the Reverend Richard
Bentley, who was writing a series of sermons on God and natural
law.[31] At the conclusion of the *Principia*, Newton asserted that "this
most beautiful system of the sun, planets, and comets, could only
proceed from the counsel and dominion of an intelligent and powerful Being."

Newton saw science as a form of worship, yet Newtonian mechanics had a dolorous effect upon traditional belief in a Christian God. Its determinism seemed to deny free will; as Voltaire wrote, "It would be very singular that all nature, all the planets, should obey eternal laws, and that there should be a little animal, five feet high, who, in contempt of these laws, could act as he pleased."[32]

Newton himself did not believe that his theory had diminished the role of the deity. As he saw it, the real miracle is existence itself, and he invoked the hand of God at the origin of the universe: "The Motions which the Planets now have could not spring from any natural Cause alone, but were impressed by an intelligent Agent," he wrote Bentley.[33] In modern scientific terminology the question he was addressing is called the problem of initial conditions. We think that the formation of the solar system can be explained in terms of the workings of natural law, but the authorship of the laws remains a mystery. If for every effect there must have been a cause, then what, or who, was responsible for the *first* cause? But to ask such questions is to leave science behind, and to enter precincts still ruled by Saint Augustine of Hippo and Isaac Newton the theologian.

7

A PLUMB
LINE TO THE
SUN

In Tahiti . . . the women are possessed of a
delicate organization, a sprightly turn of mind,
a lively, fanciful imagination, a wonderful
quickness of parts and sensibility, a sweetness
of temper, and a desire to please.
—Johann Georg Forster, 1778

The conception of the solar system that the Western
world had attained by the beginning of the eighteenth century was
accurate in its proportions but indeterminate in scale. Thanks principally to the theoretical work of Copernicus and Kepler and to the
observations of Tycho and Galileo, it had been established beyond
dispute that the earth was one of five known planets moving in
elliptical orbits around the sun. And, thanks to Newton, these
motions could be interpreted and predicted in terms of a mathematically cogent dynamical scheme that embraced terrestrial as well
as extraterrestrial physics. But, though the relative distances of the
sun and planets were understood, their absolute distances were not.

Copernicus had measured the proportions of the solar system
to within 5 percent of the correct values, and Kepler had come
closer still. These relative distances customarily were expressed in
terms of the distance from the earth to the sun, a quantity known
as the astronomical unit. But nobody knew what the distance to

the sun might be; in other words, the value of the astronomical unit had not been determined. Here was a clear challenge. Since the proportions of the system already were known, if the distance to the sun or to any one planet could be ascertained, the distances of all the other planets would follow. And, since the apparent diameters of the planets could by now be measured rather well, by using a micrometer eyepiece attached to a good telescope, the sizes of the planets could be ascertained as soon as their distances had been measured. Beyond that lay the exciting prospect that, by using the astronomical unit as a baseline, it might be possible to triangulate nearby stars and measure their distances as well. Accomplishing this feat constituted one of the heroic endeavors of eighteenth-century astronomy.

Traditional estimates of the distance from the earth to the sun were of little help. Beginning with Hipparchus in the second century B.C. and ranging down through Ptolemy, Copernicus, and Tycho, astronomers had assumed as a rule of thumb that the astronomical unit was equal to about twelve hundred times the radius of the earth—in modern figures, some 4.8 million miles. Such a distance seemed appropriately vast; to borrow a conceit from the thirteenth century, had Adam started walking on the day of the creation (usually set at 4004 B.C.) he would have required six hundred years to reach the sun, and would have arrived, footsore, at the planet Jupiter in the twentieth century. Nevertheless, an astronomical unit of twelve hundred earth radii was twenty times smaller

Estimated Mean Distances of the Planets from the Sun (In Astronomical Units)

	Copernicus	Kepler	Twentieth Century
Mercury	0.3763	0.389	0.387
Venus	0.7193	0.724	0.723
Earth	1.0000	1.000	1.000
Mars	1.5198	1.523	1.524
Jupiter	5.2192	5.200	5.202
Saturn	9.1743	9.510	9.539

than the real distance. Kepler and later observers suspected that it was an underestimate—Kepler guessed that the value was more like thirty-five hundred earth radii, nearly three times the previous estimates—but these early observers lacked observational instruments adequate to test their hunches.

Two ways of obtaining distance data were available. One, micrometry, was theoretically crude but practically accessible. The other, triangulation, was perfect in theory but difficult to accomplish in practice.

Micrometry consisted of using a micrometer—an eyepiece equipped with an adjustable knife blade—to measure the apparent diameter of a planet as seen through a telescope. The astronomer then estimated the distance of a planet by comparing its apparent diameter with what he guessed to be its actual diameter. Obviously, the result could be no better than the guess as to the planet's diameter. A few astronomers guessed very well indeed: Christian Huygens in 1659 assumed that the diameter of Mars was about 60 percent that of the earth (the correct figure is 53 percent), then measured the apparent size of the disk of Mars through a telescope and calculated a value for the astronomical unit of one hundred million miles. This came astonishingly close to the truth—the mean distance separating the earth from the sun is ninety-three million miles—but it depended entirely upon the accuracy of Huygens's hunch about the size of Mars, and that, as Huygens himself was the first to concede, was "a slippery basis" upon which to base so important a result.[1] The issue was not who made the luckiest guess, but who could obtain observational data that would establish the value of the astronomical unit to everyone's satisfaction. This micrometry alone could not do.

Triangulation, called parallax (from the Greek *parallaxis*, for the value of an angle), was the sounder method. If a planet were observed simultaneously by two observers stationed thousands of miles apart—one in France, say, and the other in Mexico—its position against the background stars would appear to be slightly altered for the astronomer in Mexico as compared to the one in France, owing to their different perspectives on it. If both this angle and the baseline distance separating the two astronomers could be measured, the distance to the planet could be calculated through the straightforward application of euclidean geometry.

That triangulation was theoretically sound had been appreci-

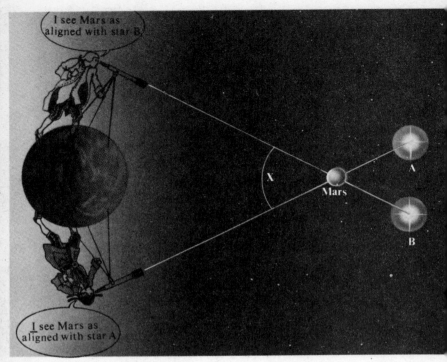

The parallax of Mars was first obtained by simultaneously observing the planet's apparent location from two widely separated places on Earth. The difference in perspective made it possible to measure the value of angle X, which yields the distance from Earth to Mars. The angle, however, is small: Were the earth the size depicted in this illustration, Mars would be five hundred feet away.

ated since ancient times. The difficulty lay in execution. First, one had to know the exact distance between two widely separated observers; this required reasonably accurate intercontinental maps. Second, the observations had to be carried out at the same time, to avoid errors introduced by the motions of the planets and by the rotation of the earth on its axis; this required accurate clocks and a way of synchronizing them. Third, the position of the planet against the stars had to be plotted precisely, because any triangle drawn between a planet and two points on Earth is going to be a very long, thin triangle indeed. Still, the thing could be done, given sufficient exactitude in the measurement of terrestrial space and time.

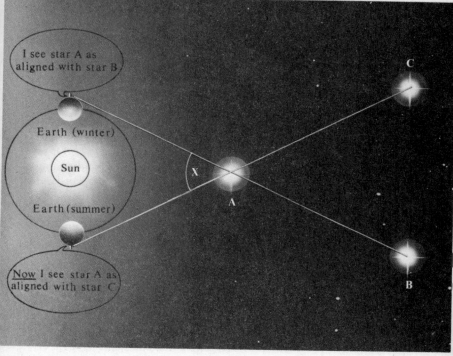

The parallax of stars can best be measured by using as a baseline not the earth but the earth's orbit around the sun. Even with so large a baseline, however, angle X is *extremely* small: Were the earth's orbit the size here depicted, the nearest star would be more than two miles away.

Fortunately for science, rapid progress was being made in both cartography and chronometry. The agency responsible, however, was less the pursuit of pure knowledge than the accumulation of the booty of empire. The wealth of the world flowed into eighteenth-century Europe in ships: From their holds came the Indian rosewood of the dining tables where Newton and Halley were entertained, the African gold inlay on the plates, the turkey with corn they were served as the main course, the chocolate for dessert, and the tobacco they smoked afterward. But blue-water navigation was as hazardous as it was inexact, and sailors who ventured far beyond the sight of land were forever groping their way in the unknown—they were "at sea," as we still say today—with results that ranged from delay to disaster. Many a cargo of silver, sugar, or hardwood had been conveyed across the Atlantic or Indian oceans only to be dashed against the rocks of Land's End or the Cape of Good Hope. The situation had improved little in the century that had passed since the geographer Richard Hakluyt wrote of navigators that "no kind of men in any profession in the commonwealth pass their years in so great and continual hazard of life. . . . Of so many, so few grow to gray hairs."[2] The definitive catastrophe came in 1707, when Sir Cloudesley Shovell, four ships of his fleet, and fully two thousand of his men were lost on the rocks of the Scilly Islands of southwest England, this on a night when his navigators had reckoned that the fleet was in safe waters hundreds of miles to the west. Clearly, something had to be done.

The problem had to do with the determination of longitude. It had long been possible for a ship's navigator to find his latitude—his location in a north-south direction—by measuring the altitude above the horizon of the pole star or of the sun at noon. The instrument employed for this purpose was the astrolabe (from the Greek for "to take a star"), a disk made of copper or tin, five to seven inches in diameter, fitted with a movable sighting arm. At local noon on any clear day aboard a ship of the line, three officers could be seen helping to shoot the sun—one holding the astrolabe steady, another sighting it, and a third reading the elevation—while deckhands stood by to catch the navigator when he fell or to retrieve the astrolabe if it were dropped and went scuttling across the rolling deck. The efficiency of the astrolabe had been improving, through the endeavors of Newton, Halley, John Hadley, Thomas Godfrey, and others, who made the instrument less cumbersome by reducing it first to a quarter of a circle (the "quadrant"), then to a sixth (the "sextant"), by employing mirrors to fold its optics so that the observer could see sun and horizon superimposed, and by adding filters and a telescope for greater accuracy. But, although these improvements helped navigators refine their calculations of latitude, they did not help them determine their longitude—their position in the east-west direction. Here the question was as much one of time as of space.

As the earth turns, the stars troop across the sky at a rate of fifteen degrees per hour. This means that if you know the time, the sky will tell you where you are. But knowledge of the exact time was just what navigators of Newton's day lacked. On land, time was kept by pendulum clocks, but pendulums do not work at sea; the rolling of the boat wrecks their performance. A typical ship's clock in the early eighteenth century was accurate to no better than five or ten minutes per day, which translated into a miscalculation of fully five hundred miles in longitude after only ten days at sea. It was just such an error that had dashed Cloudesley Shovell's fleet on the rocks of the Scilly Islands.

The problem of determining longitude at sea had defied resolution for so long that many regarded it as unsolvable. The mathematician in Cervantes's *The Dog's Dialog* muses crazily that he has "spent twenty-two years searching for the fixed point"—*el punto fijo*, the correct longitude—"and here it leaves me, and there I have it, and when it seems I really have it and it cannot possibly escape me, then, when I am not looking, I find myself so far away again that I am astonished. The same thing happens with squaring the circle."[3] Sebastian Cabot on his deathbed claimed that God had revealed the answer to him, but added, alas, that He had also sworn him to secrecy.

Still, the longitude problem was obviously imperative, and more than a few inventors took it on, encouraged by the large cash prizes proffered by the governments of seafaring states like Spain, Portugal, Venice, Holland, and England. The richest of these was a prize of twenty thousand pounds, offered by the British Board of Longitude to anyone who could devise a practicable method of determining longitude on a transatlantic crossing to within one-half a degree, which equals sixty-three nautical miles at the latitude of London. John Harrison, an uneducated carpenter turned clockmaker, pursued the prize for much of his working life. He constructed a succession of "watches" (the term, meaning a portable clock, comes from the shipboard practice of dividing up the day into six watches of four hours each) of increasingly subtle and rugged design, checking them for accuracy by observing the disappearance of designated stars behind a neighbor's chimney each night. His masterpiece, a marine chronometer that took him nineteen years to complete, was transported to Port Royal, Jamaica, aboard H.M.S. *Deptford* in 1761–1762, was there tested against sightings of the sun, and was found to have lost only 5.1 seconds in eighty days—a performance that many of today's timepieces could not match. Nonetheless, it took Harrison years of lobbying to collect a portion of the prize, and he never got it all; twenty thousand pounds was a lot of money.

The astronomers and geographers, however, did not have to wait as long as did the mariners to improve their measurements of earthly space and time. Maps were constantly improving: Although pendulum clocks were not yet reliable at sea, they could be synchronized on land, by observing transits and eclipses of the satellites of Jupiter. (The Dutch had awarded Galileo a gold chain for proposing this ingenious idea, though they could not make it work on board ship, since any telescopic magnification sufficient to resolve Jupiter's moons also magnified the rocking of the boat too much for the planet to be kept in view.) In France, cartographers led by Giovanni Cassini and Jean Picard employed Galileo's method to cage the continent in a cat's cradle of surveyor's triangles, producing an accurate map that enabled Picard to determine the circumference of the earth to within 126 miles of the correct value.*

Equipped with better maps and clocks, astronomers tried to triangulate the neighboring planets Mars and Venus. In 1672, an international expedition led by the young French astronomer Jean Richer sailed to Cayenne, on the South American seacoast three hundred miles north of the equator. There he observed Mars during its closest approach to Earth at the same time that his colleagues, their clocks synchronized to Richer's, sighted Mars from their post at the French Academy. Cassini sorted through the data and derived a value for the astronomical unit of eighty-seven million miles. This approximated the correct figure of ninety-three million miles, but given the many residual inaccuracies of the instruments and techniques of the time, Cassini's like Huygens's earlier estimate necessarily was regarded as but an educated guess.

Venus comes closer to Earth than does Mars, and so should be still more accessible to triangulation, but when closest it is lost in the glare of the sun. Twice in a long while, however, in pairs of events separated by just over a century, Venus passes directly in front of the sun. During these transits, as they are called, the planet appears as a black circle silhouetted against the blazing solar

---

*The map revealed that France was smaller than had been thought, prompting the Sun King to remark that the scholars of the French Academy of Sciences had cost him more territory than had been lost to all France's enemies in war.

disk. Edmond Halley, who had observed a transit of Mercury during his expedition to St. Helena, realized that the distance to Venus
might be determined by timing, from widely separated stations,
exactly when the planet appeared and disappeared from the face of
the sun. The edge of the sun would serve as a clearly defined
backdrop, the planet as a kind of surveyor's stake out in space.

Halley knew that he would not live to observe a transit of
Venus. There had been a pair of transits in 1631 and 1639, a
generation before he was born; the next pair were due in 1761 and
1769, by which time he would have been over a hundred years
old.* (Halley must have been getting used to this sort of thing; he
didn't live to see the return of Halley's Comet, either.) And so it
was with the insistence of a man striving to project his words beyond
the grave that Halley, in a paper published in 1716 "which," he
wrote, "I prophecy will be immortal," outlined the procedure for
the benefit of astronomers yet unborn:

> We therefore recommend again and again, to the curious in-
> vestigators of the stars to whom, when our lives are over, these
> observations are entrusted, that they, mindful of our advice,
> apply themselves to the undertaking of these observations vig-
> orously. And for them we desire and pray for all good luck,
> especially that they not be deprived of this coveted spectacle
> by the unfortunate obscuration of cloudy heavens, and that
> the immensities of the celestial spheres, compelled to more
> precise boundaries, may at last yield to their glory and eternal
> fame.[4]

Previous observations of transits had been rare and rather hap-
hazard. Pierre Gassendi in Paris managed to observe a transit of
Mercury in 1631 that Kepler had predicted; he stamped on the floor
to alert his young assistant to measure the altitude of the sun, but
the boy, growing impatient after three days of waiting for the great
event, had wandered off. Gassendi's solitary published observation
was useless for triangulation, though it did reveal that the disk of
Mercury was much smaller than had been thought: "I could hardly
be persuaded that it was Mercury, so much was I preoccupied by
the expectation of a greater size,"[5] Gassendi wrote. This supported

---

*The most recent transits of Venus were in 1874 and 1882; the next will occur
on June 7, 2004, and June 5, 2012.

the contention of Galileo that the solar system was considerably larger than had been estimated by Ptolemy and the other geocentrists.

As for Venus, its transit on December 6–7, 1631, was visible only from the New World and appears to have been viewed by not a single human being, and the transit of November 24, 1639, was observed by but two people, the English astronomer and clergyman Jeremiah Horrocks and his friend William Crabtree. Alarmingly for Horrocks, who was a clergyman, the transit occurred on a Sunday, when he was obliged to preach two sermons. He rushed home from church, peered through his telescope at 3:15 P.M., and saw Venus, "the object of my most sanguine wishes . . . just wholly entered upon the Sun's disk."[6] Venus, like Mercury, looked smaller than had been predicted—Kepler thought Venus would cover one quarter of the sun, an enormous overestimate—and so to behold its tiny apparent size helped improve human appreciation of interplanetary distances. But Horrocks had no way to measure the apparent diameter of the disk precisely, and, since he was but one observer, he could not have triangulated Venus even if he had possessed an accurate clock. Crabtree, for his part, was so overwhelmed by the sight of an entire world dwarfed by the sun that he made no coherent notes at all, prompting Horrocks to protest that "we astronomers have a certain . . . disposition [to be] distractedly delighted with light and trifling circumstances."[7]

But the world had changed by the time the transits of Venus of 1761 and 1769 came due. Astronomy had become an organized science, conducted by professionals, sponsored by scientific societies, and supported by government funds. Now at last, it was felt, science had the resources to sound the dimensions of the solar system. Halley's implorations were remembered, and the transits were scrutinized by scores of observers equipped with micrometers, accurate clocks, and brass telescopes mounted on hardwood tripods at sites as far away as Siberia, South Africa, Mexico, and the South Pacific.

And, to an extent, the transit observers succeeded, though not without suffering sufficient tribulations to remind them that while the motions of the planets may be sublime the affairs of this world are marbled with chaos. The astronomer Charles Mason and the surveyor Jeremiah Dixon, later of the Mason-Dixon Line, were attacked by a French frigate while making their way to Africa (this was during the Seven Years' War) with a loss of eleven dead and thirty-seven wounded; they reached Cape Town under military escort and observed the 1761 transit, only to find that they differed by many seconds in their estimate of the time when Venus had entered and left the disk of the sun. William Wales timed the transit from Hudson Bay, Canada, after enduring mosquitoes, horseflies, and a winter sufficiently severe that, as he noted with empirical exactitude, a half-pint of brandy left unattended iced over in only five minutes. Jean-Baptiste Chappe d'Auteroche, dispatched by the French Academy into the depths of Russia, raced across the frozen Volga and through Siberian forests in horse-drawn sleds, arrived at Tobolsk six days prior to the transit, posted guards to repel angry mobs who blamed him for causing spring floods by interfering with the sun, and managed to observe the transit. He died eight years later in Baja California after timing the 1769 transit, of an epidemic that spared but one member of his party, who dutifully returned his data to Paris. Alexandre-Gui Pingré was rained out for most of the transit in Madagascar, lost his ship to the British and was returned to Lisbon under British guns; a humanist as well as a scientist, he took comfort in the ship's rations of spirits: "Liquor," he wrote, "gives us the necessary strength for determining the distance of . . . the sun."[8]

Least fortunate of all was Guillaume le Gentil, who sailed from France on March 26, 1760, planning to observe the transit the following year from the east coast of India. Monsoons blew his ship off course, and transit day found him becalmed in the middle of the Indian Ocean, unable to make any useful observations. Determined to redeem the expedition by observing the second transit, Le Gentil booked passage to India, built an observatory atop an obsolete powder magazine in Pondicherry, and waited. The sky remained marvelously clear throughout May, only to cloud over on June 4, the morning of the transit, then clear again as soon as the transit was over. Wrote Le Gentil:

> I was more than two weeks in a singular dejection and almost did not have the courage to take up my pen to continue my journal; and several times it fell from my hands, when the moment came to report to France the fate of my operations. . . . This is the fate which often awaits astronomers. I had gone more than ten thousand leagues; it seemed that I had

crossed such a great expanse of seas, exiling myself from my
native land, only to be the spectator of a fatal cloud which
came to place itself before the sun at the precise moment of
my observation, to carry off from me the fruits of my pains
and of my fatigues.[9]

Worse lay ahead. Stricken with dysentery, Le Gentil remained
in India for another nine months, bedridden. He then booked passage home aboard a Spanish warship that was demasted in a hur-
ricane off the Cape of Good Hope and blown off course north of
the Azores before finally limping into port at Cadiz. Le Gentil
crossed the Pyrenees and at last set foot on French soil, after eleven
years, six months, and thirteen days of absence. Upon his return
to Paris he learned that he had been declared dead, his estate looted,
and its remains divided up among his heirs and creditors. He re-
nounced astronomy, married, and retired to write his memoirs.
Cassini, eulogizing Le Gentil, praised his character but allowed
that "in his sea voyages he had contracted a little unsociability and
brusqueness."[10]
   The most elaborate of the transit expeditions, mounted by the
Royal Society, departed aboard the ninety-eight-foot bark H.M.S.
Endeavour from Plymouth on August 26, 1768, with a deputation
of scientists led by Joseph Banks, a wealthy botanist and future
president of the Royal Society. *Endeavour* was equipped with crates
full of clocks, telescopes, and meteorological equipment, as well as
a barrel of nails for trading with the Tahitians, who had a passion
for anything made of metal. The commander was Captain James
Cook, an expert navigator, marine surveyor, and mathematician
who had taught himself astronomy so well that, by observing the
solar eclipse of 1766, he had been able to determine his longitude
in Newfoundland to within two nautical miles. An empiricist in
the social as well as the physical sciences, Cook found by experimenting with diet that he could ward off scurvy by feeding his
men sauerkraut, which he shrewdly popularized among the hands
by at first restricting it to the officers' mess. The voyage was un-
eventful by the standards of the day: The expeditionaries took on
three thousand gallons of wine and a thousand pounds of onions
at Madeira, were fired upon in the Falklands by a half-mad viceroy
who understood the transit to involve "the North Star passing
through the South Pole," and lost four men—a veteran seaman who

drowned, a young marine who jumped overboard in shame after having stolen a bit of sealskin, and Banks's two servants, who got drunk in a snowstorm in Tierra del Fuego and froze to death. After seven and a half months *Endeavour* reached Tahiti, then as now a synonym for paradise.

Cook issued strict orders to his men against unauthorized trading of metal objects with Tahitian females, who adorned their thighs with intricate tattoos of arrows and stars and saw nothing wrong in trading sexual favors for a nail or two. Cook recalled with concern that the crew of an earlier ship to reach Tahiti, the *Dolphin*, had in their enthusiasm for the Tahitian girls extracted so many nails from the ship that they nearly pulled it apart. When two of Cook's marines deserted, married Tahitians, and fled to the mountains, Cook had them brought back and clapped in irons; he was a humane man, but he intended to return to England. His orders notwithstanding, though, nails and other metal objects kept vanishing from the ship.

Under Cook's and Banks's direction a fortress observatory was erected on Tahiti at what has ever since been known as Point Venus, and from there the transit of June 3, 1769, was observed under clear blue skies.

Timing the transit, however, proved difficult. The trouble was that Venus has a thick atmosphere, which refracts and diffuses the sunlight passing through it. As a result the disk of the planet, rather than snapping crisply into view as does the disk of airless Mercury when it is in transit, seems instead to adhere to the edge of the sun, like a raindrop hanging from a branch. "We very distinctly saw an Atmosphere or dusky shade round the body of the Planet which very much disturbed the times of the Contacts," Cook noted in his journal.[11] Consequently, Cook and astronomer Charles Green, observing through identical telescopes, differed in their estimates of the entry and exit times of Venus' disk by as much as twenty seconds.

But despite these difficulties, the data gathered by Cook's and the other scientific expeditions yielded estimates of the distance from the earth to the sun that came within 10 percent of the correct value. The astronomical unit subsequently was measured even more accurately by scientists who drew imaginary triangles, ever more refined, to Venus during its nineteenth-century transits, to Mars when it was in opposition in 1877, and to dozens of asteroids (or "minor planets") as these previously useless chunks of rock drifted past the earth.

The immensity of the solar system, nearly a hundred times the Ptolemaic estimate of the size of the entire universe, now stood revealed, and scientists could with assurance turn their attention to the depths of interstellar space, and take on the still more ambitious task of measuring the distances of stars.

Here, too, some ground had been cleared by educated guesswork. One early approach to measuring stellar distances consisted of assuming that a given star was intrinsically just as bright as the sun, then measuring its apparent brightness (or *magnitude*) and estimating its distance by applying the law, known since Kepler's day, that the apparent brightness of any object in space diminishes by the square of its distance. (This was analogous to earlier attempts to approximate the distances of planets by assuming that they were roughly the same size as the earth.) In the late seventeenth century, Christian Huygens observed the sun from a darkened room through pinholes of various sizes until he obtained an image that seemed equal in brightness to that of Sirius, the brightest star. Since the appropriate pinhole admitted 1 part in 27,664 of the sun's light, Huygens concluded that Sirius was 27,664 times farther away than the sun—an underestimate by some twenty times, but an enormous distance nonetheless. A somewhat more refined approach, proposed by James Gregory in 1668 and detailed by Isaac Newton in a draft of the *Principia*, was to use Saturn, the outermost known planet, as a sort of reflecting mirror to gauge the intensity of sunlight. By guessing at Saturn's reflectivity and assuming the stars to be of similar brightness to the sun, Newton concluded that the brightest stars are (to convert his figures into modern terms) about sixteen light-years away. The flaw here was that stars differ tremendously in their intrinsic luminosity; most of the bright stars we see in the sky are tens of times more luminous than the sun, and are, therefore, much more distant than we would guess by assuming that they resemble the sun.

The more promising strategy was to triangulate the stars. This could be accomplished by using, not the earth, but the *orbit* of the earth, as the baseline. The idea was to chart the position of a nearby star on two evenings six months apart, when the earth was at opposite extremities of its orbit, then look for a change in position produced by the change in our angle of sight on the nearby star against the more distant stars in the background. This method, known as stellar parallax, became theoretically practicable once the radius of the earth's orbit—the astronomical unit—had been measured. Before it could be employed successfully, however, some of the subtleties of the earth's motion had to be better understood.

The hero of this dry but vital business was the British astronomer James Bradley, Halley's successor as Astronomer Royal. Raised on parallax, Bradley had triangulated Mars while still in his twenties, in the company of his uncle the amateur astronomer James Pound and Halley himself. Their observations indicated that the astronomical unit was equal to some 93 million to 125 million miles.

Eight years later, in 1725, Bradley and another amateur astronomer, Samuel Molyneux, installed a precision telescope in the chimney of Molyneux's home. This "zenith telescope" pointed straight up, to the part of the sky where distortions of starlight induced by the earth's atmosphere are at a minimum. It was used to observe but a single star, Gamma Draconis, which passed through the zenith at London's latitude. Bradley and Molyneux reasoned that as the months went by the apparent position of Gamma Draconis would slowly shift, owing to the changing perspective introduced by the earth's motion. The extent of this shift was to be measured by means of a plumb bob that would indicate how much the telescope's aim had to be altered to bring the star back into the crosshairs. (Hooke, in the previous century, had used a zenith telescope to observe the same star, but the crudity of his instruments prevented his reaching any useful conclusion.)

The new assault on the parallax of Gamma Draconis proved more successful, but in an entirely unexpected way. As the months passed and Bradley's observations of the star accumulated, he was surprised to find that the largest variation in its position occurred not annually, but daily. Intrigued, Bradley installed a second telescope, one capable of greater latitude of motion and, therefore, of observing more stars, and mounted it on the roof of his aunt's home. (She obligingly permitted holes to be cut through the floors so that the measuring instruments could be placed in the cool, stable air of the basement.) By 1728, Bradley had observed more than two hundred stars and had found, to his amazement, that every one of them behaved in the same way: Each seemed to crawl slightly

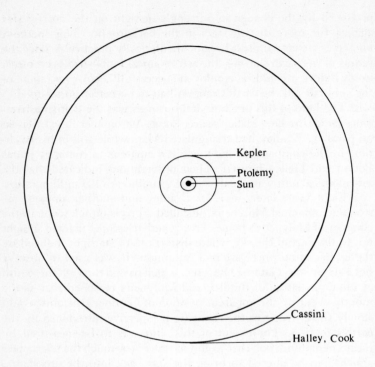

Estimates of the size of Earth's orbit, A.D. 100–1769

northward, then southward, every twenty-four hours. Bradley had
no idea why.

As often happens, the answer came to him not while he was
at work in his observatory but while he was relaxing. While on a
boat in the Thames, Bradley found himself gazing at a wind vane
mounted atop the mast. It pointed into the wind and therefore
seemed to change direction whenever the boat turned. What was
changing, of course, was the orientation, not of the wind, but of
the boat.

It occurred to Bradley that the earth is like a boat adrift in
winds of starlight—that, as the earth moves through the starlight,
its motion alters the apparent positions of the stars. Think of the
earth as a woman walking briskly through the rain; her motion

makes the raindrops seem to slant toward her, so she tilts her umbrella forward to compensate. Similarly, the earth's motion makes starlight seem to slant, altering the apparent position of the stars hour by hour. Bradley had discovered what is called the *aberration* of starlight.

Twenty years later Bradley detected another subtlety in the earth's motion, a *nutation*, or wobble, in the direction of its axis of rotation. These complications frustrated his efforts to detect the parallax of Gamma Draconis, but they paved the way for future parallax measurements—and, not incidentally, provided direct proof of the old Copernican hypothesis that the earth spins on its axis and orbits the sun.

But, since the stars are very far away, their triangulation called for instruments more precise than were available in Bradley's day. Were the earth's orbit represented by a serving platter one foot in diameter, a triangle drawn from the edges of the plate to the very nearest star would be twenty-six miles long, and its sides would be almost indistinguishable from parallel lines. The job facing the parallax astronomers was to detect the angle of convergence of such a triangle, and much thinner ones as well, and to measure the angles precisely enough to say where the lines would meet, for at that point stood the location of the star in three-dimensional space.

Bradley did not live to see the day when so great a degree of exactitude became attainable. But telescopes and their mountings kept improving, and in December 1838, Friedrich Wilhelm Bessel, a mathematician and astronomer who worked at the observatory of Königsberg with a precision telescope constructed by the master optician Joseph Fraunhofer of Munich, announced that after eighteen months of observations he had succeeded in measuring the parallax of the star 61 Cygni. Bessel's measurement yielded a distance to 61 Cygni that came within 10 percent of the modern value of 10.9 light-years. Soon thereafter Thomas Henderson at the Cape of Good Hope obtained the parallax of Alpha Centauri, and Friedrich Struve in Russia found the parallax of the bright blue-water star Vega.

The angles, as expected, were tiny. The parallax of Alpha Centauri, which is the nearest star to the sun and therefore has the largest parallax, is only 0.3 seconds of arc, or one ten-thousandth of a degree. Clearly, interstellar space is built on an almost inconceivably gigantic scale. Light from our neighbor Alpha Centauri, traveling at 186,000 miles per second, takes four years and fifteen weeks to reach us (which is to say that Alpha Centauri is 4.3 light-years away), while 61 Cygni, the inconspicuous star scrutinized by Bessel, lies 11 light-years from the earth. But the vastness of the distances, which had long been inferred from the supposition that the stars are suns, made less of an impression than did the fact that such distances actually could be measured by human beings. Tri-angles born in the mind of Aristarchus of Samos had been extended out into the previously soundless depths of interstellar space, throw-ing back the conceptual horizons of cosmological thought. The sky was no longer the limit.

And yet, the more that came to be understood about the distant stars, the more intimate they seemed, as connections were identified linking the earth and the stars. One such insight in particular would have interested Captain Cook. It has to do with the iron that made the nails that the Tahitians found so alluring.

When the nuclear chemistry that powers the stars began to be deciphered by twentieth-century astrophysicists, it emerged that iron plays a central role in the evolution of stars. Stars burn by fusing the nuclei of the light atoms of hydrogen, the nucleus of which consists of but a single proton, and helium, which consists of two protons and two neutrons. In doing so stars release energy, which is how they shine, but they also build heavier atoms out of the lighter ones. As the process continues, each star forges atoms of carbon, oxygen, neon, sodium, magnesium and silicon, then nickel, cobalt, and, finally, iron. At iron the building stops; a nor-mal, first-generation star lacks the energy required to make any heavier nuclei. The Sumerian name for iron, which means "metal from heaven," is literally true: Iron is a working star's proudest product.

When a star runs out of fuel, it can become unstable and explode, spewing much of its substance, now rich in iron and other heavy elements, into space. As time passes, this expanding bubble of gas is intermixed with passing interstellar clouds. The sun and its planets congealed from one such cloud. Time passed, human beings appeared, miners in the north of England dug the iron from the earth, and ironmongers pounded it into nails that longshoremen loaded in barrels into the holds of H.M.S *Endeavour*. Off the nails went to Tahiti, continuing a journey that had begun in the bowels of stars that died before the sun was born. The nails that Cook's men traded with the Tahitian dancing girls, while on an expedition to measure the distance of the sun, were, themselves, the shards of ancient suns.

8

DEEP SPACE

The infinitude of the creation is great enough
to make a world, or a Milky Way of worlds,
look in comparison with it what a flower or
an insect does in comparison with the earth.
                              —Immanuel Kant

I have looked farther into space than ever [a]
human being did before me.
                              —William Herschel

Bright *nebulae* (from the Latin for "fuzzy") are diffuse
patches of glowing material found scattered here and there among
the stars. Most can be seen only with a telescope. Although they
resemble one another superficially, the bright nebulae actually comprise three very different classes of objects. Some, misnamed "planetary" because they are spherical in shape and bear a passing
resemblance to planets, are shells of gas thrown off by old, unstable
stars; a typical planetary nebula is about one light-year in diameter
and has one-fifth the mass of the sun. Others, the reflection and
emission nebulae, are clouds of gas and dust illuminated by nearby
stars; in many cases, the stars doing the illuminating have themselves recently condensed from the surrounding cloud. These nebulae measure hundreds of light-years in diameter and can harbor

the mass of a million or more suns. They represent the bright, congealed parts of the still more extensive dark nebulae that wend their way throughout much of the disk of the Milky Way galaxy —though this was not recognized at first, since the dark nebulae are too inconspicuous to call attention to themselves. Finally there are the elliptical and spiral nebulae. These are galaxies in their own right, millions of light-years away. A major galaxy can measure over one hundred thousand light-years in diameter and contain hundreds of billions of stars.

In much the same way that human beings could not investigate interstellar space until we understood that the sun is one among many stars, so the realization that we live in a universe of galaxies, scattered across immense gulfs of space, required that we first understand the nature of the nebulae. This involved comprehending not only the appearance of the nebulae but also their chemical composition, an effort that spawned the sciences of spectroscopy and astrophysics.

Science is said to proceed on two legs, one of theory (or, loosely, of deduction) and the other of observation and experiment (or induction). Its progress, however, is less often a commanding stride than a kind of halting stagger—more like the path of the wandering minstrel than the straight-ruled trajectory of a military marching band. The development of science is influenced by intellectual fashions, is frequently dependent upon the growth of technology, and, in any case, seldom can be planned far in advance, since its destination is usually unknown. In the case of the exploration of intergalactic space, the first step was taken by armchair theorists—by the philosopher Immanuel Kant and the mathematician Johann Lambert—followed by the observations of the prescient amateur astronomer William Herschel.

When Kant first wrote on cosmology he was not yet *Kant*, the intellectual titan whose unification of empiricism and rationalism was to illuminate and enliven philosophy throughout the world. The year was 1750, and he was but twenty-six years old. The death of his father four years earlier had obliged him to interrupt his education, and he was working as a private tutor in East Prussia. He had earned a bachelor's degree (paying his tuition out of his earnings from gambling at billiards and cards) but five more years would pass before he was awarded his doctorate. He had not yet ruined his writing style by trying to satisfy the formal requirements

established by the philosophy faculty at the University of Königsberg, where, at the age of forty-six, he would finally be appointed
professor of logic and metaphysics. He was a witty, outgoing man
and attractive to women, though he could never bring himself to
marry. A creature of habit, he ate one meal a day, always with
friends, consulted a barometer and thermometer by his bedside
each morning in order to determine how to dress, and took his
evening walk so punctually that neighbors literally set their clocks
by his appearance on the street. He taught mathematics and physics, revered Lucretius and Newton, and read everything from theological history to the actuarial tables.

One day Kant read, in a Hamburg journal, a review of a book
titled *An Original Theory or New Hypothesis of the Universe*, by an
English surveyor and natural philosopher named Thomas Wright.
Wright in his piety had taught himself astronomy the better to
appreciate the grandeur of God's creation, and his books and lectures, freighted with moral and theological lessons, were popular
in society circles. In the course of a variegated career, Wright proposed a number of models of the universe, many of them contra-
dictory and all burdened with such concerns as the location of the
throne of God, which he put at the center of the cosmos, and hell,
which he relegated to the outer darkness.

The cosmological speculations of such a thinker might not
normally have commanded the attentions of a Kant, but the summary of Wright's book that Kant read distorted Wright's theories,
and, in the process, improved upon them. The result was one of
journalism's signal contributions to cosmology, the inadvertent pro-
motion of a nonexistent hypothesis that Kant then turned into this
world's first glimpse of the universe of galaxies.

Wright, following the same erroneous route that had misled
Plato, Aristotle, Ptolemy, and Copernicus, assumed the universe
to be spherical. But where his pre-Copernican predecessors had
put the sun at the center of the universe, Wright suggested that
the sun instead belongs to the celestial sphere. What he had done,
really, was to resurrect the starry sphere of Aristotle and Ptolemy, but with the sun as one of its stars. Wright's cosmos was
hollow, like an orange with the pulp sucked out and with the sun
and other stars in the skin. Wright noted that the appearance of
the Milky Way as a band of stars in the sky might be explained as
our view of this starry shell from our location within it: When we

look along a line tangential to the sphere we see many stars—the Milky Way—and when we look along the sphere's radius we see relatively few stars.

The synopsis that Kant read in the newspaper stressed this last point—happily, the most felicitous part of Wright's theory—and was vague about the rest. Consequently, Kant got the mistaken impression that Wright's universe consisted of a flattened disk of stars, like a thumbnail-sized slice cut tangentially from the skin of an orange. Kant therefore supposed (as he thought Wright had also) that the stars of the Milky Way are arrayed across a disk-shaped volume of space. So excited was Kant about this idea that he wrote a book on it. He stated its thesis this way:

> Just as the planets in their system are found very nearly in a common plane, the fixed stars are also related in their positions, as nearly as possible, to a certain plane which must be conceived as drawn through the whole heavens, and by their being very closely massed in it they present that streak of light which is called the Milky Way. I have become persuaded that because this zone, illuminated by innumerable suns, has almost exactly the form of a great circle, our sun must be situated quite near this great plane. In exploring the causes of this arrangement, I have found the view to be very probable that the so-called fixed stars may really be slowly moving, wandering stars of a higher order.[1]

From this precarious foothold Kant made a cat's leap to the universe of galaxies. He knew from reading of the observations of the French astronomer Pierre-Louis de Maupertius that elliptical nebulae had been found here and there in the sky. One of these, the Andromeda nebula, could be seen with the unaided eye; others were discernible through telescopes. Kant realized that if the universe were composed of many disk-shaped aggregations of stars—galaxies, as we would say today—then the elliptical nebulae could be other galaxies of stars like our Milky Way. "I come now to that part of my theory which gives it its greatest charm, by the sublime idea which it presents of the plan of the creation," he wrote.

> If a system of fixed stars which are related in their positions to the common plane, as we have delineated the Milky Way to be, be so far removed from us that the individual stars of

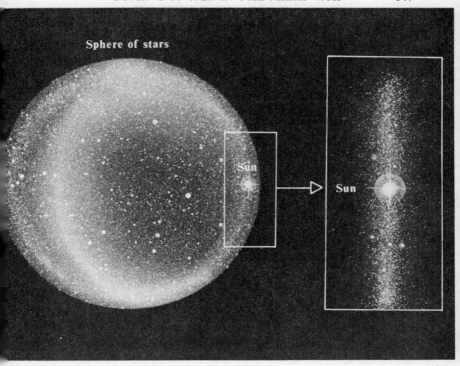

Wright envisioned the universe as a bubble, and proposed that the appearance of the Milky Way in the sky represented our view of its starry skin. Kant was unaware of the first part of his argument, but seized upon the second, correctly conceiving of the sun as belonging to a flattened system of stars—a galaxy.

which it consists are no longer sensibly distinguishable even by the telescope; if its distance has the same ratio to the distance of the stars of the Milky Way as that of the latter has to the distance of the sun; in short, if such a world of fixed stars is beheld at such an immense distance from the eye of the spectator situated outside of it, then this world [i.e., the Milky Way galaxy] will appear under a small angle as a patch of space whose figure will be circular if its plane is presented directly to the eye, and elliptical if it is seen from the side or obliquely. The feebleness of its light, its figure, and the apparent size of its diameter will clearly distinguish such a phenomenon when it is presented from all the stars that are seen single.[2]

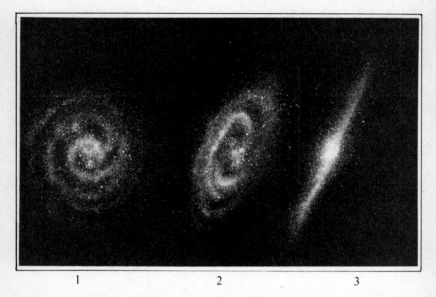

1   2   3

Kant realized that disk-shaped galaxies, viewed at random angles, would produce the appearance of seemingly round, oval, and linear "nebulae."

The elliptical nebulae, Kant wrote, present us with just such apparitions. The nebulae are "systems of many stars" lying "at immense distances."[3] Here for the first time was a portrait of the universe as consisting of galaxies adrift in the vastness of cosmological space.

Kant's book, titled *Universal Natural History and Theory of the Heavens*, was published—if that is the word—in 1755, but its publisher immediately went bankrupt, the books were seized to satisfy his debts, and the world, consequently, heard little of it. Kant dedicated it to Frederick the Great, but many better-known artists and philosophers were dedicating their works to this singularly enlightened monarch—Johann Sebastian Bach, for one, had recently composed his *Musical Offering* in Frederick's honor—and the king never saw Kant's book.

Frederick did, however, come upon the idea of a universe of galaxies, by another and even less likely avenue. His acquaintanceship with it began one evening in March 1764 when he entered a darkened room, nearly all its candles extinguished, to interview, for membership in the Berlin Academy of Sciences, a candidate whose appearance and manner were so off-putting that the friends who had arranged the meeting had feared that Frederick would never admit him if he could see him clearly.

The man in the dark was Johann Heinrich Lambert, and his friends had ample grounds for their concern. Lambert's appearance was unsettling: His forehead was so high that most of his face stood above, not below, the eyebrows, and he dressed uniquely, in a scarlet tailcoat, turquoise vest, black trousers and white stockings, an outfit to which, on special occasions, he added a broad ribbon tied in two bows, one adorning his pigtail and the other his chest. Though his eyes were piercing he seldom looked directly at anyone, preferring, instead, to strike a profile. If an interrogator tried to step around to get a look at him, Lambert would turn slowly on his heel, maintaining the profile, a human moon.

"Would you do me the favor," said Frederick to the darkling Lambert, "of telling me in what sciences you are specialized?"

"In all of them," Lambert replied, addressing a point in space ninety degrees away from the king.

"Are you also a skillful mathematician?" asked Frederick.

"Yes."

"Which professor taught you mathematics?"

"I myself."

"Are you therefore another Pascal?" asked Frederick, referring to the great mathematician of the previous century.

"Yes, Your Majesty," replied the voice in the dark.[4]

Frederick turned away, barely able to contain his laughter, and left the room. That night at dinner he remarked that he had just met the biggest blockhead in the world. But Lambert, when consoled by his friends on the outcome of the interview, serenely assured them that he would get the appointment, since should Frederick "not name me, it would be a blot in his own history."[5] And, indeed, following a review of his publications, Lambert was appointed to the Academy.

Among his works was a collection of essays titled *Cosmological Letters*, which this solitary man, so freakish-looking that children followed him through the streets as they might a fakir in a loincloth, had written as a series of letters to an imaginary friend. In it, Lambert proposed that the sun lies toward one edge of a disk-shaped system of stars, the Milky Way, and that there are "innumerable other Milky Ways."[6] He indicated that he had arrived at this theory while gazing for long hours at the night sky:

> I sat at the window and as the objects on Earth put aside all their charm to draw attention, there still remained for me the starry sky as, of all showplaces, the most worthy of contemplation. . . . I take on wings of light and soar through all spaces of the heavens. I never come far enough and the desire always grows to go still farther. In such reflections did I present to myself the Milky Way. . . . This luminous arch, which stretches all around the firmament and decorates the world like a ring studded with gems, roused in me astonishment and wonderment.[7]

The galactic rhapsodies of Kant and Lambert helped awaken the human mind to the potential richness and reach of the universe. But rapture in itself, no matter how insightfully founded, is of course an inadequate foundation upon which to ground a scientific cosmology. To determine whether the universe is in fact comprised of galaxies would require actually mapping the universe in three dimensions, by means of observations more exacting, if no less enchanting, than Lambert's meditative stargazing.

The point man of this observational campaign was William Herschel, the first astronomer to make acute, systematic observations of the universe beyond the solar system, where lies the vast majority of everything there is. Herschel was born in Hanover on November 15, 1738, the son of a musician with an active intellect who taught his six children to think for themselves, inciting heated dinner-table discussions of science and philosophy and taking them outdoors on clear nights to teach them the constellations. The Seven Years' War found the eighteen-year-old Herschel playing oboe in his father's unit, the Hanoverian Guards' band. Mars hates music, and the band was superfluous in battle. "Nobody had time to look after the musicians," Herschel recalled, in his deadpan way. "They did not seem to be wanted."[8] For a time he wandered through the carnage in a state of abstraction worthy of Buster Keaton in *The General*. Then one day, when French troops got within firing range of the muddy field where the band was encamped, Herschel's father advised his son on the better part of valor, and the boy obediently walked out of the war. "Nobody seemed to mind," he noted.[9]

He fled to England, where the king at the time was the polit-

ically disinterested but indisputably Hanoverian George II, and
there flourished. Herschel's English was excellent, his manner re-
freshingly direct and personable; "I have the good luck to make
friends everywhere," he wrote home. [10] He continued his education
by reading constantly; many years later he would tell his son John
that once while reading on horseback he suddenly found himself
standing on the road, book still securely in hand, the horse having
tossed him in a perfect somersault though the air. His mind was
sufficiently powerful to impress the likes of David Hume, yet he
wore his learning lightly enough to thrive in London society. His
musical fortunes benefited from the example set by his distinguished
countryman George Frederick Handel, and by age thirty Herschel
had been appointed organist of the chapel at Bath, a genteel post
where he could expect to abide in comfort all the rest of his days.

Instead, he felt dissatisfied. Music was not enough; he knew
he was no Handel, and was not content with mere facility. "It is
a pity that music is not a hundred times more difficult as a science,"
he wrote. ". . . My love of activity makes it absolutely necessary
that I should be busy, for I grow sick by idleness; it kills me almost
to do nothing."[11]

He found deliverance by following Kepler's and Galileo's path
across the bridge that leads from music to astronomy. Like many
amateur astronomers before and since, he began by reading books
of popular science. He was particularly impressed by James Ferguson's *Astronomy Explained Upon Sir Isaac Newton's Principles* and
Robert Smith's *A Compleat System of Opticks*.

Ferguson had begun his study of astronomy when as an uneducated shepherd boy he used to lie on his back in the fields of
Scotland at night and measure the angles between stars with beads
positioned on a thread. He taught himself to read, became a teacher
and popular lecturer, wrote two well-received books on astronomy,
and ultimately was elected to the Royal Society. It was in Ferguson's book that Herschel first read about the nebulae. Some nebulae
appeared to be starless; as Ferguson wrote, "There are several little
whitish spots in the Heavens, which appear magnified and more
luminous when seen through a telescope; yet without any stars in
them. One of these is in Andromeda's girdle." Other nebulae were
tangled in stars. "They look like dim stars to the naked eye," wrote
Ferguson, "but through a telescope they appear [to be] broad illuminated parts of the sky; in some of which is one star, in others

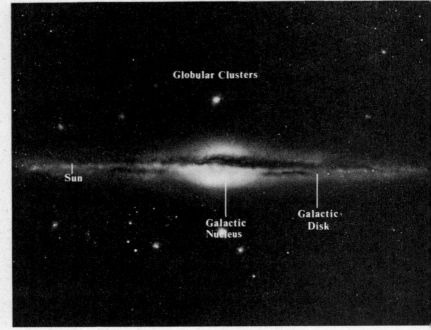

Globular Clusters

Sun

Galactic
Nucleus

Galactic
Disk

The Milky Way galaxy, seen edge-on, is disklike in shape, with an elliptical central region. The disk is surrounded by a halo of globular star clusters and old stars.

more. . . . The most remarkable of all the cloudy stars is that in the middle of Orion's sword."[12]

In Smith's book, Herschel read that although the stars—and, presumably, the nebulae—are distant, the immense spaces they inhabit may be penetrated by using large telescopes: More stars can be seen, Smith wrote, "as the aperture is more enlarged to take in more light."[13] Herschel took this lesson utterly to heart. His career was one long epitomization of the principle that telescopes enable us to see *into* space, and that the larger the telescope, the farther we can see.

Herschel began by purchasing a refracting telescope, but he soon found, as Newton had, that it suffered from chromatic aberration, meaning that it tended to introduce false colors. This defect eventually would be overcome by the development of compound apochromatic lenses, but at the time Herschel got into as-

The spiral arms of the Milky Way galaxy are produced by the light of millions of fiercely burning young, giant stars.

tronomy the only way to avoid it in refracting telescopes was to make them with very long focal lengths. This situation had driven observers to extremes. John Flamsteed erected a refractor 90 feet long at the Royal Greenwich Observatory, and Cassini in Paris studied Saturn through a series of ever more ambitiously constructed telescopes with focal lengths of 17, 34, 100, and 136 feet. Since a rigid tube of such a length could scarcely be constructed, much less mounted successfully, the tube often was done away with, and the objective lens was instead mounted on the highest available platform, such as the roof of a tall public building or, in the case of James Pound of England, on a maypole in Wanstead Park. The observer stood several city blocks away, eyepiece in hand, and sighted on the distant lens, awaiting the few precious moments when the planet Jupiter or the binary star Epsilon Lyrae would swim across his field of view. An astronomer blessed with great patience could, occasionally, make useful observations with such a contraption—Bradley in 1722 managed to measure the angular diameter of Venus using a tubeless refractor 212 feet long—but most found such reedy spyglasses so unwieldy that the cure was worse than the disease. Herschel constructed refractors with focal lengths of 4, 12, 15, and 30 feet, then gave up on them. "The great trouble occasioned by such long tubes, which I found it almost impossible to manage, induced me to turn my thoughts to reflectors," he wrote. He rented a small reflecting telescope of the sort invented by Newton, and found it "so much more convenient than my long glasses that I soon resolved to try whether I could not make myself such another."[14]

This decision marked the beginning of extragalactic astronomy, and the end of Herschel's leisure. Soon he was at work in every free hour, casting metal mirrors and laboriously grinding them to the precise concave figure required to bring starlight to a sharp focus. His sister Caroline—who had joined him in England in hopes of singing with the orchestra, but found instead that their lives were being given over to astronomy and their home turned into an optics shop—helped out by reading to him and feeding him sandwiches while he ground and polished mirrors for up to sixteen hours at a stretch. With a delicacy of touch that he attributed to his boyhood training as a violinist, Herschel fashioned hardwood telescope tubes as elegant as cellos and topped them off with magnifying eyepieces made of cocus, the wood used in oboes like the one he had played as a boy. Less than ten years after he opened his first astronomy books, he could boast confidently "that I absolutely have the best telescopes that were ever made."[15]

Herschel's skill as an observer was equally refined; he had a way with telescopes. "Seeing is in some respect an art, which must be learnt," he wrote.

> I have tried to improve telescopes and practiced continually to see with them. These instruments have played me so many tricks that I have at last found them out in many of their humours and have made them confess to me what they would have concealed, if I had not with such perseverance and patience courted them.[16]

With the ardor of a man possessed, Herschel stayed at the telescope on virtually every clear night of the year, all night long, taking only a few minutes off every three or four hours to warm himself—or, as happened one night when the temperature dropped to 11 degrees Fahrenheit, to fetch a tool to break through the ice that had glazed over his inkwell. He rushed to the telescope to observe during intermissions in the concerts he conducted at Bath. When skies were cloudy he and Caroline waited up, hoping for a change in the weather. "If it had not been for the intervention of a cloudy or moonlit night I know not when he or I either would have got any sleep," wrote Caroline in her diary.[17] When they moved to Datchet, to a dank house so near the Thames that the yard was often flooded, Herschel waded through the water and climbed to the eyepiece of the telescope, staving off ague by rubbing onion on his hands and face. "He has an excellent constitution," wrote Caroline, "and thinks about nothing else in the world but the celestial bodies."[18]

Herschel's preferred method of observing consisted of "sweeping" the sky. Wearing a black hood to keep any stray light from dazzling his dilated, dark-adapted eyes, he would move the telescope across a segment of sky, pausing to note the locations of interesting objects, then move the telescope slightly in the perpendicular and sweep back along an adjacent path. Ten to thirty such oscillations he called a sweep, and he registered each in what he called his "Book of Sweeps." This was making a virtue of necessity; his telescope lacked the equatorial mountings and clock drives that are employed today to compensate for the earth's rotation and to hold a single object effortlessly in view. Its great advantage was that it encouraged Herschel to memorize whole swathes of sky; the most significant northern hemisphere star map of the later eighteenth century may well have existed not on the pages of a celestial atlas but in Herschel's mind.

It was to this familiarity with the sky that Herschel owed his discovery, on the night of March 13, 1781, of the planet Uranus. Uranus had been glimpsed dozens of times before, by Bradley, Flamsteed, and others, but always had been mistaken for a star. Herschel, his mind an encyclopedia of the night sky, realized as soon as he saw it that no star belonged there. At first he mistook the little green dot for a comet, but the Astronomer Royal, Nevil Maskelyne, calculated its orbit and determined that it must be a planet, one far beyond Saturn. In a single stroke, Herschel had doubled the radius of the known solar system. The resulting fame brought him a fellowship in the Royal Society, a pension, and an appointment as astronomer to King George III—who was being blamed for losing the American Revolution and was suffering a mental breakdown at the time, and must have felt grateful for a little good news.

Herschel received a royal grant of four thousand pounds to build and operate what would be the world's largest telescope. Out of his own funds he had already managed to build a reflector twenty feet long, with a mirror eighteen and a half inches in diameter, but there were clear signs that he had pushed his private efforts about as far as they could go. Most ominous was the episode of the horse-dung mold. Herschel had wanted to cast a mirror fully three feet in diameter, with three times the light-gathering power of the eighteen-inch. No foundry would take on the unprecedented project, so Herschel resolved to do it himself, in the basement of his house at 19 New King Street in Bath. He constructed an inexpensive mold out of what the uncomplaining Caroline described as "an immense quantity" of horse dung. She, William, and their brother Alex took turns pounding the dung, assisted by their friend William Watson of the Royal Society. Finally came the day to, as Herschel put it, "cast the great mirror." At first all went well, but then the mold cracked under the intense heat and molten metal flowed out across the floor, exploding flagstones and sending them caroming off the ceiling. The party fled into the garden, pursued by a rapidly expanding pool of liquid metal. Herschel took refuge on a pile of bricks and there collapsed. He had reached the practical limits of amateur telescope making.

The largest telescope in the world was built, therefore, with the king's money, by a team of workmen under Herschel's direction. It had a forty-eight-inch mirror that weighed a ton, housed in a tube forty feet long. To reach the eyepiece, Herschel had to climb a scaffolding that rose fifty feet into the air. Oliver Wendell Holmes described the instrument as "a mighty bewilderment of slanted masts, spars and ladders and ropes, from the midst of which a vast tube . . . lifted its mighty muzzle defiantly towards the sky."[19] At its dedication, the king took the archbishop of Canterbury by the arm with the words, "Come, my Lord Bishop, I will show you the way to Heaven."[20]

With the forty-eight-inch reflector, Herschel discovered Enceladus and Mimas, the sixth and seventh satellites of Saturn, but ultimately the heroic telescope proved to be a disappointment. Training it on a given piece of sky was a taxing process that involved shouting instructions down to a team of laborers stationed in the rigging below, and its mirror tended to warp and mist over with changes in temperature and humidity. Herschel soon returned to working with smaller telescopes he had built by hand.

The nebulae continued to fascinate him. In 1781 he received a copy of Charles Messier's new catalog of these glowing islands of light, promptly set to work observing them, and found that "most of the nebulae . . . yielded to the force of my light and power, and were resolved into stars." He concluded, prematurely, that *all* nebulae were but star clusters, and could be resolved into their constituent stars once large enough telescopes were employed in observing them. His confidence in this comprehensive but erroneous hypothesis was shaken by his subsequent investigations of what he labeled the "planetary" nebulae—the ones now known to be shells of gas ejected by stars. When Herschel observed planetary nebulae in which the central star was too dim to be seen, he assumed that they were globular star clusters. But then, on the night of November 13, 1790, he came upon a planetary nebulae in Taurus with a clearly visible central star. He appreciated its significance immediately. "A most singular Phaenomenon!" he wrote in his journal. "A star of about the eighth magnitude, with a faint luminous atmosphere. . . . The star is perfectly in the center and the atmosphere is so diluted, faint and equal throughout, that there can be no surmise of its consisting of stars; nor can there be a doubt of the evident connection between the atmosphere and the star." He decided that some nebulae must, after all, be composed not of stars but of "a shining fluid" of unknown constitution. "Perhaps it has been too hastily surmised that all milky nebulosity, of which there is so much in the heavens, is owing to starlight only," he wrote, modifying his earlier hypothesis. "What a field of novelty is here opened to our conceptions!" he exclaimed, more delighted by the variety of the sky than bothered at having been wrong.[21]

Herschel could be astonishingly acute. He called the Orion Nebula, a knot of congealing gas sixteen hundred light-years from Earth, "the chaotic material of future suns," which is exactly what it is.[22] He argued that the sun belongs to a vast cluster of stars— a galaxy, as we would say today—and he tried to map its boundaries, by counting stars of various magnitudes in various directions in the sky. This effort failed, owing both to the fact that apparent magnitude is not a reliable index to the distance of stars and to the presence of dark, obscuring nebulae in the Milky Way that Herschel mistook for empty space. Nonetheless the inspiring fact remains that an oboe player with a handmade telescope undertook, in the eighteenth century, a scientifically defensible project aimed at charting the extent of the entire Milky Way galaxy.

Herschel studied other galaxies, too, notably the great nebula in Andromeda, which he assumed, correctly, to glow with "the united luster of millions of stars." He even noted that the central part of Andromeda was of "a faint red color." The central region of this giant galaxy is, indeed, warmer in hue than the surrounding disk—it consists of old red and yellow stars, while young blue stars predominate in the surrounding disk—but it seems incredible that this distinction, which was not fully established until the twentieth century, could have been detected by the eye of an eighteenth-century astronomer. And yet, Herschel being Herschel, one sometimes wonders.

In any case, Herschel's legacy has less to do with the extent to which his conclusions were right or wrong than with his prophetically modern approach to deep-space astronomy. At a time when most astronomers were peering at planets through the narrow fields of refracting telescopes, Herschel was harvesting great swaths of ancient light from distant nebulae and galaxies. While they were refining their estimates of distances within the solar system to the second decimal place, he was endeavoring to chart the starshoals of intergalactic space. While they were using estimates of the velocity of light to adjust their calculations of the orbits of the satellites of Jupiter, he was, he realized, seeing so far into space as to be viewing the universe as it looked millions of years in the past. Herschel's use of large reflecting telescopes to discern what he called "the construction of the heavens" may have been technologically precipitate, but it presaged the methods of the twentieth-century astronomers who were to realize his dreams. Cosmology for Kant and Lambert had been principally an indoor discipline; Herschel took it outdoors.

Sustained by his love for what he called "this magnificent collection of stars" in which we live, Herschel kept working until the end. "Lina, there is a great comet," he wrote his sister Caroline

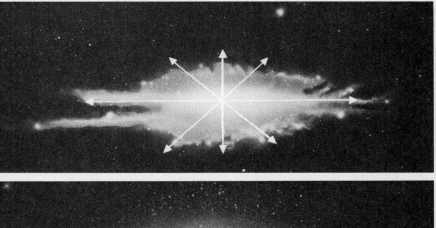

William Herschel sought to chart our galaxy by counting stars of given apparent magnitudes in all quarters of the sky (*top*). The resulting chart (*bottom*), though extremely rough, hinted at the existence of the galactic plane.

on July 4, 1819. "I want you to assist me. Come to dine and spend the day here. If you can come soon after one o'clock we shall have time to prepare maps and telescopes. I saw its situation last night, it has a long tail."[23] He was eighty years old at the time, and he was still at work when he died, two years later.

9

ISLAND
UNIVERSES

The light of the fixed stars is of the same
nature [as] the light of the sun.
—Newton

Observations always involve theory.
—Edwin Hubble

Two schools of thought about the nature of the
elliptical nebulae held sway in the nineteenth century.

One, the "island universe" theory of Kant and Lambert—the
phrase is Kant's—maintained that our sun is one among many stars
in a galaxy, the Milky Way, and that there are many other galaxies,
which we see across great gulfs of space as the spiral and elliptical
nebulae. The other, the "nebular hypothesis," maintained that the
spiral and elliptical nebulae are whirlpools of gas condensing to
form stars, and that they are nearby and relatively small. The
nebular hypothesis also had originated with Kant, but was usually
called "Laplacian," after the French mathematician Pierre-Simon
de Laplace, who had published a detailed account of how the sun
and its planets might have congealed from a whirling nebula. Both
theories were to some extent correct—some nebulae are, indeed,
star-forming gas clouds, while the elliptical and spiral nebulae are
galaxies of stars—but there was an understandable tendency to

assume that a single theory would explain all types of nebulae, and this assumption bred confusion.

The observational evidence seemed to favor the nebular hypothesis. Most spectacular was the discovery by William Parsons, the third earl of Rosse, that some elliptical nebulae display a spiral structure. Lord Rosse, who employed a six-foot reflecting telescope that was at the time the largest in the world, actually was seeing spiral galaxies, but his observations were thought instead to support the nebular hypothesis, with its vision of stars condensing from whirlpools of gas. This impression was strengthened when photographs taken by Isaac Roberts in England in the 1880s revealed that most elliptical nebulae are spirals; when Roberts's photographs were exhibited, at the Royal Astronomical Society in London in 1888, learned spectators were said to have gasped in recognition at the photographic evidence of "the nebular hypothesis made visible."[1] The hypothesis gained even more ground when time-exposure photographs made by James Keeler at Lick Observatory in California in the 1890s indicated that there are a great many spiral nebulae; Keeler estimated that over one hundred *thousand* spiral nebulae lay within the range of the Lick telescope. Hundreds of thousands of new solar systems seemed plausible, given the multitude of suns that bedeck the Milky Way, but it strained credulity to imagine that there could be hundreds of thousands of galaxies, each home to billions of stars.

The riddle ultimately was solved, not by the telescope or the camera alone, but by combining both with the spectroscope, which was to reveal what the stars and nebulae are made of—something that the philosopher Auguste Comte, as late as 1844, could cite as an example of knowledge forever denied the human mind.

The development of spectroscopy dates from 1666, when Newton noted that white sunlight directed through a prism produces a rainbow of colors. In 1802, the English physicist William Wollaston found that if he placed a thin slit in front of the prism the spectrum displayed a series of parallel dark lines, like the cracks between piano keys. But Wollaston set the experiment aside, and the elevation of spectroscopy to the status of an exact science was left to a skinny, impoverished teenager with a persistent cough, who when Wollaston made his discovery was in hospital, recuperating from injuries he had suffered in the collapse of the optics shop where he worked in the Munich slums. His name was Joseph Fraun-hofer, and his fortunes were about to improve.

Optics in the early nineteenth century was a growth industry. Napoleon Bonaparte's passion for maps and spyglasses had set sur-veyors and generals to writing orders for portable telescopes and theodolites, and the research of William Herschel and his son John, who charted the southern skies from an observatory at the Cape of Good Hope, had inspired interest in large telescopes among both enthusiasts who wanted to view the wonders of deep space for themselves and skeptics who were out to test the Herschels' claims. A new breed of artisans prospered—the opticians, bitterly com-petitive, fiercely innovative, as hard as the brass and glass they worked with and as eccentric as the scientists and engineers they served. Emblematic of the breed was Jesse Ramsden of London, a perfectionist who toiled over his projects until he got them right, no matter how long it took; the eight-foot altitude-measuring circle that he crafted for Dunsink Observatory in Dublin, admittedly a masterpiece of precision, was delivered twenty-three years after the contract deadline.*

If the opticians expected to be treated like artists, that is just what many of them were. Alvan Clark, the great American tele-scope-maker, prospered as a portrait painter before he switched careers and built what are still regarded as the finest refracting telescopes in the world; keen-sighted, Clark was said to be able to fire six rifle bullets "through a distant board with such precision that one would say only a single shot had been fired," and to detect tiny bubbles and ripples in glass that were invisible to lesser mortals.[2]

Fraunhofer was born into the steerage class of this flourishing profession. The eleventh son of an indigent master glazier, he had been orphaned at age eleven and apprenticed to one Philipp Weich-selberger, a dull-witted Munich glasscutter who kept him over-worked, underpaid, underfed, and uneducated. On July 21, 1801, the dilapidated building that comprised Weichselberger's house and shop collapsed, and Fraunhofer, the only survivor, was at length pulled from its wreckage. His rescue made news, and his plight

*Equally unpunctual in his social commitments, Ramsden once arrived for a party at Buckingham Palace at the hour and day inscribed on an invitation sent him by the king, but one year late.

attracted the attention of Maximilian Joseph, the elector of Bavaria, who visited the injured boy in hospital and was impressed by his intelligence and cheerful disposition. The elector made Fraunhofer a present of eighteen ducats, enough to buy a glass-working machine, books, and release from what was left of his apprenticeship. Once free, Fraunhofer never looked back. He had an instinct for the essential, and his spirited research into the basic characteristics of various kinds of glass soon established him as the world's foremost maker of telescope lenses.

Fraunhofer started out using spectral lines as sources of monochromatic light for his experiments in improving the color correction of his lenses, but soon became fascinated by the lines themselves. "I saw with the telescope," he wrote, "an almost countless number of strong and weak vertical lines which are darker than the rest of the color-image. Some appeared to be perfectly black."[3] He mapped hundreds of such lines in the spectrum of the sun, and found identical patterns in the spectra of the moon and planets—as one would expect, since these bodies shine by reflected sunlight. But when he turned his telescope on other stars, their spectral lines looked quite different. The significance of the difference remained a mystery.

Fraunhofer died on June 7, 1826, at the age of thirty-nine, of tuberculosis, leaving the mysterious Fraunhofer lines as his legacy. In 1849, Léon Foucault in Paris and W. A. Miller in London found bright lines that coincided with Fraunhofer's dark lines. Today these are known respectively as the emission and absorption lines, and they play a role in spectroscopy as potent as that of fossils in geology, producing information on the temperatures, compositions, and motions of gaseous nebulae and stars.

In the years 1855 through 1863, the physicists Gustav Kirchhoff and Robert Bunsen (the inventor of the Bunsen burner) determined that distinct sequences of Fraunhofer lines were produced by various chemical elements. One evening they saw, from the window of their laboratory in Heidelberg, a fire raging in the port city of Mannheim ten miles to the west. Using their spectroscope, they detected the telltale lines of barium and strontium in the flames. This set Bunsen to wondering whether they might be able to detect chemical elements in the spectrum of the sun as well. "But," he added, "people would think we were mad to dream of such a thing."[4]

Kirchhoff was mad enough to try, and by 1861 he had identified sodium, calcium, magnesium, iron, chromium, nickel, barium, copper, and zinc in the sun. A link had been found between the physics of earth and the stars, and a path blazed to the new sciences of spectroscopy and astrophysics.

In London, a wealthy amateur astronomer named William Huggins learned of Kirchhoff's and Bunsen's finding that Fraunhofer lines were generated by known chemical elements in the sun, and saw at once that their methods might be applied to the stars and nebulae. "This news came to me like the coming upon a spring of water in a dry and thirsty land," he wrote.[5] Huggins fitted a spectroscope to the Clark telescope at his private observatory, on Upper Tulse Hill in London. By carefully studying each spectrum until he could make sense of their many overlapping lines, he succeeded in identifying iron, sodium, calcium, magnesium, and bismuth in the spectra of the bright stars Aldebaran and Betelgeuse. This was the first conclusive evidence that other stars are made of the same substances that we find here in the solar system.

With mounting excitement, Huggins turned his telescope to a nebula. His journal for the year 1864 records the feeling "of excited suspense, mingled with a degree of awe, with which, after a few moments hesitation, I put my eye to the spectroscope. Was I not about to look into a secret place of creation?" He was not disappointed:

> I looked into the spectroscope. No spectrum such as I expected! A single bright line only! . . . The riddle of the nebulae was solved. The answer, which had come to us in the light itself, read: Not an aggregation of stars, but a luminous gas. Stars after the order of our own sun, and of the brighter stars, would give a different spectrum; the light of this nebula had clearly been emitted by a luminous gas.[6]

Because this first nebula Huggins observed with his spectroscope happened to be gaseous, he was led to the erroneous conclusion that all nebulae, the ellipticals and spirals included, were gaseous and that none was composed of stars.

But life is seldom simple, and misleading evidence for the nebular hypothesis continued to accumulate. The positions of hundreds of spiral nebulae were charted, and they were found to be most numerous in the parts of the sky that lie well away from the Milky Way—to "avoid" the Milky Way, in astronomical jargon. The avoidance effect suggested that the spiral nebulae were associated with our galaxy. (Actually, avoidance results from the fact that dark clouds along the plane of *our* galaxy obscure our view of the *other* galaxies, so that we see mostly those that lie away from the galactic plane.) The nebular hypothesis was strengthened on the theoretical front as well, when the astrophysicist James Jeans demonstrated, with considerable mathematical rigor, that a collapsing cloud of gas would tend to assume a disk shape much like that of the spiral nebulae. Jeans even managed to coax his model into generating spiral arms like those seen in the astrophotographs.

By now the nebular hypothesis was so successful that a bandwagon syndrome took over and astronomers began seeing what they thought they ought to see. One announced that he had measured the parallax of the Andromeda spiral. (Parallax is detectable only out to a few hundred light-years; the Andromeda galaxy is over two *million* light-years away.) Another found that by examining older photographs he could detect signs of circular motion in spiral nebulae. (In reality, galaxies are so large that to see a galaxy turn by as much as the second hand on a clock moves in one second would require taking two photographs separated by an interval of fully five million years.)

As the twentieth century began, then, several of the most stupefying aspects of the closed, pre-Copernican cosmology had been resurrected on a galactic scale. The sun was widely thought to be located at or near the center of a stellar system—the Milky Way—which embraced every star and nebula in the telescopic sky, and which, therefore, constituted nothing less than the entire observable universe. Beyond our galaxy might lie an infinite void, but this question remained as purely academic as had been the nature of space beyond the outer sphere of stars in Aristotle's model.

But there *is* a self-correcting mechanism to science, and by the turn of the century it had begun to assert itself. The first cracks in the facade of the nebular hypothesis appeared on the theoretical side, when a fatal defect was identified in the Jeans theory of how the solar system had condensed. Were the hypothesis correct, the mathematicians calculated, the sun should have retained most of the angular momentum of the solar system, and be spinning very rapidly; instead, the solar "day" lasts a leisurely twenty-six days at the sun's equator, and the planets harbor 98 percent of the angular momentum of the solar system.* The observational evidence began to turn against the nebular hypothesis as well. Huggins took a spectrum of the Andromeda nebula in 1888 but found it hard to interpret. Nine years later, Julius Scheiner in Germany published a spectrum of the Andromeda nebula, noting that the spectrum was not gaseous but starlike. Undoubtedly, at least some spiral nebulae were made of stars.

Exploding stars then came to the astronomers' aid, as they had centuries earlier for Tycho, Kepler, and Galileo. Two or three supergiant stars explode in an average major galaxy every century, with such brilliance that they can be seen across the reaches of intergalactic space. Since thousands of galaxies (or elliptical and spiral nebulae, as they were then being called) lay within the reach of existing telescopes and cameras, it was only a matter of time before supernovae began to be detected in photographs of other galaxies. The first such extragalactic supernova to be noticed, in Andromeda in 1885, happened to be near the center of the spiral, and so could be explained away as the sputtering of a Laplacian protosun. But then, in 1917, George Ritchey, an optician at Mount Wilson, and Heber Curtis, an astronomer at Lick, announced that they had found several novae in old file photographs of spirals. Other astronomers started ransacking their plate files, and found scores more. The novae were not central, but occurred primarily in the spinal arms. This was extremely damaging to the notion that all nebulae were gaseous: Dozens of exploding stars in galaxies full of stars made sense; in Laplacian gas disks, they did not. As Curtis commented, "The novae in spirals furnish weighty evidence in favor of the well known 'island universe' theory."[7]

The stage was set for the discovery of galaxies. What remained

---

*By the 1980s, theoretical astrophysicists using computer models had derived a general theory of the origin of the solar system that, though more sophisticated than those of Kant, Laplace, or Jeans, resembles them at least superficially. The new model envisions the sun congealing from a nebular cloud, the remnants of which formed a flattened disk that condensed, as it cooled, into a multitude of little chunks of material, or "planetesimals," which in turn collided to form the planets. Indirect confirmation of the theory came when an orbiting infrared telescope detected cold, planetesimallike disks around Vega and several other bright, young stars. The details of the theory, however, are quantitatively difficult, and still have not been worked out. It is one of the humbling truths of contemporary science that, while we theorize about the origin of the entire universe, we do not yet fully understand how our own little planetary system began.

was the most expansive surveying project in the history of our planet—to chart the location of the solar system in the Milky Way, and to determine the distances of the other galaxies, if such they were, beyond.

The champion of this cause was the founder of observational astrophysics, George Ellery Hale. Hale's early career reenacted the progression of spectroscopy from the sun to the stars. He became enchanted by the sun as a boy growing up in the Chicago suburbs, built a backyard observatory where he observed solar spectra, and by the age of twenty-four had invented the spectrohelioscope, a device that made it possible to examine the solar atmosphere in one wavelength of light at a time. Captivated by the realization that, as he kept repeating all his life, "the sun is a star," he then turned his attention to the depths of space. He was responsible for building four telescopes, each in its day the world's largest—the 40-inch refractor at Yerkes Observatory in Wisconsin and, in southern California, the 60- and 100-inch reflectors at Mount Wilson and the 200-inch reflector at Palomar. Mount Wilson in particular stood as a monument to Hale's dual passions in spectroscopy: There, solar telescopes recorded the spectra of the sun by day, while by night giant reflecting telescopes were employed to probe the multitude of other suns scattered through the Milky Way and beyond.

Hardworking even by the hard-boiled standards of the opticians and astronomers of the day, Hale rode mules up the rocky, twisting road from Pasadena to Mount Wilson's peak, and when no mules were available simply ran up the side of the mountain. He did a lifetime's worth of research of his own and managed simultaneously to act as the observatory director, raising funds for ever larger telescopes and recruiting some of the world's leading astronomers to Mount Wilson. One of the cleverest of his recruits was Harlow Shapley.

Shapley had studied at Princeton Observatory under Henry Norris Russell, where he specialized in Eclipsing Binaries. These are double stars, so close together in the sky that they look like single stars even through the most powerful telescopes, that happen to be oriented in space in such a way that they periodically eclipse each other. The resulting variations in the total brightness of the system bear a superficial resemblance to genuine variable stars, which change their brightness owing to internal pulsations. In this fashion Shapley came to study variable stars as well. The knowledge he gained in this somewhat backhanded way came in handy, for a class of variable stars—the Cepheid variables—were to provide astronomy with a means of measuring distances across interstellar space and even intergalactic. Thanks to the Cepheids, Shapley was to earn a place in history as the first human being to establish the location of the sun in the Milky Way galaxy.

Cepheids—as Shapley was the first to propose—pulsate, varying in brightness as they change in size. Astrophysically speaking, they are giant stars, three or more times the mass of the sun, undergoing a period of instability that occurs when they run low on hydrogen fuel and begin burning helium. The wonderful thing about them is that the period of each Cepheid—i.e., the time it takes to go through a cycle of variation in brightness—is directly related to its intrinsic brightness (i.e., its absolute magnitude). Once the absolute magnitude of any star is known, it is a simple matter to compute its distance: All the astronomer has to do is measure its *apparent* magnitude and then apply the formula that brightness decreases by the square of the distance. If, for instance, we have two Cepheid variables with the same period, we may assume that they have about the same absolute magnitude. If the apparent magnitude of one is four times that of the other, we conclude (barring complications such as the interference of an intervening interstellar cloud) that the dimmer star is twice as far away.

The relationship between the periodicity and the absolute magnitude of Cepheid variable stars was discovered in 1912 by Henrietta Swan Leavitt, one of a number of women hired at meager wages to work as "computers" in the Harvard College Observatory office in Cambridge, Massachusetts. Leavitt spent her days examining photographic plates taken through the twenty-four-inch refracting telescope at the Harvard station in Arequipa, Peru. One of her tasks was to identify variable stars. This involved comparing thousands of pinpoint star images on plates taken on different dates, looking for changes in brightness. It was painstaking toil, considered too menial to claim the time of a full-fledged astronomer. Leavitt spent thousands of hours at it, and in doing so acquired an unusual degree of familiarity with the southern sky.

She happened to be assigned to a region that includes the Magellanic Clouds. So named because they attracted the attention of Magellan and his crew on their voyage around the world, the Magellanic Clouds are two large, shaggy patches of softly glowing light that resemble detached swatches of the Milky Way. We know today, as Leavitt and her contemporaries did not, that the Clouds are nearby galaxies, and that the stars in each Cloud therefore all lie at about the same distance from us, like fireflies in a bottle viewed from across a field at night. This means that any significant difference in the apparent magnitudes of stars in a Magellanic Cloud must result from genuine differences in their absolute magnitudes and not from the effect of differing distances. Thanks to this happy circumstance, Leavitt in studying Cepheid variable stars in the Magellanic Clouds was able to notice a correlation between their brightness and their period of variability—the brighter the Cepheid, the longer its cycle of variation. The period-luminosity function Leavitt discovered was to become the cornerstone of measuring distance in the Milky Way and beyond.

Shapley, out to chart the Milky Way galaxy, seized on the Cepheids with great enthusiasm. Using the big sixty-inch Mount Wilson telescope, he photographed globular star clusters—spectacular assemblages of hundreds of thousands to millions of stars —identified Cepheid variable stars in each of them, then employed the Cepheids to calibrate the distances of the clusters. "The results are continual pleasure," he wrote the astronomer Jacobus Kapteyn in 1917. "Give me time enough and I shall get something out of the problem yet."[8] The payoff came sooner than Shapley had hoped, and within a matter of months he could write, to the astrophysicist Arthur Stanley Eddington: "Now, with startling suddenness and definiteness, they [the globular clusters] seem to have elucidated the whole sidereal structure."[9]

Shapley had found that the globular clusters are distributed across a spherical expanse of space, as if they were part of an enormous metaglobular cluster, and that the center of this sphere is nowhere near the sun, but lies far away to the south, past the stars of Sagittarius. In a superbly daring intuitive leap, Shapley then conjectured—accurately, as it turned out—that the center of the realm of the globular clusters was also the center of the galaxy itself. As Shapley put it, "The globular clusters are a sort of framework—a vague skeleton of the whole Galaxy—the . . . best indicators of its extent and orientation." If so, the sun lies far from the center of things: "The solar system can no longer maintain a central position," Shapley asserted.[10]

Shapley's triumph was marred only by problems with his calculation of distances. The diameter of the Milky Way galaxy previously had been reckoned—by various investigators, Shapley among them—at some fifteen to twenty thousand light-years. Now, with his Cepheid variable work in hand, Shapley concluded that the correct figure was three hundred thousand light-years—more than ten times larger than the dimensions entertained by his contemporaries, and three times the most generous estimates we have today.*

Various errors contributed to Shapley's inflated picture of the Milky Way galaxy. Like many of his contemporaries, he underestimated the extent to which clouds of interstellar gas and dust dim the images of distant stars, making them appear farther away than they really are. Moreover, he assumed that the Cepheid variable stars he observed in globular clusters were essentially identical to those Henrietta Leavitt had found in the Magellanic Clouds; actually, as Walter Baade and other astrophysicists were to find, the cluster variables are less massive and intrinsically less bright, and therefore by implication less distant, than a straightforward comparison of their periods with those of their younger cousins would imply. Inaccuracies of this sort are routine on the cutting edge of science, but they had the dolorous effect of misleading Shapley into thinking that the Milky Way, rather than being but one galaxy among many, was a system of unique grandeur. He began to think of the Milky Way as more or less the entire universe, and to regard the spiral nebulae but its subjects or its satellites.

For these and perhaps for subtler psychological reasons as well, Shapley came to take a proprietary interest in defending what he called "the enormous, all-comprehending" dimensions of the galaxy that he had charted. He called this view his "big galaxy" hypothesis. Those who agreed with him tended to think of the word "big" in terms of its Norse etymology, from *bugge*, meaning "important." Those who disagreed preferred to emphasize the word's Latin etymology, from *buccae*, for "puffed up."

Among the dissidents was Heber Curtis of Lick Observatory, an advocate of the "island universe" theory. Shapley reacted to

---

*Modern estimates put the diameter of the Milky Way disk at seventy to one hundred thousand light-years. There probably are dim stars much farther out, however, as well as stray halo stars and "tramp" globular clusters orbiting the galaxy at distances of over three hundred thousand light-years from the galactic center.

Curtis's arguments with the abhorrence of a patient contemplating
surgery: Curtis, he noted, "must shrink my galactic system enormously to have much luck with island universes."[11] The issue was
formally debated by Shapley and Curtis under the auspices of the
National Academy of Sciences, in Washington, D.C., on April 26,
1920. Shapley was generally judged the loser, but, as is usually the
case in science, the debate settled little and the last word belonged
not to men but to the sky.

The hypothesis defended by Curtis, that the spiral nebulae
were galaxies of stars, would be confirmed if a spiral could be
unambiguously resolved into stars. That vital step was accomplished in 1924 by Shapley's colleague and nemesis, Edwin Hubble.
A tall, elegant, and overbearing man with a highly evolved opinion
of his potential place in history, Hubble made everything he did
look effortless—he had been a track star, a boxer, a Rhodes scholar,
and an attorney before turning astronomer—and one of the things
he did most effortlessly was to infuriate Shapley. Hubble took
scores of photographs of M33 and its neighbor M31, the Andromeda
spiral, and found there what he later called "dense swarms of images
in no way differing from those of ordinary stars."[12]

Whether the pinpoints of light on Hubble's photographic plate
really *were* stars, however, was open to contention; Shapley and
other advocates of the island universe hypothesis dismissed them
as curds in a Laplacian nebula. Here, again, Henrietta Leavitt's
Cepheid variable stars provided the needed mileposts. Cepheids are
bright enough to be discernible across intergalactic distances. Using
the new one-hundred-inch telescope at Mount Wilson, Hubble photographed the spirals again and again, comparing the plates to find
stars that had varied in brightness. His efforts soon bore fruit, and
on February 19, 1924, he wrote Shapley, who by then had left
Mount Wilson to become director of Harvard College Observatory,
a laconic note containing one of the most momentous findings in
the history of science: "You will be interested to hear that I have
found a Cepheid variable in the Andromeda Nebula."[13]

Hubble deduced that Andromeda lies about one million light-
years away, an estimate half the distance of later ones but clearly
sufficient to establish that the spiral was well beyond even Shapley's
"big galaxy." Shapley replied sourly that he found Hubble's letter
to be "the most entertaining piece of literature I have seen for a
long time."[14] Later he complained that Hubble had given him in-

sufficient credit for his priority in using Cepheid variables to chart distances. But the game was over. Hubble's paper announcing that he had found Cepheids in spirals—read (in his Olympian absence) at a joint meeting of the American Astronomical Society and the American Association for the Advancement of Science in Washington, D.C., on New Year's Day, 1925—initiated the final decline of the nebular hypothesis, the ascendancy of the island universe hypothesis, and humankind's realization that we live in one among many galaxies.

Hubble went on to identify not only Cepheids but novae and giant stars in Andromeda and other galaxies. These studies helped allay his fear that the laws of physics might break down beyond our home galaxy, rendering his distance measurements invalid. Newton, too, had wondered whether "God is able . . . to vary the Laws of Nature, and make Worlds of several sorts in several Parts of the Universe."[15] Hubble, in his short paper announcing the finding of Cepheids in M31, took time to caution that his results depended upon the assumption that "the nature of Cepheid variation is uniform throughout the observable portion of the universe." When he found Cepheids and other familiar stars in the galaxy NGC 6822, he wrote with evident relief that "the principle of the uniformity of nature thus seems to rule undisturbed in this remote region of space."[16]

Some astronomers have a gift for making lovely, sharp photographs of galaxies with large telescopes. Hubble was not one of them, though he was adept at extracting essential data from the generally flawed plates he did obtain. Nor was he especially skillful at taking spectra, but in this he was soon aided by one Milton Humason, a resourceful young man with an inquiring mind who started out on Mount Wilson as a muleteer and observatory janitor, began assisting the astronomers in their work at the telescope, and eventually became an expert observational astronomer in his own right. Throughout the 1930s and 1940s, Hubble and Humason pushed back the frontiers of the observable universe, charting and cataloging ever more distant galaxies. Eventually, Hubble was taking photographs that were strewn with the images of more remote galaxies than foreground stars.

In 1952, the year before Hubble died, Walter Baade announced at a meeting of the International Astronomical Union in Rome that he had discovered an error in the calibration of the Cepheid period-luminosity value, the correction of which doubled the cosmic distance scale. Further refinements in the distance scale were attained by Hubble's former assistant Allan Sandage, later in collaboration with the Swiss astronomer Gustav Tammann, and it became possible for astronomers to measure, with some confidence, the distance to galaxies hundreds of millions to billions of light-years away.

At these distances, time commands a significance equal to that of space. Inasmuch as it takes time for the light from a distant galaxy to pass through space, we see the galaxy as it was long ago: The galaxies of the Coma cluster, for instance, appear to us as they looked seven hundred million years ago, when the first jellyfish were just appearing on Earth. Owing to this phenomenon, called lookback time, telescopes probe not only out into space but back into the past. It should, therefore, be possible to determine, by looking far into deep space, whether the universe was once different than it is today. Evidence that this is indeed the case came in the 1960s, when Sandage and radio astronomer Thomas Matthews discovered quasars, and Maarten Schmidt determined that they were extraordinarily far away. Quasars appear to be the nuclei of young galaxies, at distances of a billion light-years and more. There is nothing quite like them in the universe today. And so the exploration of space opened the pages of cosmic history.

The work of charting our place in the universe goes on, and today we can say with some confidence that the sun is a typical yellow star that lies in the disk of a major spiral galaxy, about two thirds of the way out from the galactic center. The disk contains not only stars and their planets but also vast, rarefied lakes of hydrogen and helium gas, denser knots of gas where atoms have been able to find one another and bind together as molecules, and giant thunderheads of soot given off by smoky stars. Waves generated by harmonics in the gravitational interaction of the myriad stars move across the disk in a graceful, spiral pattern, plowing the interstellar material into globules dense enough to collapse under the attraction of their own gravitational force. In this way new stars are formed, and it is the light of the most massive and shortest-lasting of the young stars that illuminates the spiral arms, making them visible. The spiral arms, then, are not objects but processes —as transitory, by the bounteous spatiotemporal standards of the Milky Way, as the back-blowing veils of froth that whitecap the waves of earthly oceans.

Beyond the Milky Way lie more galaxies. Some, like the Large Magellanic Cloud and the Andromeda galaxy, are spirals. Others are ellipticals, their stars hung in pristine, cloudless space. Others are dim dwarfs, some not much larger than globular clusters. Most belong, in turn, to clusters of galaxies. The Milky Way is one of a few dozen galaxies comprising a gravitationally bound association that astronomers call the Local Group. That group in turn lies near one extremity of a lanky archipelago of galaxies called the Virgo Supercluster. If we could fly the sixty million or so light-years from here to the center of the supercluster, we would encounter many sights worth seeing along our way—the giant cannibal galaxy Centaurus A, an elliptical busily gobbling up a spiral that blundered into it; the distended spiral M51, with its one outflung arm stretching after a departing companion galaxy; the furiously glowing spiral M106, with its bright yellow nucleus and its shoals of blue-white stars; and, at the supercluster core, the giant elliptical Virgo A, wreathed in thousands of globular star clusters, harboring some three trillion stars, and adorned by a blue-white plasma jet that has been spat from its core with the velocity of a bolt of lightning.

Beyond Virgo lie the Perseus, Coma, and Hercules clusters, and beyond them so many more clusters and superclusters of galaxies that it takes volumes just to catalog them. There is structure even on these enormous scales; the superclusters appear to be arrayed into gigantic cosmic domains that resemble the cells of a sponge. Beyond *that*, light from faraway galaxies, riding the contours of curved space, becomes as dappled as the moon's reflection on a pond in a gentle breeze. Out there, awaiting some future Hubble or Herschel, lie many a tale of things past, or passing, or to come.

10

EINSTEIN'S SKY

I want to know how God created this world.
I am not interested in this or that phenomenon, in the spectrum of this or that element.
I want to know His thoughts, the rest are
details.

—Einstein

Once the validity of this mode of thought has
been recognized, the final results appear almost simple; any intelligent undergraduate can
understand them without much trouble. But
the years of searching in the dark for a truth
that one feels, but cannot express; the intense
desire and the alternations of confidence and
misgiving, until one breaks through to clarity
and understanding, are only known to him
who has himself experienced them.

—Einstein

In much the same way that Newton's account of
gravitation and inertia advanced physics to the point that it could
embrace a moving Earth in a heliocentric solar system, Einstein's

relativity enabled physics to deal with the much higher velocities, greater distances and more furious energies encountered in the wider universe of the galaxies. If Newton's domain was that of the stars and planets, Einstein's extended from the centers of stars to the geometry of the cosmos as a whole.

To bring about so great an expansion of the scope of science, Einstein was obliged to abandon Newton's conceptions of space and time. Newtonian space and time were inflexible and inalterable; they formed a changeless proscenium arch within which all events took place and against which all could be unambiguously measured. "Absolute space, in its own nature, without relation to anything external, remains always similar and immovable," Newton wrote. ". . . Absolute, true and mathematical time, of itself, and from its own nature, flows equally without relation to anything external."[1] Einstein determined that this assumption was both superfluous and misleading. The special theory of relativity revealed that the rate at which time flows and the length of distances gauged across space vary, according to the relative velocities of those measuring them. The general theory of relativity went on to portray space as curved, and derived from spatial curvature the phenomena that Newtonian dynamics had attributed to the force of gravity.

Einstein grew up in an age when the classical conception of space, if not of time, was already coming unraveled. In order to explain how "absolute" space could have any reality—and, more to the point, how light and gravitational force could be conveyed across the empty space separating the stars and planets—Newton and his followers had postulated that space is pervaded by an invisible substance, an *aether*. The word was borrowed from Aristotle's term for the celestial element of which the stars and planets were made, and like its namesake this new, updated aether was wonderful stuff. Lucid and friction-free, static and unchanging, it not only permitted the unimpeded motion of the planets and stars but actually wafted right through them—like a breeze through a grove of trees, as the English physicist Thomas Young put it.*

The appealing idea that space is pervaded by an aether began to run into trouble once it became possible to make precise meas-

---

*If, instead, the earth dragged the aether along with it, like a ship gathering up seaweed as it plows through the Sargasso Sea, then the aberration of starlight that Bradley had first observed (the effect of moving through starlight like a woman running in the rain) would not occur.

urements of the velocity of light. That light travels at a finite velocity had been appreciated since the 1670s, when the Danish astronomer Olaus Römer detected periodic variations in the time when Io, innermost of the four bright moons of Jupiter, went into eclipse: The eclipses came earlier than expected when Jupiter was relatively close to the earth and later when Jupiter was farther away. Römer realized that the discrepancy must be caused by the time it takes light to travel across the changing distance from Jupiter to Earth. From what was then known of the absolute distance of Jupiter, he was able to calculate the velocity of light to within about 30 percent of the accurate value (which is 186,272 miles per second).

Galileo had once tried to determine the velocity of light. He stationed two men with shuttered lanterns on hilltops about one mile apart, then timed the interval that elapsed between the instant when the first man opened his shutter and the second, responding to this signal, opened *his* shutter, sending a light beam back to the first. Römer's finding made it clear why Galileo had failed; the interval he had attempted to measure (without a clock!) was less than a hundred thousandth of a second.

Römer's result also suggested a way of measuring the velocity of the earth relative to absolute space: If light were propagated by a stationary aether, the absolute motion of the earth relative to the aether could be detected by measuring variations in the observed velocity of light. Imagine that the earth were a sailboat on an aether lake, and think of the light coming from two stars on opposite sides of the sky as ripples spreading from two stones dropped in the lake, one ahead of the boat and one behind. If we were standing on the deck of the boat and we measured the velocity of each set of ripples, we would find that those radiating from the stone dropped ahead would appear to be moving faster than those coming from behind. By measuring the difference in the observed velocity of the ripples coming from ahead and behind, we could calculate the speed of the boat. Similarly, it was assumed that the velocity of the earth's motion could be determined by observing differences in the velocity of light waves coming through the stationary aether from stars ahead and behind.*

To measure this "aether drift"—as it was called, though what

---

*I have adopted this metaphor from one employed by Einstein's colleague Banesh Hoffmann.[2]

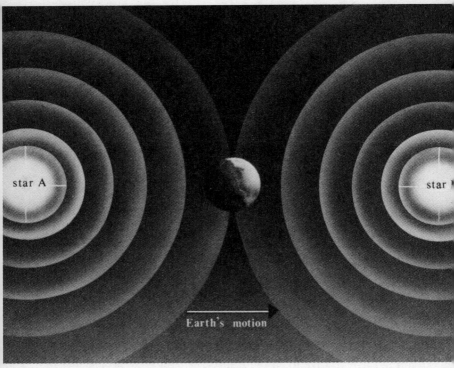

star A

Earth's motion

star B

Aether drift theory held that if the velocity of light was constant relative to a stationary, all-pervading aether, then when the earth in its orbit was moving away from star *A* and toward star *B*, the observed speed of the light coming from star *B* would be higher than that of the light coming from star *A*.

was thought to be drifting was not the aether but the earth—would of course be a delicate matter, since the velocity of the earth amounts to but a tiny fraction of the velocity of light. But by the latter part of the nineteenth century, technology had advanced to a sufficient degree of precision to make the task feasible. The critical experiment was conducted in the 1880s by the physicist Albert Michelson (who devoted his career to the study of light, he said, "because it's so much fun") and the chemist Edward Morley.

The Michelson-Morley apparatus, set up in a basement laboratory at Western Reserve University in Cleveland, Ohio, was based on the principle of interferometry. A beam of light was split and the two resulting light beams were reflected at right angles, then recombined and brought to a focus at an eyepiece. The idea was that the earth's motion through the stationary aether would show up as a change in the interference pattern produced when one of the light beams, the one that had to travel into the aether wind, was retarded relative to the other beam. As Michelson explained the principle to his young daughter Dorothy, "Two beams of light race against each other, like two swimmers, one struggling upstream and back, while the other, covering the same distance, just crosses and returns. The second swimmer will always win, *if there is any current in the river.*"[3] Since we know the earth is moving, there had to be *some* current—provided that, as Michelson and most other physicists then believed, there was such a thing as an aether that delineated the frame of reference of absolute Newtonian space.

To minimize exterior vibrations, the interferometer floated on a pool of mercury. To alter its orientation relative to the motion of the earth, it rotated on its mercury pool. Michelson spent days peering through the slowly moving eyepiece of the interferometer, looking for the telltale change in the interference patterns that would betray the earth's motion through the aether. To his intense disappointment, he saw no such change at all. The conclusion was as inescapable as it was repugnant to Michelson: There was no detectable "aether drift."

At first, few theorists were prepared to abandon the aether hypothesis, and several tried to reconcile it with the null outcome of the Michelson-Morley experiment. Their efforts gave rise to the bizarre idea that the experimental apparatus—and, indeed, the entire earth—contracted in the direction of its motion by just enough to cancel the effects of their velocity through the aether. "The only way out of it that I can see," said the Irish physicist George FitzGerald, "is that the equality of [light] paths must be inaccurate."[4] In other words, the two beams of light *seemed* to be of equal length, because their length was distorted by the very motion of the earth they were intended to detect. As FitzGerald put it, "The block of stone [holding the apparatus] must be distorted, put out of shape by its motion . . . the stone would have to shorten in the direction of motion and swell out in the other two directions."[5] The Dutch physicist Hendrik Antoon Lorentz independently arrived at the same hypothesis, and worked it out in mathematical detail.

This, the "Lorentz contraction," was to emerge in a different form as a key element in the special theory of relativity. The French physicist Henri Poincaré, one of the few leading scientists to take

the Lorentz contraction seriously, came close to developing it into
a form that was mathematically equivalent to Einstein's theory;
Poincaré spoke presciently of "a principle of relativity" that would
prescribe that no object could exceed the velocity of light.[6] But
most researchers found it odd to the point of desperation to suggest
that the velocity of the earth causes the entire planet to contract,
like an orange squashed between a titan's hands, and Lorentz himself soon set the idea aside. "I think he must have been held back
by fears," the physicist Paul Dirac speculated, years later. ". . . I
do not suppose that one can ever have great hopes without their
being combined with great fears."[7]

Enter Einstein. He was born in 1879, in Ulm, where Kepler
had once wandered in search of a printer, the manuscript of the
Rudolphine Tables under his arm. A strong-willed but dreamy
boy, Einstein did not begin speaking until he was three years old,
and he forever retained something of the brooding intensity owned
by the silent child. Intuitively antiauthoritarian, he rebelled against
outside discipline, a habit that infuriated many of his teachers.
(Years later he would joke that "to punish me for my contempt for
authority, Fate made me an authority myself.")[8]

At the age of sixteen Einstein escaped from the confines of the
Luitpold Gymnasium in Munich—where his Greek instructor told
him, "You will never amount to anything," thus unwittingly earning himself a place in history—by persuading a doctor to write a
note stating that the school regimen was pushing him to the brink
of a nervous breakdown.[9] He failed his college entrance examination, spent a year in preparatory school, and was graduated from
the Federal Polytechnic Institute in Zurich in 1900 with respectable
but unexceptional marks, having habitually cut classes to play the
violin, languish in the cafés, and idle on Lake Zurich aboard rented
sailboats with his fiancée, Mileva Maric, one of the few female
students at the Polytechnic.

Unable to get a job as a scientist or even as a high school
science teacher, Einstein advertised himself as a tutor in mathematics and physics, appending the invitation, "Trial lessons free."[10]
The few who responded found him to be a bewildering teacher,
cheerful and bright but inclined to romp down arcane avenues of
inspiration with a fleetness that left them far behind. Eventually,
Einstein found steady employment, as a "technical expert, third
class," in the Swiss patent office in Bern. He married Mileva in

1903 and they had a son, the first of two, in 1904. (Their first child, a daughter, was born out of wedlock and is thought to have died in infancy, perhaps of scarlet fever; no letters between Einstein and Mileva on this point have been found.) His hopes of getting a raise, the better to support his wife and family, were rewarded when in 1906 he was promoted to technical expert, *second* class.

With his mane of black hair, his limpid, penetrating gaze and his devotion to literature and music and philosophy, Einstein in those days resembled a poet as much as a scientist. Nor was he especially well informed about the progress of physics: His efforts to keep abreast of the scientific literature were impaired by the fact that the technical library generally was closed when he got off work. His technical writings, though occasionally interesting, were in general limited to the sort of speculations about infinity and entropy that may be found in the notebooks of a thousand other postgraduates.

Einstein was indifferent to convention and quick to laugh, a natural enemy of pomp and ceremony. When a friend prevailed upon him to attend festivities at the university in Geneva honoring the 350th anniversary of its founding by Calvin, he marched among the berobed professors in the academic procession wearing an old straw hat and rumpled suit, having no more suitable clothing, and recalled that at the banquet afterward he "said to a Genevan patrician who sat next to me, 'Do you know what Calvin would have done if he were still here? . . . He would have had us all burned because of sinful gluttony.' The man uttered not another word."[11] He was, in short, a Bohemian and a rebel and a high-spirited young man, but nobody's candidate for scientific distinction.

Yet in 1905, Einstein's thoughts began to crystallize, and in that year alone he wrote four epochal papers that transformed the scientific landscape. The first, published three days after his twenty-sixth birthday, would help to lay the foundations of quantum physics. Another was to alter the course of atomic theory and statistical mechanics. The other two enunciated what came to be known as the special theory of relativity. When Max Planck, the editorial director of the German *Annals of Physics,* looked up from reading the first relativity paper, he knew at once that the world had changed. The age of Newton was over, and a new science had arisen to replace it.

In retrospect all is clear, and veins of scientific genius may be found running through the musings of the young Einstein. He had been a quietly religious child, who at the age of eleven composed little hymns in honor of God that he sang on his way to school. But at about age twelve, as he recalled many years later, he

> experienced a second wonder of a totally different nature, in a little book dealing with Euclidian plane geometry, which came into my hands at the beginning of a school year. Here were assertions, as for example the intersection of the three altitudes of a triangle in one point, which—though by no means evident—could nevertheless be proved with such certainty that any doubt appeared to be out of the question. This lucidity and certainty made an indescribable impression upon me.[12]

Einstein later saw this conversion, from traditional religion to what he called his "holy" Euclid text, as involving two ways of striving for the same deliverance:

> It is quite clear to me that the religious paradise of youth, which was thus lost, was a first attempt to free myself from the chains of the "merely-personal," from an existence which is dominated by wishes, hopes and primitive feelings. Out yonder there was this huge world, which exists independently of us human beings and which stands before us like a great, eternal riddle, at least partially accessible to our inspection and thinking. The contemplation of this world beckoned like a liberation. . . . The road to this paradise was not as comfortable and alluring as the road to the religious paradise; but it has proved itself as trustworthy, and I have never regretted having chosen it.[13]

Even Einstein's lack of precocity looked in hindsight like a gift in disguise. Einstein felt that he had "developed so slowly that I only began to wonder about space and time when I was already grown up. In consequence I probed deeper into the problem than an ordinary child would have done." Whatever the cause, he certainly possessed unusual powers of concentration: Like Newton, who credited his insights into deep problems to his habit of "thinking of them without ceasing," Einstein was implacably tenacious

in pursuing a line of thought once it had captured his attention.*
And, like Galileo, he combined a taste for fundamental philosophical questions with an appreciation of the importance of testing his
ideas empirically: "Direct observation of facts," he said, "has always
had for me a kind of magical attraction."[15]

The intellectual odyssey that led Einstein to the special theory
of relativity—and from there to the general theory, which was to
deliver theoretical cosmology from its infancy—began when he was
no more than five years old. He was sick in bed, and his father
showed him a pocket compass to keep him amused. He asked what
made the compass needle point north, and was told that the earth
is enshrouded in a magnetic field to which the needle responds. He
was astonished. It seemed, he recalled many years later, "a miracle"
that an invisible, intangible field could govern the behavior of the
very real compass needle. "Something deeply hidden," he thought,
"had to be behind things."[16]

He learned what that something might be a few years later,
when he read a textbook description of James Clerk Maxwell's
theory of the electromagnetic field. Maxwell had built his field
theory on the experimental work of the English scientist Michael
Faraday. The two men, as Einstein later noted, were related to
each other much as were Galileo and Newton—"the former of each
pair grasping the relations intuitively, and the second one formulating those relations exactly and applying them quantitatively."[17]
Faraday, a blacksmith's son, had been an apprentice to a London
bookbinder. He read popular-science books in his spare time, and
when a friend took him to hear a series of public lectures by the
chemist Sir Humphry Davy, Faraday took notes on the lectures,
printed them, bound them in leather, and sent them to Davy, who
responded by hiring him as a laboratory assistant at the Royal
Institution of Great Britain. There Faraday remained for the next

---

*The American mathematician Ernst Straus was treated to an example of Einstein's
tenacity one afternoon while working as his assistant at the Institute for Advanced
Study in Princeton in the 1940s. "We had finished the preparation of a paper and
we were looking for a paper clip," Straus writes. "After opening a lot of drawers
we finally found one which turned out to be too badly bent for use. So we were
looking for a tool to straighten it. Opening a lot more drawers we came on a whole
box of unused paper clips. Einstein immediately started to shape one of them into
a tool to straighten the bent one. When I asked him what he was doing, he said,
'Once I am set on a goal, it becomes difficult to deflect me.' "[14]

forty-six years, eventually succeeding Davy as the institution director. He was an Edisonian figure, his white hair parted in the
middle over wide-set eyes and slab-flat cheekbones, his shoulders
stooped with work and his large hands buried in laboratory apparatus, though he smiled as habitually as Edison scowled.

In the course of more than fifteen thousand experiments, Faraday found that electricity and magnetism are conveyed by means
of invisible lines of force arrayed in space—i.e., by fields. (Students
today who sprinkle iron filings on a paper resting on a horseshoe
magnet to watch the filings trace out the magnetic field lines are
replicating an old Faraday experiment.) His gift to science amounted
to a fundamental shift in emphasis, from the visible apparatus, the
magnet or electrical coil, to the invisible field that surrounds it and
conveys the electrical or magnetic force. Here began field theory,
which today explores processes ranging from the subatomic to the
intergalactic scale and portrays the entire material world as but a
grand illusion, spun on the loom of the force fields. Einstein was
to be its Bach.

But though Faraday established the existence of electrical and
magnetic fields, he lacked the mathematical acumen required to
write a quantitative description of them. This was left for Maxwell.
Thin-boned as a bird, with trusting, farsighted eyes and a choir-
boy's fragile countenance, Maxwell was at home in mathematical
castles inaccessible to Faraday. A methodical thinker, he first studied electricity and magnetism by reading Faraday's papers—this
on Lord Kelvin's advice, in order to introduce himself to the fields
through Faraday's eyes—and only thereafter subjected them to the
arc lamp of his mathematical skills. The result, Maxwell wrote
Kelvin in 1854, was to "have been rewarded of late by finding the
whole mass of confusion beginning to clear up under the influence
of a few simple ideas."[18]

This was the beginning of the abstraction of the field concept,
a step that would spell the end of purely mechanistic science and
lead to the nonvisualizable but far more flexible mathematical flights
of relativity and quantum physics. Faraday read the papers Maxwell
sent him with the bemusement of a tone-deaf man listening to
Beethoven's quartets, understanding that they were beautiful without being able to appreciate just how. "I was almost frightened
when I saw such mathematical force made to bear upon the subject,
and then wondered to see that the subject stood it so well," Faraday

wrote Maxwell. In another letter he asked, touchingly and tentatively:

> When a mathematician engaged in investigating physical actions and results has arrived at his conclusions, may they not be expressed in common language as fully, clearly, and definitely as in mathematical formulae? If so, would it not be a great boon to such as I to express them so?—translating them out of their hieroglyphics.[19]

Maxwell obligingly rendered some of his explanations of field theory into the mechanical cogwheels and sprocket formulations that Faraday could understand, but it was when stripped to bare equations that his theory flew. With fuguelike balance and power, Maxwell's equations demonstrated that electricity and magnetism are aspects of a single force, electromagnetism, and that light itself is a variety of this force.* Thus were united what had been the separate studies of electricity, magnetism, and optics.

When the young Einstein encountered Maxwell's equations they struck him, he said, "like a revelation." Here was a precise and symmetrical account of the invisible field that governed the compass needle. It animated space, could "weave a web across the sky" as Maxwell had put it, and its differential equations etched the outlines of that web with exquisite balance and precision.

"What made this theory appear revolutionary," Einstein recalled, "was the transition from forces at a distance to fields as fundamental variables."[20] It was no longer necessary to invoke the idea of an aether transmitting light across space; the electromagnetic field in itself could do the job. This had not been appreciated by the older, classical physicists, Maxwell himself among them. Theirs was a putatively hardheaded, mechanical world view, in which the

---

*Maxwell found that the speed with which electromagnetic fields are propagated is equal to the ratio between the electrical force exerted by two electrical charges when at rest and the magnetic force they exert when in motion. As this turned out to be nothing other than the velocity of light, Maxwell concluded that light itself is an electromagnetic field. Since popular accounts of the special theory of relativity sometimes convey the mistaken impression that the velocity of light is an arbitrary speed limit, like that set by legislatures for public highways, it is helpful to keep in mind Maxwell's finding—that the velocity of light results from a fundamental constant in the equations that describe the behavior of electromagnetic fields.

fields taken by themselves appeared too insubstantial to be real. It was by their consensus that the aether hypothesis had glided on, a ghost ship alive with Saint Elmo's fire, well after Maxwell's equations and the Michelson-Morley experiment had emptied the wind from its sails. Einstein, caring little for tradition, abandoned the aether and focused his attention on the field.

Yet if one adhered to both Maxwell's equations and Newton's absolute space, the result was a paradox. This the giants of physics understood; it was one of their reasons for underestimating the importance of Maxwell's field equations. Einstein, ignorant of their wisdom, discovered the paradox for himself, at the age of sixteen. He was at the time enrolled in a preparatory school at Aarau, in the Swiss Oberland, where he enjoyed taking walks along the river oxbows. (Years later he would write a paper defining how rivers meander.) One day, Einstein asked himself what he would see if he were to chase a beam of light at the velocity of light. The answer, according to classical physics, was that "I should observe such a beam of light as a spatially oscillatory electromagnetic field at rest. However, there seems to be no such thing, whether on the basis of experience or according to Maxwell's equations."[21] Velocity was inherent to light; it was, after all, by way of its velocity that light had revealed to Maxwell its identity as an electromagnetic field. Yet if we live in an absolute, Newtonian space demarcated by the aether, it should be possible to catch up with a light beam and thus rob it of its speed. Something had to give, in either Newton's physics or Maxwell's.

Einstein was acquainted with another electrodynamic paradox as well, one that had turned up literally in his own backyard, in the iron and copper dynamos that his father and his uncle Jakob had built in an electrical shop behind the family home in the Munich suburbs. The principle of the dynamo, established by Faraday, was that the field created by a whirling magnet will generate an electrical current in a surrounding web of wire. This finding had tremendous practical potential: The energy of a steam engine or a flowing stream could be harnessed to produce electricity that could then be exported via electrical lines to power machinery and illuminate city streets miles away. Although the Einstein family never managed to make much of a living from it, dynamo design was on the forefront of contemporary technology, and giant steam-driven dynamos were being commissioned and built at considerable expense.* Yet their performance could not be predicted with exactitude so long as the behavior of the electromagnetic field within the dynamo was so poorly understood. Under existing theory, the moving field was to be explained according to one set of rules if viewed from the perspective of a dynamo's rotating magnet, and another if viewed from the stationary electrical coil. Every dynamo housed a whirling mystery.

The situation was economically embarrassing for the industrialists. It bothered Einstein as unaesthetic. "The thought that one is dealing here with two fundamentally different cases was for me unbearable," he recalled. "The difference between these two cases could not be a real difference but rather, in my conviction, only a difference in the choice of the reference point."[23]

Such questions were still on Einstein's mind when he completed his year of preparatory school, but if he hoped to find guidance in solving them at the Polytechnic Institute, he was soon disappointed. His physics professor, the capable but conservative Heinrich Friedrich Weber, was fascinated by dynamos, owed his chair to the philanthropy of the dynamo builder Werner von Siemens, and was sufficiently devoted to the study of electricity that he repeatedly submitted himself to electrical shocks of one thousand volts and more of alternating current—this as part of an effort to determine how much voltage a human being could endure.† Yet Weber, steeped in the traditions of classical physics, never lectured

---

*A sense of the allure of the dynamo was preserved by the American historian Henry Adams in his *The Education of Henry Adams*. Describing his visit to the "great hall of dynamos" at the Paris Exposition in 1900, he writes, "To Adams the dynamo became a symbol of infinity. As he grew accustomed to the great gallery of machines, he began to feel the forty-foot dynamos as a moral force, much as the early Christians felt the Cross. The planet itself seemed less impressive, in its old-fashioned, deliberate, annual or daily revolution, than this huge wheel."[22]

†Research like Herr Weber's was being applied with dispatch to the execution of convicts and the punishment of malingering conscripts. The first electrocution of a criminal in the United States occurred in 1890, less than ten years after the first public power station in America started operating; the method was purportedly humane, but it took the victim fifteen long minutes to die. Shell-shocked German soldiers in the trenches of the First World War were administered jolts of electricity and then sent back to the front; if they returned, they were given still more severe shocks, in a closed circuit of fear and pain that drove some to suicide.[24]

on Maxwell or Faraday. Einstein soon lost interest and started cutting Herr Weber's classes. He read physics on his own and conducted experiments in the Polytechnic's superb laboratories. One of his experiments resulted in an explosion that badly injured Einstein's hand and nearly wrecked the lab.

Professor Weber responded by doing what he could—which was a great deal—to prevent Einstein's getting a job after graduation. Thus stigmatized, Einstein went nowhere. The distasteful experience of cramming for the comprehensive final examinations had in any event left him unable to think about science for a full year, and he spent his time reading philosophy and playing the violin. When he did resume the study of physics, it was with little encouragement from the outside world. He submitted a thesis on the kinetics of gases to the University of Zurich, but no doctorate was forthcoming. He wrote a few scientific papers, but they were almost worthless. And yet, though regretting that he was a disappointment to his parents, Einstein remained serenely self-confident. "I have a few splendid ideas," he wrote to his friend Marcel Grossman, "which now only need proper incubation."[25]

It was with the help of Grossman's father that Einstein got the patent office job, and while we may shake our heads at the spectacle of so great a man in so slight a position, Einstein remembered it as "my best time of all."[26] He enjoyed contemplating the mechanical gadgets that came before him for review, found that writing critiques of patent applications helped him learn to express himself succinctly, and reveled in the companionship of his friend Michele Angelo Besso, with whom he discussed philosophy, physics, and everything under the sun. "I could not have found a better sounding-board in all of Europe," he said.[27]

At Besso's urging, Einstein read the works of the Austrian physicist and philosopher Ernst Mach, one of the few leading scientific thinkers to critique the mechanical paradigm on which rested belief in a Newtonian space pervaded by an aether. *"The simplest mechanical principles are of a very complicated character,"* Mach wrote (his italics). *". . . They can by no means be regarded as mathematically established truths but only as principles that not only admit of constant control by experience but actually require it."*[28] A scalding critic of Newtonian space in general and of the aether hypothesis in particular, Mach sought to replace such "metaphysical obscurities," as he called them, with more economical precepts anchored firmly in the sense data of observation. Space, Mach argued, is not a thing, but an expression of interrelationships among events. "*All* masses and *all* velocities, and consequently *all* forces, are relative," he wrote.[29] Einstein agreed, and was encouraged to attempt to write a theory that built space and time out of events alone, as Mach prescribed. He never entirely succeeded in satisfying Mach's criteria—it may be that no workable theory can—but the effort helped impel him toward relativity.

The emergence of the special theory of relativity was as unconventional as its author. The 1905 paper that first enunciated the theory resembles the work of a crank; it contains no citations whatever from the scientific literature, and mentions the aid of but one individual, Besso, who was not a scientist. (At the time, Einstein knew no scientists.) Einstein's first lecture in Zurich explaining the theory was delivered not in a university but in the Carpenters' Union hall; he went on for over an hour, then suddenly interrupted himself to ask the time, explaining that he did not own a watch. Yet here began the reformation of the concepts of space and time.

With the special theory of relativity, Einstein had at last resolved the paradox that had occurred to him at age sixteen, that Maxwell's equations failed if one could chase a beam of light at the velocity of light. His did so by concluding that one *cannot* accelerate to the velocity of light—that, indeed, the velocity of light is *the same* for all observers, regardless of their relative motion. If, for instance, a physicist were to board a spaceship and fly off toward the star Vega at 50 percent the velocity of light, and while on board measure the velocity of the light coming from Vega, he would find that velocity to be exactly the same as would his colleagues back on Earth.

To quantify this strange state of affairs, Einstein was obliged to employ the Lorentz contractions. (At the time he knew little of Lorentz, whom he was later to esteem as "the greatest and noblest man of our times . . . a living work of art.")[30] In Einstein's hands, the Lorentz equations specify that as an observer increases in velocity, his dimensions, as well as those of his spaceship and any measuring devices aboard, will shrink along the direction of their motion by just the amount required to make the measurement of light's velocity always come out the same. This, then, was why

Michelson and Morley had found no trace of an "aether drift." In
fact the aether is superfluous, as is Newton's absolute space and
time, for there is no need for any unmoving frame of reference:
"To the concept of absolute rest there correspond no properties of
the phenomena, neither in mechanics, nor in electrodynamics."[31]
What matters are observable events, and no event can be observed
until the light (or radio waves, or other form of electromagnetic
radiation) that brings news of it reaches the observer. Einstein had
replaced Newton's space with a network of light beams; *theirs* was
the absolute grid, within which space itself became supple.

Observers in motion experience a slowing in the passage of
time, as well: An astronaut traveling at 90 percent of the velocity
of light would age only half as fast as her colleagues back home,
so that at, say, a twentieth class reunion of interstellar astronauts,
those who had served the most aboard relativistic spacecraft would
be the youngest. Mass, too, is rendered plastic within the framework of the light beams; objects approaching the speed of light
increase in mass. The effects of relativistic time dilation, mass increase, and change in dimension are minute at ordinary velocities
like that of the earth in its orbit or the sun through space (which
is why it had not been noticed sooner) but become pronounced as
speeds increase, and go to infinity at the speed of light. If the earth
could be accelerated to the velocity of light (a feat that would require
infinite energy to achieve) it would contract into a two-dimensional
wafer of infinite mass, on which time would come to a stop—which
is one way of saying that acceleration to light speed is impossible.

Nor are these effects illusory, or merely psychological: They
are as real as the stone that Dr. Johnson kicked in his famous
refutation of Bishop Berkeley, and have been confirmed in scores
of experiments. The relativistic increase in the mass of particles
moving at nearly the velocity of light is not only observed in particle
accelerators, but is what gives the speeding particles most of their
punch. Relativistic time dilation has been tested by flying atomic
clocks around the world in commercial aircraft; the clocks were
found to run slow by just the tiny amount the theory specifies. A
NASA ground controller once threatened to dock astronauts in
space a fraction of a penny of their flight pay, to compensate for
the decrease in the passage of time they experienced as a result of
their velocity in orbit.

These and other implications of special relativity initially struck the lay public, and many scientists as well, as uncommonly strange.* But if Einstein's approach was radical, his intention was essentially conservative. As is implied by the title of his original relativity paper, "On the Electrodynamics of Moving Bodies," his aim was to redeem the laws of electrodynamics so that they could be shown to work in every imaginable situation, not just in a quiet laboratory in Zurich but in whirling dynamos and on moving worlds hurtling past one another at staggering speeds. The term *relativity*, coined not by Einstein but by Poincaré and applied to the theory by the physicist Max Planck, is somewhat misleading in this sense; Einstein, stressing its conservative function, had preferred to call it *Invariantentheorie*—"invariance theory."

Relativity nonetheless cast its net wide, embracing the study not only of light and space and time, but of matter as well. The theory derives its catholic impact from the fact that electromagnetism is implicated not only in the propagation of light but also in the architecture of matter: Electromagnetism is the force that holds electrons in their orbits around nuclear particles to make atoms, binds atoms together to form molecules, and ties molecules together to form objects. Every tangible thing, from stars and planets to this page and the eye that reads it, carries electromagnetism in the fiber of its being. To alter one's conception of electromagnetism is, therefore, to reconsider the very nature of matter. Einstein caught sight of this connection only three months after the first account of special relativity had appeared, and published a paper titled, "Does the Inertia Content of a Body Depend Upon Its Energy Content?" The answer was yes, and ours has been a sadder and wiser world ever since.

In the first paper, as we have seen, Einstein demonstrated that the inertial mass of an object increases when it absorbs energy. It follows that its mass *decreases* when it *radiates* energy. This holds true, not only for a spaceship gliding toward the stars, but for an object at rest as well: A camera loses a (very) little mass when the

---

*Einstein shared this fate with Newton, whose ideas were routinely characterized as incomprehensible. A student who saw Isaac Newton passing in his carriage is said to have remarked, "There goes the man that writ a book that neither he nor anybody else understands."

flash goes off, and the people whose picture is being taken become a little more massive in the exchange. Mass and energy are interchangeable, with electromagnetic energy doing the bartering between them.

Einstein, contemplating this fact, concluded that energy and inertial mass are the same, and he expressed their identity in the equation

$$m = \frac{E}{c^2}$$

in which m is the mass of an object, E is its energy content, and c is the velocity of light. In composing this singularly economical little equation, which unifies the concepts of energy and matter and relates both to the velocity of light, Einstein initially was concerned with mass. If instead we solve for energy, it takes on a more familiar and more ominous form, as

$$E = mc^2$$

Viewed from this perspective, the theory says that matter is frozen energy. This of course is the key to nuclear power and nuclear weapons, though Einstein did not consider these applications at the time and rejected them as impractical once they were proposed by others. In the hands of the astrophysicists, the equation would be used to discern the thermonuclear processes that power the sun and stars.

But for all its protean achievements, special relativity was silent with regard to gravitation, the other known large-scale force in the universe. The special theory has to do with *inertial* mass, the resistance objects offer to change in their state of motion—their "clout," or "heft," so to speak. Gravitation acts upon objects according to their *gravitational* mass—i.e., their "weight." Inertial mass is what you feel when you slide a suitcase along a polished floor; gravitational mass is what you feel when you lift the suitcase. There would appear to be distinct differences between the two: Gravitational mass manifests itself only in the presence of gravitational force, while inertial mass is a permanent property of matter. Take the suitcase on a spaceship and, once in orbit, it will weigh nothing (i.e., its gravitational mass will measure zero), but its inertial mass will remain the same: You'll have to work just as hard to wrest it around the cabin, and once in motion it will have the same momentum as if it were sliding across a floor on Earth.*

Yet for some reason, the inertial and gravitational mass of any given object are equivalent. Put the suitcase on the airport scale and find that it weighs thirty pounds: That is a result of its gravitational mass. Now set it on a sheet of smooth, glazed ice or another relatively friction-free surface, attach a spring scale to the handle, and pull it until you get it accelerating at the same rate at which it would fall (i.e., 16 feet per second, on Earth), and the scale will register, again, thirty pounds: That is a result of its *inertial* mass. Experiments have been performed to a high degree of precision on all sorts of materials, in many different weights, and the gravitational mass of each object repeatedly turns out to be exactly equal to its inertial mass.†

The equality of inertial and gravitational mass had been an integral if inconspicuous part of classical physics for centuries. It can be seen, for instance, to explain Galileo's discovery that cannonballs and boccie balls fall at the same velocity despite their differing weight: They do so because the cannonball, though it has greater gravitational mass and ought (naïvely) to fall faster, also has a greater inertial mass, which makes it accelerate more slowly; since these two quantities are equivalent they cancel out, and the cannonball consequently falls no faster than the boccie ball. But in Newtonian mechanics the equivalence principle was treated as a

---

*I was treated to an inadvertent demonstration of this effect one day, aboard a DC-3 in a violent storm over the Bahamas, when a doctor's iron scale, standing about four feet tall, tore loose from its moorings in the aft end of the cabin. The plane then plunged into a downdraft, rendering everything momentarily weightless, and the scale rose into the air and drifted toward me. I fended it off with my feet, thus briefly experiencing its inertial mass absent its gravitational mass. The fact that the menacing object happened to be a weightless device for measuring weight invested the lesson with a certain ironic intensity.

†The definitive experiments were conducted by Baron Roland von Eötvös in Budapest in 1889 and 1922. Eötvös suspended objects of various compositions from threads and looked for deviations in these plumb lines caused by differences between their gravitational mass (which was being pulled straight down) and their inertial mass (which was being pulled sideways, by the rotation of the earth). "In no case," he wrote, "could we discover any detectable deviation from the law of proportionality of gravitation and inertia." This remains the case today, although one recent reenactment of the experiment did produce subtle anomalies that could not immediately be accounted for.

mere coincidence. Einstein was intrigued. Here, he thought, "must lie the key to a deeper understanding of inertia and gravitation."[32] His inquiry set him on his way up the craggy road toward the general theory of relativity.

Einstein's first insight into the question came one day in 1907, in what he later called "the happiest thought of my life." The memory of the moment remained vivid decades later:

> I was sitting in a chair in the patent office at Bern, when all of a sudden a thought occurred to me: "If a person falls freely he will not feel his own weight." I was startled. This simple thought made a deep impression on me. It impelled me toward a theory of gravitation.[33]

To appreciate why this seemingly straightforward picture should have so excited Einstein, imagine that you awaken to find yourself floating, weightless, in a sealed, windowless elevator car. A diabolical set of instructions, printed on the wall, informs you that there are two identical such elevator cars—one adrift in deep space, where it is subject to no significant gravitational influence, and the other trapped in the sun's gravitational field, plunging rapidly toward its doom. You will be rescued only if you can *prove* (not guess) in which car you are riding—the one floating in zero gravity, or the one falling in a strong gravitational field. What Einstein realized that day in the patent office was that you *cannot* tell the difference, neither through your senses nor by conducting experiments. The fact that you are weightless does not mean that you are free from gravitation; you might be in free fall. (The "weightlessness" experienced by astronauts in orbit is precisely of this sort: Though trapped by the earth's gravitational field they feel no weight—i.e., no effect of gravitation—because they and their spaceship are constantly falling.) The gravitational field, therefore, has only a *relative* existence. One is reminded of the joke about the man who falls from the roof of a tall building and, seeing a friend looking aghast out a window on the way down, calls out encouragingly, "I'm okay so far!" His point was Einstein's—that the gravitational field does not exist for him, so long as he remains in his inertial framework. (The sidewalk, alas, is in an inertial framework of its own.)

The same ambiguity applies in the opposite situation: Suppose that when you awaken you find yourself standing in the elevator car, at your normal weight. This time the instructions say that you are either (1) aboard a elevator stopped on the ground floor of an office building on Earth, or (2) adrift in zero-gravity space, in an elevator attached by a cable to a spaceship that is pulling it away at a steady acceleration, pressing you to the floor with a force equal to that of Earth's gravitation—at one "G," as the jet pilots say. Here again, you cannot prove which is the case.

Einstein reasoned that if the effects of gravitation are mimicked by acceleration, gravitation itself might be regarded as a kind of acceleration. But acceleration through what reference frame? It could not be ordinary three-dimensional space; the passengers in the elevator in the New York skyscraper, after all, are not flying through space relative to the earth.

The search for an answer required brought Einstein to consider the concept of a four-dimensional space-time continuum. Within its framework, gravitation *is* acceleration, the acceleration of objects as they glide along "world lines"—paths of least action traced over the slopes of a three-dimensional space that is curved in the fourth dimension.

A forerunner here was Hermann Minkowski, who had been Einstein's mathematics professor at the Polytechnic Institute. Minkowski remembered Einstein as a "lazy dog" who seldom came to class, but he was quick to appreciate the importance of Einstein's work, though initially he viewed it as but an improvement on Lorentz. In 1908 Minkowski published a paper on Lorentz's theory that cleared away much of the mathematical deadwood that had cluttered Einstein's original formulation of special relativity. It demonstrated that time could be treated as a dimension in a four-dimensional universe. "Henceforth space by itself, and time by itself, are doomed to fade away into mere shadows, and only a kind of union of the two will preserve an independent reality," Minkowski predicted.[34] His words proved prophetic, and the special theory of relativity has been viewed in terms of a "space-time continuum" ever since. Einstein initially dismissed Minkowski's formulation as excessively pedantic, joking that he scarcely recognized his own theory once the mathematicians got hold of it. But he came to realize that if he were to explore the connection between weight and inertia, he would do well to travel farther up the trail Minkowski had blazed.

Minkowski's space-time continuum, though suitable for special relativity, would not support what was to become general relativity. Its space was "flat"—i.e., euclidean. If gravitation were to be interpreted as a form of acceleration, that acceleration would have to occur along the undulations of curved space. So it was that Einstein was led, however reluctantly, into the forbidding territory of the noneuclidean geometries.

Euclidean geometry, as every high school math student is taught, has different characteristics depending upon whether it is worked in two dimensions ("plane" geometry) or three ("solid" geometry). On a plane, the sum of the angles of a triangle is 180 degrees, but if we add a third dimension we can envision surfaces such as that

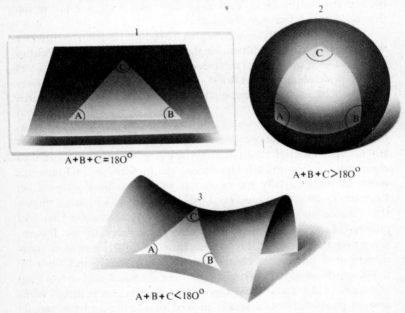

Triangles in flat two-dimensional space have interior angles that always add up to 180 degrees. But when two-space is curved into a third dimension, the angles always total either *less* than 180 degrees (if the curvature is hyperbolic, or "open") or *more* than 180 degrees (if the curvature is spherical, or "closed"). Similarly, the geometry of the three-dimensional universe may be either flat (euclidean), or open or closed (noneuclidean), when viewed in the context of Einstein's four-dimensional space-time continuum.

of a sphere, on which the angles of a triangle add up to *more* than 180 degrees, or a saddle-shaped hyperbola, on which the angles add up to *less* than 180 degrees. The shortest distance between two points on a plane is a straight line, but on a sphere or a hyperbola the shortest distances are *curved* lines. In the noneuclidean geometries a fourth dimension is added, and the rules are changed in a similarly consistent manner to allow for the possible curvature of three-dimensional space within a four-dimensional theater. Two categories of curved space then can be imagined (or at least calculated): spherical, or "closed" space, in which three-space obeys geometrical rules analogous to those of the two-space on the surface of a sphere, and hyperbolic, or "open" space, analogous to the surface of a three-dimensional hyperbola. (One can also work out a flat, euclidean four-dimensional geometry, but in that case the rules do not change, just as plane two-dimensional geometry obeys the same rules if the planes happen to be the sides of a three-dimensional cube.)

By the time Einstein came on the scene, the rules of four-dimensional geometry had been worked out—those of spherical four-space by Georg Friedrich Riemann and those of the four-dimensional hyperbolas by Nikolai Ivanovich Lobachevski and János Bolyai. The whole subject, however, was still regarded as at best difficult and arcane, and at worst almost disreputable.* The legendary mathematician Karl Friedrich Gauss had withheld his papers on noneuclidean geometry from publication, fearing ridicule by his colleagues, and Bolyai conducted his research in the field against the advice of his father, who warned him, "For God's sake, please give it up. Fear it no less than the sensual passions because it, too, may take up all your time and deprive you of your health, peace of mind and happiness in life."[36]

Einstein rushed in where Bolyai's father feared to tread. With the aid of his old classmate Marcel Grossman—"Help me, Marcel,

---

*The very term "fourth dimension" called to mind the enthusiasms of eccentrics and ecstatics like Charles Hinton, who sought to enhance his appreciation of its subtleties by manipulating 81 cubes that represented the units of a 3-by-3-by-3-by-3-unit euclidean hypercube. Hinton's career was interrupted—and his subject cast into further ignominy—when he was convicted of bigamy for living out the free-love philosophy of his father, who liked to say that "Christ was the Savior of men, but I am the savior of women, and I don't envy Him a bit!" Hinton *fils* dropped dead at a banquet of the Society of Philanthropic Inquiry in Washington, D.C., moments after delivering a toast in honor of femininity.[35]

or I'll go crazy!" he wrote—Einstein struggled through the complexities of curved space, seeking to assign the fourth dimension to time and make the whole, infernally complicated affair come out right. He had by now begun to win professional recognition, had quit the patent office to accept a series of teaching positions that culminated in a full professorship in pure research at the University of Berlin, and was doing important work in quantum mechanics and a half-dozen other fields. But he kept returning to the riddle of gravitation, trying to find patterns of beauty and simplicity among thick stacks of papers black with equations. Like a lost explorer discarding his belongings on a trek across the desert, he found it necessary to part company with some of the most cherished of his possessions—among them one of the central precepts of the special theory itself, which to his joy was ultimately to return as a local case within the broader scheme of the general theory. "In all my life I have never before labored so hard," he wrote to a friend. ". . . Compared with this problem, the original theory of relativity is child's play."[37] Nowhere in human history is there to be found a more sustained and heroic labor of the intellect than in Einstein's trek toward general relativity, nor one that has produced a greater reward.

He completed the theory in November 1915 and published it the following spring. Though its equations are complex, its central conception is startlingly simple. The force of gravitation disappears, and is replaced by the geometry of space itself: Matter curves space, and what we call gravitation is but the acceleration of objects as they slide down the toboggan runs described by their trajectories in time through the undulations of space. The planets skid along the inner walls of a depression in space created by the fat, massive sun; clusters of galaxies rest in spatial hollows like nuggets in a prospector's bowl.

In marrying gravitational physics to the geometry of curved space, general relativity emancipated cosmology from the ancient dilemma of whether the universe is infinite and unbounded or finite and bounded. An infinite universe would be not just large but *infinite*, and this posed problems. The gravitational force generated by an infinite number of stars would itself be infinite, and would, therefore, overwhelm the local action of gravity; this prospect so troubled Newton that he resorted to invoking God's infinite grace to resolve it. Moreover, the light from an infinite number of stars might be expected to turn the night sky into a blazing sheet of light; yet the night sky is dark.* The alternative, however—a finite euclidean universe with an edge to it—was equally unattractive: As Liu Chi posed the question, in China in the fourteenth century, "If heaven has a boundary, what things could be outside it?"[39] The difficulty of imagining an end to space had been enunciated as early as the fifth century B.C., by Plato's colleague Archytas the Pythagorean; Lucretius summed it up this way:

> Let us assume for the moment that the universe is limited. If a man advances so that he is at the very edge of the extreme boundary and hurls a swift spear, do you prefer that this spear, hurled with great force, go where it was sent and fly far, or do you think that something can stop it and stand in its way?[40]

General relativity resolved the matter by establishing that the universe could be both finite—i.e., could contain a finite number of stars in a finite volume of space—and unbounded. The key to this realization lay in Einstein's demonstration that, since matter warps space, the sum total of the mass in all the galaxies might be sufficient to wrap space around themselves. The result would be a closed, four-dimensionally spherical cosmos, in which any observer, anywhere in the universe, would see galaxies stretching deep into space in every direction, and would conclude, correctly, that there is no end to space. Yet the amount of space in a closed universe would nonetheless be finite: An adventurer with time to spare could eventually visit every galaxy, yet would never reach an edge of space. Just as the surface of the earth is finite but unbounded in two dimensions (we can wander wherever we like, and will not fall off the edge of the earth) so a closed four-dimensional universe is finite but unbounded to us who observe it in three dimensions.†

---

*This disturbing puzzle, known today as Olbers's paradox after the nineteenth-century German astronomer Wilhelm Olbers, was discovered independently by other astronomers, among them Halley, who lectured on it at a Royal Society meeting in 1721. Newton chaired that meeting, but for some reason never wrote about the paradox. The historian of science Michael Hoskin suggests that the old man was napping while Halley spoke.[38]

†Alternately, general relativity allows that the universe might be structured like a four-dimensional hyperbola, in which case it would be both *in*finite and unbounded. This possibility resurrects some of the difficulties that afflict all infinite-universe models, but they could perhaps be resolved, should the observational data indicate that space is indeed hyperbolically rather than spherically curved.

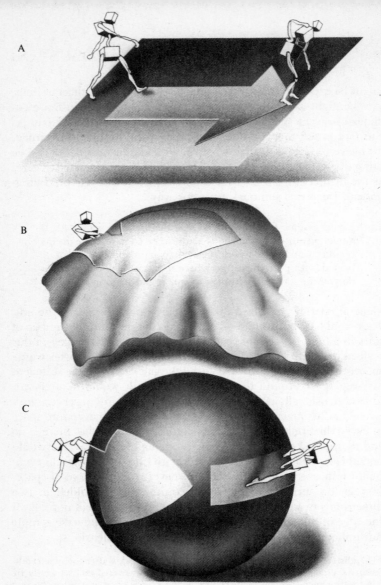

A

B

C

Two-dimensional inhabitants of a finite universe must confront the paradox of an "edge" to their cosmos. But if we add a dimension, curving the plane on which they live into a sphere, their world, though still finite, becomes unbounded. General relativity reveals a similar prospect for the four-dimensional geometry of the universe we three-dimensional creatures inhabit: hence Einstein's "closed, unbounded" universe.

202

The question of whether the universe is hyperbolic and open or spherical and closed remains unanswered, as we shall see. But, thanks to Einstein, the problem is no longer clouded by paradox. By introducing the scientific prospect of a finite, unbounded cosmos, Einstein's general theory initiated a meaningful dialogue between the human mind and the conundrums of cosmological space.

The theory was beautiful, but was it true? Einstein, having been to the mountaintop, felt supremely confident on this score. General relativity explained a precession in the orbit of the planet Mercury that had been left unaccounted for in Newtonian mechanics, and he did not doubt it would survive further tests as well. As he wrote his friend Besso, "I am fully satisfied, and I do not doubt any more the correctness of the whole system. . . . The sense of the thing is too evident."[41]

The wider scientific community, however, awaited the verdict of experiment. There would be a total solar eclipse on May 29, 1919, at which time the sun would stand against the bright stars of the Hyades cluster. The English astronomer Arthur Stanley Eddington led an expedition to a cocoa plantation on Principe Island off west Equatorial Africa to observe the eclipse and see whether the predicted curvature of space in the region of the sun would distort the apparent positions of the stars in the briefly darkened sky. It was a scene of high drama—English scientists testing the theory of a German physicist immediately after the end of the Great War. As the time of the eclipse approached, rain clouds covered the sky. But then, moments after the moon's shadow came speeding across the landscape and totality began, a hole opened up in the clouds around the sun, and the camera shutters were triggered. The results of Eddington's expedition, and of a second eclipse observation conducted at Sobral, Brazil on the same day, were presented by the Astronomer Royal at a meeting of the Royal Society in London on November 6, 1919, with Newton's portrait looking on. They were positive: The light rays coming from the stars of the Hyades were found to be offset to just the degree predicted in the theory.

When Einstein received a telegram from Lorentz announcing the outcome of the Eddington expedition, he showed it to a student, Ilse Rosenthal-Schneider, who asked, "What would you have said if there had been no confirmation?"

"I would have had to pity our dear Lord," Einstein replied. "The theory is correct."[42]*

Subsequent experiments have further vindicated Einstein's confidence. The curvature of space in the vicinity of the sun was established with much greater accuracy, by bouncing radar waves off Mercury and Venus when they lie near the sun in the sky, and the extent of curvature matched that predicted by the general theory of relativity. A light beam directed up a tower in the Jefferson Physical Laboratory at Harvard University was found to be shifted toward the red by the earth's gravitation to just the anticipated degree. Maelstroms of energy detected at the centers of violent galaxies indicate that they harbor black holes, collapsed objects wrapped in infinitely curved space that shuts them off from the rest of the universe; the existence of black holes was another prediction of the general theory. And the theory has been tested in many other ways as well—in examinations of entombed dead stars, the whirling of active stars around one another, the wanderings of interplanetary spacecraft well past Jupiter, and the slowing of light as it climbs up out of the sun's space well—and all these trials it has survived.

Too modest to be immodest, Einstein had written when publishing his completed account of general relativity that "hardly anyone who has truly understood this theory will be able to resist being captivated by its magic."[44] But, even if only those mathematicians and physicists who have mastered general relativity are in a position properly to understand it, still we can all appreciate it to some degree, if, while keeping in mind its basic concepts, we contemplate the universe of effortlessly wheeling galaxies deployed across the blossom petals of gently curving space. Einstein's epitaph could be Christopher Wren's: If you seek his monument, look around.

---

*Einstein once astonished Ernst Straus by saying of Max Planck, the father of quantum physics, "He was one of the finest people I have ever known and one of my best friends; but, you know, he didn't really understand physics." When Straus asked what he meant, Einstein replied, "During the eclipse of 1919, Planck stayed up all night to see if it would confirm the bending of light by the gravitational field of the sun. If he had really understood the way the general theory of relativity explains the equivalence of inertial and gravitational mass, he would have gone to bed the way I did."[43]

11

THE
EXPANSION
OF THE
UNIVERSE

Nature lives in motion.

—James Hutton

Eyesight should learn from reason.

—Kepler

When Einstein began to investigate the cosmological implications of the general theory of relativity, he found something strange and disturbing: The theory implied that the universe as a whole could not be static, but must be either expanding or contracting. This was a completely novel idea, and one for which there was, at the time, no observational evidence whatever: The astronomers he consulted informed Einstein that stars wander more or less randomly through space, but display no concerted motion of the sort that would suggest cosmic expansion or contraction. Faced with this disjunction between his theory and the empirical data, Einstein reluctantly concluded that there must be something wrong with the theory, and he modified its equations by adding a term that he called the cosmological constant. Symbolized by the Greek letter lambda, the new term was intended to make the radius of the universe hold steady with the passing of time.

Einstein never liked the cosmological constant. He called it
"gravely detrimental to the formal beauty of the theory," pointing
out that it was nothing more than a mathematical fiction, without
any real physical basis, one that had been introduced solely to being
the theory into accord with the observational facts. As he wrote in
1917:

> [W]e admittedly had to introduce an extension of the field
> equations of gravitation which is not justified by our actual
> knowledge of gravitation. . . . That term is necessary only for
> the purpose of making possible a quasi-static distribution of
> matter, as required by the fact of the small velocities of the
> stars.[1]

Moreover, as soon became apparent, the term did not even accomplish its avowed function of making the relativistic universe stand
still. The Russian mathematician Aleksandr Friedmann found that
Einstein in introducing the term had made an algebraic error, dividing by a quantity that could be zero. When Friedmann corrected
the error, general relativity broke free of its fetters and the relativistic universe, to Einstein's frustration, once again took on wings.

Connoisseurs of irony's serrated edge will appreciate that it
was in 1917, the very year that Einstein besmirched his general
theory of relativity by introducing the cosmological term, that
the American astronomer Vesto Slipher published a paper containing the first observational evidence that the universe is in fact
expanding.

Slipher knew nothing of general relativity. He was a nose-to-
the-grindstone staffer at Lowell Observatory, in Flagstaff, Arizona,
an isolated and idiosyncratic private institution so remote from the
theoretical physics community that it might as well have been on
the far side of the moon. His employer was Percival Lowell, of the
Boston Lowells, a loftily unconventional thinker remembered chiefly
for having charted the (illusory) canals of Mars, which he took to
be global waterways dug by a parched alien civilization desperately
importing water from the polar ice caps. Like many astronomers
of his day, Lowell thought that the spiral nebulae were Laplacian
solar systems aborning. To test this thesis, he assigned Slipher to
take spectra of a number of spirals, using a new and more efficient
spectrograph, in order to search for the rotation velocities charac-

teristic of Laplacian nebulae eddying their way into stars and planets. Slipher did, indeed, find evidence of rotation in the spirals—as Edwin Hubble would find, this was actually the motion of billions of stars orbiting in spiral galaxies—but he also found, superimposed on the rotation velocities, an enormous displacement in the spectral lines of most spirals toward the red end of the spectrum.

The only reasonable explanation for this astonishing finding was that Slipher was observing Doppler shifts. The name comes from the Austrian physicist Christian Johann Doppler, who noted in 1842 that light, sound, or other radiation coming from a moving source is received at a higher frequency if the source emanating it is approaching, and at a lower frequency if the source is receding. (It is owing to this "Doppler shift" that an automobile horn sounds higher in pitch if the car is approaching and lower if it is speeding away.) Astronomers had long been making use of Doppler shifts in spectra to measure the velocities of stars: The spectral lines of stars that were moving toward the sun would be displaced toward the blue, while those of stars moving away from the sun would be displaced toward the red, or lower frequency, end of the spectrum. Indeed, it was by virtue of just such measurements that astronomers had been able to inform Einstein that stellar motions in the Milky Way were generally random.

The velocities of the spiral nebulae implied by Slipher's redshifts, however, were much more rapid than those of the stars. Two of the first fifteen spirals Slipher observed were moving at over two million miles per hour. Even more unexpectedly, their motions were concerted: Twenty-one of the twenty-five spirals for which Slipher had accumulated spectra by 1917 were displaced toward the red, indicating that they were flying away from each other and from the earth. (The exceptions, we understand today, were nearby galaxies that are gravitationally bound to the Milky Way in the Local Group of galaxies, and therefore do not participate in the cosmic expansion.)

But, though Slipher's findings were provocative, they did not in themselves indicate that the universe is expanding. Slipher had no way of knowing the distances to the galaxies whose redshifts he obtained—nor, for that matter, of ascertaining that they were galaxies. That shoe was to be dropped, instead, by Edwin Hubble.

Hubble was thirty years old by the time he was discharged from service as a private in the American Expeditionary Force in

France and began training the big telescopes at Mount Wilson on the spiral nebulae, but he worked fast: Only five years later he was able to write Harlow Shapley that he had found Cepheid variable stars in the spirals, establishing that they are galaxies and making it possible to estimate their distances. Five years after that, in 1929, he was able to plot the distances of twenty-five galaxies against his and Slipher's measurements of their redshifts. The result was a straight line—a direct correlation between distance and velocity of recession.

Inscribed in the Hubble diagram was the signature of cosmic expansion. Imagine that the earth were expanding, and let the two-dimensional surface of the earth stand for three-dimensional cosmic space. In such a situation, every observer would find that every other city on Earth was receding from his city, each at a velocity directly proportional to its distance. If, for instance, the rate of expansion were such that the earth doubled in diameter every hour, then the distances between cities also would double each hour. Chicago and Memphis are five hundred miles apart; therefore observers in Chicago would find that Memphis was receding at a rate of five hundred miles per hour. Looking out to San Francisco, eighteen hundred miles away, the Chicagoans would find that *its* velocity of recession was fully eighteen hundred miles per hour. And that, a velocity-distance relation, is what Hubble found for the galaxies.

It was also, of course, just what the general theory of relativity had predicted, at least before being fettered by the lambda term. (Fumed Einstein, "If Hubble's expansion had been discovered at the time of the creation of the general theory of relativity, the [cosmological constant] would never have been introduced.")[2] Yet Hubble, like Slipher, was isolated by the gulf that still separated the world of the American observational astronomers from that of Einstein and the other leading theoretical physicists in Europe. Hubble knew next to nothing of general relativity; neither did his boss, George Ellery Hale, who confessed that "the complications of the theory of relativity are altogether too much for my comprehension," adding, "I fear it will always remain beyond my grasp."[3] Nor had either man heard of relativity's prediction that the universe might be expanding. Lacking a theoretical explanation for what he had observed, Hubble, alert to the fact that "observations always involve theory,"[4] was reluctant to draw conclusions about the mean-

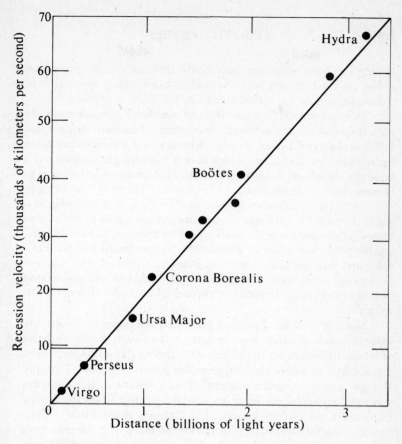

The Hubble law, that galaxies are receding from one another at velocities directly proportional to the distances separating them, holds true across the known universe. This plot includes ten major clusters of galaxies. The boxed area at the lower left represents the galaxies observed by Hubble when he discovered the law.

ing of his own discovery.* He spoke of the "redshift-distance relation," and sometimes of "velocity-shifts," but seldom of what his

*The one theory Hubble *did* know about that predicted the redshift-distance relation was that of the Dutch astronomer Willem de Sitter, who had published a model in which redshifts were generated, not by velocity in an expanding universe, but by a spurious "De Sitter effect," a bit of mathematical arcana with no known physical basis. The chilly reception afforded Hubble's efforts to link his observations to the De Sitter effect did little to encourage Hubble's already hesitant ventures into the theoretical side of cosmology.

finding has been called ever since—the discovery of the expansion of the universe. Years later, he was still describing the notion of cosmic expansion as "rather startling."[5]

As it happened, the man who put together Einstein's relativity with the redshifts of the spirals was neither an eminent theorist nor a skilled observer, but an obscure Belgian priest and mathematician named Georges Lemaître. The son of a Louvain glassmaker and a brewer's daughter, Lemaître had decided at the age of nine to become both a scientist and a clergyman: "There is," he liked to say, "no conflict between science and religion."[6] Something of a joiner, Lemaître at Eddington's suggestion made a tour of the United States, attending scientific conferences and passing out cards bearing his name and address. During the trip he learned of Slipher's redshifts, and upon his return to Brussels wrote, in 1927, a prophetic paper that erected a mathematical superstructure connecting the observed redshifts with the expanding universe of general relativity.

Nobody noticed. Lemaître published the paper in an obscure journal—an admirably humble but professionally hobbling habit of which he was never to rid himself—and he did not in any event cut the sort of figure that suggests the stamp of genius. Plumply bourgeois in appearance, a homeboy in a priest's collar, Lemaître was brushed off by the luminaries he approached when the Solvay Conference on Physics convened in Brussels that October. Even the customarily forbearing Einstein lost patience with the entreaties of this pillar of middle-class normalcy. *"Vos calculs sont corrects, mais votre physique est abominable,"* Einstein told Lemaître—"Your calculations are correct, but your physics is abominable."[7] (Einstein later reconsidered, and in Brussels in 1933 turned over a lecture to Lemaître, reassuring the nervous cleric as he stuttered through his talk by uttering sotto voce pronouncements of *"très joli, très, très joli"*—"very, very beautiful.")[8]

The situation became clearer to Lemaître, if not to his more established colleagues, with the publication of Hubble's 1929 paper on the redshift-distance relation. In January 1930, Eddington, DeSitter, and the other recognized maestros of theoretical cosmology gathered at a meeting of the Royal Astronomical Society and there labored mightily, and in vain, to erect a mathematical bridge between DeSitter's relativistic cosmology and Hubble's discovery. Lemaître read of their efforts in the February issue of *The Observatory* and wrote to Eddington, reminding him that he had already solved the problem. Eddington sent a copy of Lemaître's paper to DeSitter, and then, with the generosity and judgment that had informed his similar efforts on Einstein's behalf years earlier, set about proclaiming to the world that a little-known Belgian mathematics professor had authored the first expanding-universe cosmology. It was thus that Hubble began to learn of the potential significance of his own findings.

Meanwhile, Lemaître had started thinking about the origin of the universe. An expanding universe clearly must once have been very different than it is at present. The galaxies today are millions of light-years apart; once they must have been closer together. Indeed, in the beginning, everything must have been close to everything else. The density of the young universe might have been very high indeed—as high, perhaps, as that of an atomic nucleus. Thinking along converging tracks stretching backward in time, Lemaître began forging the first links between cosmology, the science of the very large, and nuclear physics, the science of the very small.

This extrapolation did not sit well with Lemaître's champion Eddington. "Since I cannot avoid introducing this question of a beginning," he wrote, "it has seemed to me that the most satisfactory theory would be one which made the beginning *not too unaesthetically abrupt*" (Eddington's italics).[9] Eddington imagined that the universe had begun as a stable system, perhaps akin to a star cluster, that had fallen apart in such as way as to produce the cosmic expansion. Lemaître took a more radical tack. He proposed that the universe might have begun as an infinitely small pinpoint—a "singularity," in mathematical terms—at time zero, "a day without a yesterday" when space was infinitely curved and all matter and all energy was concentrated into a single quantum of energy.[10] Lemaître called this genesiac state the "primordial atom," and its eruption the "big noise."[11] Later the astrophysicist Fred Hoyle, who disliked the whole idea even more than did Eddington, designated the creation event by an intentionally ugly name that stuck. Hoyle called it the big bang.

The gap between European theory and American observations began to close in the early 1930s, when Einstein and many more German Jews, intellectuals, and other undesirables read Hitler's handwriting on the wall and began emigrating to the United States. In 1931, Einstein visited Mount Wilson, where Hubble, puffing his pipe in Churchillian disdain of observatory protocol, gave him a tour of the dome and showed him the spectrographic evidence of the cosmic expansion that the general theory had foreseen. Two years later, back in southern California, Einstein heard Lemaître describe his theory of the primordial atom at a lecture in the library of the Mount Wilson observatory office on Santa Barbara Street in Pasadena. "In the beginning of everything we had fireworks of unimaginable beauty," Lemaître said, waxing rhapsodic. "Then there was the explosion followed by the filling of the heavens with smoke. We come too late to do more than visualize the splendor of creation's birthday." Einstein rose to his feet at the end of the talk and called Lemaître's theory "the most beautiful and satisfying interpretation I have listened to."[12]*

Insufficiently accomplished to be called a theory, Lemaître's concept of genesis as a nuclear decay event might better be described as a working hypothesis. Lemaître understood this as well as anyone, and reminded readers of his book *The Primeval Atom* that "too much importance must not be attached to this description of the primeval atom, a description which will have to be modified, perhaps, when our knowledge of atomic nuclei is more perfect."[13] And yet, however tentatively, Lemaître's approach anticipated the course of cosmology in the latter twentieth century, helped to set science on that course, and had the salutary immediate effect of inviting nuclear physicists into the cosmological arena. Some accepted the invitation, and the result was an infusion of fresh blood and brainpower into the field. Soon physicists of the caliber of Enrico Fermi, Carl Friedrich von Weizsäcker, and Edward Teller were applying their considerable talents to the question of what went on in the first moments of the big bang.

To the forefront of this effort ambled the deceptively easygoing, good-natured Russian émigré George Gamow. Witty, iconoclastic, and irreverent about the doings of humanity if not those of nature, Gamow like Einstein was one of those rare individuals who seem never to lose their childhood curiosity and sense of won-

---

*Einstein was referring to Lemaître's contention that cosmic rays, high-energy subatomic particles from space, had been generated in the primordial fireworks. This did not hold up in its specifics, but it anticipated aspects of George Gamow's subsequent prediction that the universe might be suffused by a cosmic background radiation composed of ancient photons released by the big bang.

der. One of the things he wondered about the most was how the universe began.

Gamow's chief concern, as we will see, had to do with the formation of elements early in the history of the universe. He reasoned that the stuff of the young universe might have been hot and dense enough for atomic nuclei to have been fused into various combinations, creating the elements as we know them. This line of research was to have mixed results in Gamow's hands (theoretical physics was still insufficiently mature to handle many of the calculations involved) but its portrayal of the early universe as a hot, dense, rapidly evolving plasma gave rise to one of the most potent predictions in the history of science—that of the cosmic background radiation, a ubiquitous, simmering energy left over from the big bang.

Gamow's notion was that, if the universe began hot and has been expanding and cooling ever since, its temperature today, though cold, would not be *absolutely* cold. There should be some residual heat remaining from the big bang. This energy would have been stretched out and thus lowered in frequency by the cosmic expansion: In technical terms, the photons carrying the energy of the big bang, having originated in the wavelengths of light, ought to have been redshifted by the subsequent expansion of the universe into the lower frequencies of electromagnetic energy that we call microwave radio radiation. Working in his customarily rough-hewn manner, Gamow did a back-of-the-envelope calculation and estimated that the universe today should be permeated by an ocean of photons with an ambient temperature of some 50 degrees Kelvin. His colleagues Ralph Alpher and Robert Herman then corrected an arithmetic mistake and two other errors in Gamow's paper and arrived at a revised temperature for the background radiation of "about five degrees."[14]

At the time, little attention was paid to the prediction of Gamow, Alpher, and Herman that relic radiation should be left over from the big bang. It seemed arcane and, in any case, impossible to verify; radio astronomy was in its infancy, and there was as yet no such thing on Earth as a microwave radiotelescope. A decade later, when radio astronomy had become a reality, Robert Dicke at Princeton University independently hit on the same idea, and set about building a microwave receiver to listen for the cosmic background radiation. He was still at work on it when he learned that two researchers at Bell Laboratories, Arno Penzias and Robert Wilson, were having trouble accounting for a persistent hiss in a microwave horn that Bell had built for satellite communications experiments. The temperature of this unwanted noise was 2.7 degrees. Though none of the three remembered the work of Gamow, Alpher, and Herman, this was just the value that they had predicted (once their calculations were updated to correct for subsequent improvements in the Hubble scale of the age of the universe). Penzias and Wilson won a Nobel Prize in physics for their discovery, and Lemaître, then seventy-two years old, learned of their finding in one of the last conversations of his life.

Today the assertion that we live in an expanding universe rests upon three fundamental lines of research. The first is the Hubble law: The relation between the distances of galaxies and the redshift of their light appears to pertain to the limits of present-day observation—out to hundreds of millions of light-years—and the only known consistent explanation for such a state of affairs is that the redshifts are produced by the recession velocity of galaxies in an expanding universe. The second piece of evidence is the cosmic background radiation: It traces out the "black-body" curve that would characterize the spectrum of photons released in the big bang, and it is received at equal strength from all directions, except for a small anisotropy (or "hot spot") introduced by the absolute motion of the earth within the overall cosmic framework. The third piece of data is chronological: The age of the universe inferred from the expansion velocity, some ten to twenty billion years, fits with the ages of the oldest known stars, some twelve to sixteen billion years, and with the temperature of the cosmic background radiation itself.

Whatever its other implications for human thought—and there are many—the expansion of the universe had the tremendous advantage of investing cosmology with a dimension of cosmic history. The structure of the universe, from that of atomic nuclei to the vast superclusters of galaxies that stretch across hundreds of millions of light-years of space, may now be seen to have evolved from prior structures; to explain their present disposition clearly requires that we gain a better understanding of their history. Even natural laws themselves may prove to have a mutable past. These considerations will be discussed in Part III of this book. But first we need examine how our species came to comprehend the depths of terrestrial and cosmic history. It's time for time.

PART TWO

TIME

The leading idea which is present in all our researches, and which accompanies every fresh observation, the sound which to the ear of the student of Nature seems continually echoed in every part of her works, is—Time!— Time!—Time!

—George Scrope

Change is my theme. You gods, whose power
    has wrought
All transformations, aid the poet's thought,
And make my song's unbroken sequence flow
From earth's beginnings to the days we know.

—Ovid

SERMONS IN STONES

We aspire in vain to assign limits to the works
of creation in *space*, whether we examine the
starry heavens, or that world of minute ani-
malcules which is revealed to us by the microscope. We are prepared, therefore, to find
that in *time* also the confines of the universe
lie beyond the reach of mortal ken.

—Charles Lyell

And this our life, exempt from public haunt,
Finds tongues in trees, books in the running
    brooks,
Sermons in stones, and good in everything.

—Shakespeare

The conception of time that held sway in ancient
Greece was cyclical, and as closed as the crystalline spheres in which
Aristotle imprisoned cosmic space. Plato, Aristotle, Pythagoras,
and the Stoics all espoused the view, inherited from an old Chaldean
belief, that the history of the universe consisted of a series of "great
years," each a cycle of unspecified duration that ended when the
planets all came together in conjunction, unleashing a catastrophe

from the ashes of which the next cycle began anew. This process was thought to have been going on forever: As Aristotle reasoned, with a logic as circular as the motions of the stars, it would be paradoxical to think of time as having had a beginning *in time*, and so the cosmic cycles must eternally recur.

The cyclical view of time was not without its charms. It possessed a world-weary, urbane fatalism of the sort that so often appeals to the philosophically inclined, a tincture indelibly preserved by the Islamic historian Ahmad ibn Muhammad ibn 'Abd al-Ghaffar, al-Kazwini al-Ghifari, who recited the following parable of eternal recurrence:

> I passed one day by a very ancient and wonderfully populous city, and asked one of its inhabitants how long it had been founded.
>
> "It is indeed a mighty city," he replied. "We know not how long it has existed, and our ancestors were on this subject as ignorant as ourselves."
>
> Five centuries afterwards, as I passed by the same place, I could not perceive the slightest vestige of the city. I demanded of a peasant, who was gathering herbs upon its former site, how long it had been [since the city was] destroyed.
>
> "A strange question!" he replied. "The ground here has never been different from what you now behold."
>
> "Was there not once a splendid city here?" I asked.
>
> "Never," he replied, "so far as we have seen, and never did our fathers speak to us of any such city."
>
> On my return there five hundred years afterwards, I found the sea in the same place. On its shores was a party of fishermen. I enquired how long the land had been covered by the waters.
>
> "Is this a question for a man like you?" they said. "This spot has always been what it is now."
>
> Again I returned, five hundred years afterwards, and the sea had disappeared. I inquired of a man who stood alone upon the spot how long ago this change had taken place, and he gave me the same answer as I had received before.
>
> Finally, on coming back again after an equal lapse of time, I found there a flourishing city, more populous and more rich in beautiful buildings than the city I had seen the first time, and when I would have informed myself concerning its origin,

the inhabitants answered me, "Its rise is lost in remote antiquity: We are ignorant how long it has existed, and our fathers
were on this subject as ignorant as ourselves."[1]

Taken literally, cyclical time even proffered a species of immortality: As Aristotle's pupil Eudemus of Rhodes told his students,
"If you believe the Pythagoreans, everything will eventually return in the selfsame numerical order and I shall converse with you
staff in hand and you will sit as you are sitting now, and so it will
be in everything else."[2] Whether for these or other reasons, cyclical
time is still popular today, with many cosmologists arguing for
"oscillating universe" models in which the expansion of the universe
is envisioned as eventually coming to a halt, to be followed by
cosmic collapse into the cleansing fires of the next big bang.

But for all its felicities, the old doctrine of infinite, cyclical
history had the pernicious effect of tending to discourage attempts
to gauge the genuine extent of the past. If cosmic history consisted
of an endless series of repetitions punctuated by universal destruction, then it was impossible to determine what the total age of the
universe might actually be: An infinite, cyclical past is by definition
immeasurable—is "time out of mind," as Alexander the Great used
to say. Nor did cyclical time leave much room for the concept of
evolution, the fruitful idea that there can be genuine innovation in
the world.

The Greeks knew that the world changes, and that some of
its changes are gradual. Living as they did with the sea at their feet
and the mountains at their backs, they appreciated that waves erode
the land, and were acquainted with the strange fact that seashells
and fossils of marine creatures may be found on mountaintops far
above sea level.* At least two of the realizations essential to the
modern science of geology—that mountains can be thrust up from
what was once a seabed, and that they can be worn down by wind
and water—were mentioned as early as the sixth century B.C., by
Thales of Miletus and Xenophanes of Colophon. But they tended
to regard these transformations as mere details, limited to the current cycle of a cosmos that was in the long run eternal and un-

---

*Aristotle explained fossil fish by postulating that the fish had got stranded and
died while foraging for food in subterranean caverns.

changing. "There is necessarily some change in the whole world," wrote Aristotle, "but not in the way of coming into existence or perishing, for the universe is permanent."[3]

For science to begin to assess the antiquity of the earth and the wider universe—to locate humanity's place in the depths of the past as it was to chart our location in cosmic space—it had first to break the closed circle of cyclical time and to replace it with a linear time that, though long, had a definable beginning and a finite duration. Curiously enough, this step was initiated by a development that was in most other respects a calamity for the progress of empirical inquiry—the ascent of the Christian model of the universe.

Initially, Christian cosmology diminished the scope of cosmic history, much as it shrank the spatial dimensions of the empirically accessible universe. The grand, impersonal sweep of the Greek and Islamic cycles of time were replaced, in Christian thought, by an abbreviated and anecdotal conception of the past, in which the affairs of men and God counted for more than the inhuman workings of water on stone. If history for Aristotle was like the turning of a giant wheel, for the Christians it was like a play, with a definite beginning and end, punctuated by unique, singular events like the birth of Jesus or the giving of the law to Moses.

Christian scholars estimated the age of the world by consulting scriptural chronologies of human birth and death—by adding up the "begats," as they say. This was the method of Eusebius, Chairman of the Council of Nicaea convened by the emperor Constantine in A.D. 325 to define Christian doctrine, who determined that 3,184 years had elapsed between Adam and Abraham; of Augustine of Hippo when he estimated the date the Creation at about 5500 B.C.; of Kepler, who dated it at 3993 B.C.; and of Newton, who arrived at a date just five years earlier than Kepler's. Its apotheosis came in the seventeenth century, when James Ussher, bishop of Armagh, Ireland, concluded that "the beginning of time . . . fell on the beginning of the night which preceded the 23rd day of October, in the year . . . 4004 B.C."[4]

Ussher's spurious exactitude has made him the butt of many a latter-day scholarly snigger, but, for all its absurdities, his approach—and, more generally, the Christian approach to historiography—did more to encourage scientific inquiry into the past than had the lofty pessimism of the Greeks. By promulgating the idea that the universe had a beginning in time, and that the age of the earth was therefore both finite and measurable, the Christian chronologists unwittingly set the stage for the epoch of scientific age-dating that followed.

The difference, of course, was that the scientists studied not Scriptures but stones. This was how the naturalist George Louis Leclerc expressed the geologists's creed, in 1778:

> Just as in civil history we consult warrants, study medallions, and decipher ancient inscriptions, in order to determine the epochs of the human revolutions and fix the dates of moral events, so in natural history one must dig through the archives of the world, extract ancient relics from the bowels of the earth, [and] gather together their fragments. . . . This is the only way of fixing certain points in the immensity of space, and of placing a number of milestones on the eternal path of time.[5]

To learn from the stones, however, geologists had first to be able to see them, and here the steam engine, prime mover of the Industrial Revolution, played a key role. Steam-driven pumps evacuated water from coal mines in Germany and the north of England, making it possible to dig deeper than ever before; steam-driven hoists brought the coal to the surface; coal from the mines was then transported on barges through canals, and on railroad trains pulled by steam locomotives, to fuel the steam engines of the ships and factories of the industrially developing world. Canal water and steel rails have in common that both work best when level, and the engineers who dug the canals and laid the tracks dealt with hills that blocked their way by cutting through them whenever possible. In doing so they "opened the veins of the earth," as the builders of the Great Wall of China had put it, exposing previously unseen layers of geological strata deposited over hundreds of millions of years. Budding geologists put to work in the field to help supervise these excavations found themselves presented with a gift as bounteous as a library—evidence of the long history of our planet, inscribed in the strata as if on the corrugated pages of an ancient book.

Among the first to learn to read the language of the stones were Abraham Gottlob Werner, a German mining geologist, and William Smith, an English canal surveyor and consulting engineer who helped excavate the Somersetshire Coal Canal in 1793. Werner noted that the same strata could be found in the same order at widely separated locations, indicating that the mechanism that laid them in place had operated on a large scale. This implied that local strata might hold evidence of how the planet as a whole had changed. Smith, for his part, observed that the strata—laid out, as he put it, like "slices of bread and butter"—could be identified not only by their gross composition but also by the various sorts of fossils they contained. Crisscrossing the English countryside day and night, by coach on a company pass, Smith observed that "the same strata were found always in the same order and contained the same fossils."[6] This, then, was a key to deciphering the hieroglyphics of the rocks—the realization that the world's history could be read in the sequence of fossils the rocks contained.

The fossil record, however, soon began turning up evidence of creatures no longer to be found in the world today. The absence of their living counterparts presented a challenge to advocates of the biblical account of history, who had maintained, relying upon Scriptures, that all animals were created at the same time and that none had since become extinct. For a while it was argued that living specimens of the unfamiliar species might yet survive, in distant lands to which they had migrated in the years since the strata were formed. Thomas Jefferson entertained this possibility and urged naturalists headed west to look for wooly mammoths, whose roar one pioneer had reported hearing echoing through the forests of Virginia. But as the years passed the world's wildernesses were ever more thoroughly explored, and still no sign of the mammoth or its lost cousins turned up. Meanwhile, the roster of missing species grew longer—Georges Cuvier, the French zoologist who founded the science of paleontology, had by 1801 identified twenty-three species of extinct animals in the fossil record—and the word "extinct" began tolling like a bell in the scientific literature and the university lecture halls. It has gone on tolling ever since; and today it is understood that 99 percent of all the species that have lived on the earth have since died out.

Almost as troublesome to Christian interpreters of the earth's history was the bewildering variety of *living* species being discovered by biologists in their laboratories and by naturalists exploring the jungles of Africa, South America, and Southeast Asia. Some, like the giant subtropical beetles that bit the young Darwin, were noxious; their benefit to humanity, for whom God was said to have

made the world, was not immediately evident. Many were so mi-
nuscule that they could be detected only with a microscope; their
role in God's plan had not been anticipated. Others were instinc-
tively unsettling—none more so than the orangutan, whose name
derives from the Malay for "wild man" and whose warm, almost
intimate gaze, coming as it does from only a puddle or two across
the primate gene pool, seemed to mock human pretensions of
uniqueness. None of these creatures was thought to have shown
up on the passenger's roster of Noah's arc. What were they doing
here?

The religious orthodoxy took temporary refuge in the concept
of a "Great Chain of Being." This precept held that the hierarchy
of living beings, from the lowliest microorganisms to the apes and
great whales, had been created by God simultaneously, and that
all, together, formed one marvelous structure, a magic mountain
with humans at—or near—its apex. The importance of the Great
Chain of Being in eighteenth-century thought is difficult to overestimate; it figured in the framing of most of the scientific hypotheses of the time. The Chain, however, was no stronger than
its weakest link; its very completeness was itself proof of the perfection of God, and there could, therefore, be no "missing link."
(The term, later adopted by the evolutionists, began here.) As John
Locke wrote:

> In all the visible corporeal world we see no chasms or
> gaps. All quite down from us the descent is by easy steps, and
> a continued series that in each remove differ very little one
> from the other. There are fishes that have wings and are not
> strangers to the airy region, and there are some birds that are
> inhabitants of the water, whose blood is as cold as fishes. . . .
> When we consider the infinite power and wisdom of the Maker,
> we have reason to think that it is suitable to the magnificent
> harmony of the universe, and the great design and infinite
> goodness of the architect, that the species of creatures should
> also, by gentle degrees, ascend upwards from us towards his
> infinite perfection, as we see they gradually descend from us
> downwards.[7]

Here resided the horror that the prospect of extinction elicited
in the thoughts of the pious. "It is contrary to the common course
of providence to suffer any of his creatures to be annihilated," wrote

the Quaker naturalist Peter Collinson, as in awe he contemplated the mighty teeth of the extinct mastodon and the weighty bones of the equally extinct Irish elk.[8] The seventeenth-century naturalist John Ray noted that evidence of "the destruction of any one species," would amount to "a dismembring of the Universe, and rendring it imperfect."[9]

Yet still the death knell tolled, as the geologists' spades and the railroad builders' steam shovels continued to turn up the remains of an ever increasing variety of organisms that clearly once had lived but were to be found no more. There was fossil evidence of flowers never known to have bloomed in human sight, bizarre fishes and birds that no one had ever seen swim or fly, and exotic creatures that could not fail to capture the popular as well as the scientific imagination—the saber-toothed tiger, the "dawn horse," the giant armadillo, the woolly rhino, and the dinosaurs—all gone forever. And, since fossils of many of these creatures were found in climates where they could not have thrived (fish on mountaintops, polar bears in the tropics) the earth must have undergone profound and wide-reaching changes since the time when the extinct species had lived. How could all this have happened in the short span of Bishop Ussher's six thousand years?

The most promising answer for the fundamentalists lay in what came to be called catastrophism, the hypothesis that such major geological changes as had occurred had come about suddenly, as the result of cataclysmic, almost supernatural upheavals that had leveled mountains, raised seabeds toward the sky, and doomed whole species to extinction almost overnight. Catastrophism accounted for the extinction of species in the fossil record without violating biblical chronology, and it enjoyed strong support from the biblical story of the Flood, which the catastrophists came to regard as but the most dramatic among several disasters visited upon the world by a wrathful God. As to the question of whether all these cataclysms had been divinely ordained there was some disparity of opinion, with Cuvier like many geologists in the early nineteenth century proposing that while the Lord had wrought the Flood and earlier disasters, the ones since might be ascribed to conventionally causal agencies.

From a scientific standpoint, the most pernicious implication of catastrophism was that it severed the past from the present, much as Aristotle's astrophysics had divorced the aethereal from the mundane. By relegating major geological changes to the action of pre-ternaturally powerful forces that had manifested themselves only in the early history of the earth, catastrophism barred the extrapolation into history of scientific laws gleaned from the world today. "Never," wrote the Scottish geologist Charles Lyell, "was there a dogma more calculated to foster indolence, and to blunt the keen edge of curiosity, than this assumption of the discordance between the former and the existing causes of change."[10]

Lyell held a contrary view, called uniformitarianism. He maintained that *all* geological and biological change was due to ordinary, natural causes that had operated in much the same way throughout the earth's long history. The extinction of species, by uniformitarian lights, was brought about by events very much akin to those we see in action around us today—the slow erosion of rock and soil by wind and water, gradual changes in climate, and the occasional raising and lowering of mountains.

This steady-state view of the earth's history had first been advanced by the Scottish chemist James Hutton. The Herschel of geological history, Hutton was a farsighted visionary who saw the imprint of aeons etched in common rocks: "The ruins of an older world are visible in the present structure of our planet," he wrote. He illustrated his thesis with a cutaway drawing that depicted, above ground, a placid English countryside scene—an enclosed carriage drawn by two horses standing by a fence in the woods—while below stretched a frieze of strata, and, beneath that, a twisted and jumbled tableau of metamorphic rock, the frozen image of a tumultuous and ever changing world.

As change in a noncatastrophic world must on the whole proceed slowly, the uniformitarian hypothesis required that the earth be very old. There was some theoretical evidence that this might be the case: Georges Buffon, the French naturalist, had argued from astronomical premises that the earth began as a molten ball that slowly cooled, and that its age might therefore amount to as much as five hundred thousand years. Now Hutton, concerned not with the origin of the earth but with the geological processes to which it was currently being subjected, arrived at an even grander estimation of the extent of antiquity. "We find," he wrote, "no vestige of a beginning,—no prospect of an end."[11] This was daring, perhaps, but also reckless; an *infinite* past is a great deal more problematical than a very long past. (Darwin was to make a similar

Buried evidence of geological upheaval was depicted in cutaway drawings like this one in Hutton's *Theory of the Earth*. (After Hutton, 1795.)

mistake, arguing for an infinitely old Earth until his mathematician friends took him aside and explained that infinity is strong and dangerous medicine and not just a big number.)

The inaugural fortunes of uniformitarianism suffered, moreover, from the liabilities of Hutton's literary style; his *Theory of the Earth*, published in 1795, was written in a syntax as jumbled as the strata it described. The situation improved somewhat when John Playfair took the trouble to elucidate his friend Hutton's views, in his *Illustrations of the Huttonian Theory of the Earth*, but the real breakthrough came a generation later, when the uniformitarian hypothesis was taken up by Lyell. Born in 1797, the year of Hutton's death, Lyell was an energetic young man, blessed with poor eyesight, who peered at the world around him with myopic intensity. While still an undergraduate studying geology at Oxford, Lyell took a holiday trip to a spot on the seashore that he had visited as a child, and he noticed, as many another bather had not, that erosion had slightly altered the shape of the coastline near Norwich. He began to conceive of the planet as a seething, changing entity, writhing in its own good time like a living organism.

Much of the prior debate over the age of the world had been conducted from easy chairs, by the likes of the English divine Thomas Burnet, who boasted that he based his efforts to reconcile scientific and biblical accounts of history on but three sources, "Scripture, Reason, and ancient Tradition."[12] Lyell spent his days wandering to and fro upon the earth, and in his sixties was still scrambling up mountainsides and down dry washes, making notes all the while. Mount Etna in Sicily, the traditional abode of Vulcan's forge, had long been a favorite subject of studious scholars who had viewed it, if at all, from afar. Lyell climbed its slopes of freshly frozen lava, and deduced, from his measurements of the sheer bulk of the ten-thousand-foot mountain, that it had been built up from a great many lava flows, the accumulation of which "must have required an immense series of ages anterior to our historical periods for its growth."[13] In Chile, Lyell estimated that a single earthquake could elevate the coastal mountains by as much as three feet, and speculated that "a repetition of two thousand shocks, of equal violence, might produce a mountain chain one hundred miles long and six thousand feet high."[14] The identification of warm-water seashells in northern Italy and of the bodies of mammoth frozen in Siberian ice, he noted, indicate that the European climate was once "sufficiently mild to afford food for numerous herds of elephants and rhinoceroses, *of species distinct from those now living*" (Lyell's italics).[15]

A lucid and vivid writer, Lyell was as adept at demolishing the arguments of the catastrophists as he was at comprehending the construction of mountain ranges. "Geologists have been ever prone to represent Nature as having been prodigal of violence and parsimonious of time," he wrote, but, he noted, the fracturing and weathering of rock taken by the catastrophists to represent the violence of the early Earth could as easily have been imposed by the ravages of time.[16] A voracious student of biology as well as geology (his father had been a botanist, and Lyell *fils* had studied entomology) he drew upon the life sciences as well. The catastro-phists relegate extinction to brief cataclysms, he wrote, but

> if we then turn to the present state of the animate creation, and inquire whether it has now become fixed and stationary, we discover that, on the contrary, it is in a state of continual flux—that there are many causes in action which tend to the extinction of species, and which are conclusive against the doctrine of their unlimited durability.[17]

It is true, as the catastrophists point out, that the fossil record is fragmented and broken. But, Lyell argued, disastrous events were not required to break it:

> Forests may be as dense and lofty as those of Brazil, and may swarm with quadrupeds, birds, and insects, yet at the end of ten thousand years one layer of black mould, a few inches thick, may be the sole representative of those myriads of trees, leaves, flowers, and fruits, those innumerable bones and skel-etons of birds, quadrupeds, and reptiles, which tenanted the fertile region. Should this land be at length submerged, the waves of the sea may wash away in a few hours the scanty covering of mould.[18]

If Lyell was right in concluding that "the causes which produced the former revolutions [i.e., dramatic changes] of the globe" were the same as "those now in everyday operation," then the age of the earth must be reckoned not in thousands but in millions of years.[19]

But the choice, Lyell argued, was not solely between a young and an old Earth. Nor was it a clear-cut case of catastrophism versus uniformitarianism (each of which would in any case prove to contain seeds of the truth). The real choice was between a closed science, resigned to turn a blind eye on any evidence that contradicted the existing consensus, and an open science that dared to follow the evidence toward unknown inferences. If the first path, as Lyell wrote, was "calculated . . . to blunt the keen edge of curiosity," the second "cherishes a sanguine hope that the resources to be derived from observation and experiment, or from the study of nature such as she now is, are very far from being exhausted."[20]

In December 1831 the young man who would journey farthest along the road leading into the depths of time was packing his bags to depart on a voyage around the world, with a copy of the first volume of Lyell's *Principles of Geology* in his portable library. The book, published the previous year, had been recommended by his friend and teacher John Henslow. A catastrophist like virtually every other geologist at the time, Henslow advised his former student to enjoy Lyell's writing, but cautioned him on no account to take its radical views seriously. Charles Darwin cheerfully agreed, packed the book, and set sail on the *Beagle*.

13

THE AGE OF
THE EARTH

The antiquity of time is the youth of the world.
—Francis Bacon

What we take for the history of nature is only
the very incomplete history of an instant.
—Denis Diderot

Lyell's book turned Darwin's voyage into a trip
through time. Darwin began reading it almost immediately, in his
bunk, while suffering through the first of the many attacks of sea-
sickness that were to plague him during the next five years—the
Beagle, a stout, beamy brig ninety feet long by twenty-four feet
wide, was otherwise comfortable, but her hull was rounded and
she rolled. He started applying what he called "the wonderful su-
periority of Lyell's manner of treating geology"[1] as soon as the
expedition made landfall in the Cape Verde islands.

To construct an empirically based theory like Darwin's account
of evolution requires not only observational data but an organizing
hypothesis as well. Darwin drew his hypothesis—that the world
is old, and is continuing to change today much as it did in the
past—largely from Lyell. "The great merit of the *Principles*," he
wrote, "was that it altered the whole tone of one's mind, and,
therefore, that when seeing a thing never seen by Lyell, one yet

saw it partially through his eyes." Later Darwin allowed that "I feel as if my books came half out of Sir Charles Lyell's brain."[2]

The observations Darwin himself was well suited to provide. "Nothing escaped him," wrote Dr. Edward Eickstead Lane, who often walked with Darwin at Moor Park.

> No object in nature, whether Flower, or Bird, or Insect of any kind, could avoid his loving recognition. He knew about them all . . . could give you endless information . . . in a manner so full of point and pith and living interest, and so full of charm, that you could not but be supremely delighted, nor fail to feel . . . that you were enjoying a vast intellectual treat to be never forgotten."[3]

During the *Beagle* expedition Darwin saw the world as few have seen it, in rich diversity and detail, from horseback and muleback and on foot, in cave explorations and excursions across pack ice and blazing sand from Patagonia to Australia to the Keeling Islands of the Indian Ocean. He noted everything, absorbed everything, and collected so many samples of plants and animals that his shipmates wondered aloud whether he was out to sink the *Beagle*.

In Chile, Darwin found marine fossils on mountaintops twelve thousand feet high and witnessed an earthquake that raised the ground three feet in a matter of minutes—Lyellian evidence that the more or less uniform operation of geological processes can produce changes as dramatic as those ascribed by the theologians ancient catastrophes. At first he was cautious about jumping to conclusions: Reporting his findings in a letter to his teacher Henslow he wrote that "I am afraid you will tell me to learn my A.B.C.— to know quartz from Feldspar—before I indulge in such speculations."[4] But by the time the *Beagle* reached the South Pacific, Darwin had four years of rigorous fieldwork under his belt, and had begun to feel more confident of his ability to interpret observations in terms of hypotheses.

There he ventured an ingenious theory of his own, concerning the origin of coral atolls. On a hot fall day in 1834, while the *Beagle* was making headway from the Galapagos Islands toward Tahiti, he climbed the mainmast and saw the bone-white atolls of the Dangerous Archipelago scattered across the sea like so many lacy hoops. He was impressed by their appearance of frailty: "These low hollow coral islands bear no proportion to the vast ocean out of which they abruptly rise," he wrote, "and it seems wonderful, that such weak invaders are not overwhelmed, by the all-powerful and never-tiring waves of that great sea, miscalled the Pacific."⁵

Darwin theorized that the atolls marked the sites of vanished volcanos.* A new volcano can burst through the sea floor and, in successive eruptions, build itself up into a mountainous island that towers above the sea. When the lava stops flowing and things quiet down, a live coral reef can form on the flanks of the volcano, just below sea level. Here begins Darwin's contribution: Eventually, he said, the inactive volcano may begin to sink, owing either to erosion or to the slow collapse of the ocean floor. As the old island gradually subsides, live coral continues to build up atop the dead and dying coral below. Eventually the original island vanishes beneath the waves, leaving a ring of coral behind. "The reef constructing corals," Darwin wrote, "have indeed reared and preserved wonderful memorials of the subterranean oscillations of level; we see in each barrier-reef a proof that the land has there subsided, and in each atoll a monument over an island now lost."⁶

The beauty of this theory, from a uniformitarian standpoint, was that the process *had* to be gradual. Living coral requires sunlight; as Darwin noted, it "cannot live at a greater depth than from twenty to thirty fathoms," or some 120 to 180 feet.⁷ Had the islands sunk rapidly, as catastrophism demanded, the coral would have been plunged into the dark depths of the sea before new coral had time to grow on top of it, and no atoll would have been left behind.

When Darwin returned home after five years aboard the *Beagle*, his father upon first laying eyes on him "turned round to my sisters and exclaimed, 'Why the shape of his head is quite altered.' "⁸ This was something of a family in-joke; phrenology and physiognomy were Victorian passions that Robert Darwin shared with Robert Fitz-Roy, captain of the *Beagle*, who had at first refused to hire on the young Darwin owing to what Fitz-Roy took to be the inauspicious shape of his nose. But the elder Darwin was a sensitive

---

*Though Darwin, echoing Newton, characterized much of his research as purely inductive—"I worked on true Baconian principles," he said of his account of evolution, "and without any theory collected facts on a wholesale scale"—this has always been a difficult claim to justify scrupulously, and Darwin formulated his theory of coral atoll formation while still in South America, before he ever laid eyes on a real atoll.

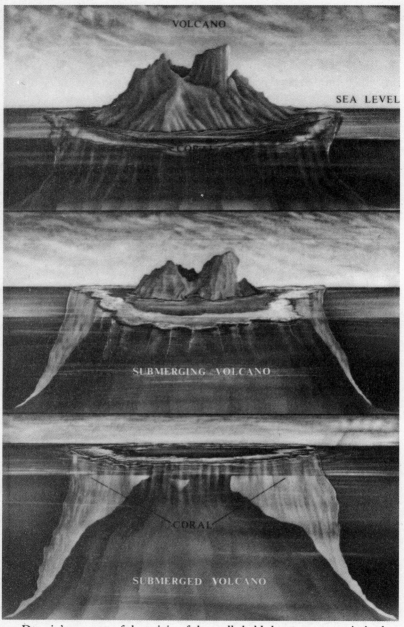

Darwin's account of the origin of the atolls held that as a mountain in the sea subsides, live coral continues to build up along what once were its coasts, until the ring of coral is all that remains.

observer, and his remark reflected his awareness that a lot had changed inside his son's skull as well. This was cause for celebration, for the young Darwin had been an idle and seemingly vacant lad with a passion for riding, hunting, gambling, drinking, and collecting twigs and stones. "You care for nothing but shooting, dogs, and rat-catching, and you will be a disgrace to yourself and all your family," his father had complained, to the delight of many a future biographer.[9] Darwin had dropped out of medical school, disappointing his father, who was a respected physician, and had failed to distinguish himself even in the undemanding theological studies to which he had been dispatched with the intention of preparing him for sinecure as a country parson.

The changes that led to a berth on the *Beagle* commenced at Cambridge. There Darwin became acquainted with Adam Sedgwick, one of the world's most accomplished field geologists, took courses in botany from Henslow, who combined an acutely rational mind with a buoyant outlook worthy of Linnaeus, and began to realize that he might, through science, combine his powers of observation with his love for the outdoors and his propensity for collecting. "I discovered," he wrote years later, "though unconsciously and insensibly, that the pleasure of observing and reasoning was a much higher one than that of skill and sport. The primeval instincts of the barbarian slowly yielded to the acquired tastes of the civilized man."[10]

When Darwin left England he was still a creationist. He did "not then in the least doubt the strict and literal truth of every word in the Bible," he recalled, and he believed, as did most of the geologists and biologists of his day, that all species of life had been created simultaneously and individually.[11] He returned home with doubts on this score. He had seen firsthand evidence that the earth is embroiled in continuing change, and he wondered whether species might change, too, and whether their mutability might cause new species to come into existence.

Evolution in itself was not a new idea. As a boy Darwin had read with interest his grandfather Erasmus Darwin's book *Zoonomia*, an evolutionary treatise full of robust exclamations over the notion that all life could have evolved from a single ancestor:

Perhaps millions of ages before the commencement of the history of mankind, would it be too bold to imagine, that all

warm-blooded animals have arisen from one living filament,
which THE GREAT FIRST CAUSE endued with ani-
mality . . . ? What a magnificent idea of the infinite power of
THE GREAT ARCHITECT! THE CAUSE OF CAUSES!
PARENT OF PARENTS![12]

Darwin was familiar, too, with the evolutionary views of the French
biologist Jean-Baptiste de Lamarck, who maintained that traits acquired by individuals through experience could be passed on to
their offspring. In a Lamarckian world, horses who grew strong
through racing bequeathed their fleetness to their young, while
giraffes, by stretching their necks to reach the leaves on trees, made
the next generation of giraffes more long-necked. Lamarckism was
replete with moral overtones gratifying to the Victorians, since it
implied that parents who worked hard and avoided vice would have
children who were genetically disposed toward hard work and clean
living. But it foundered on the question of how new species had
arisen. It pointed the way to ever better horses and giraffes, but
not to the origin of species, and thus left unanswered the question
of why different species are found in the fossil record than are living
today.

The Darwin's contribution was not simply to argue that life had
evolved—he did not even like to use the word "evolution"—but
rather to identify the evolutionary mechanism by which new species
come into existence. That was why he titled his book *The Origin of
Species*. His theory can be outlined in terms of three premises and
a conclusion.

The first premise has to do with *variation*. It notes that each
individual member of any given species is different—that each, as
we would say today, has a distinct genetic makeup. Darwin understood this very well. He grew up at a time when animal breeding
and plant hybridization was booming in England—his father-in-
law, Josiah Wedgwood, the ceramics manufacturer, was a noted
sheep breeder, and his father was a pigeon fancier—and he learned
from the husbanders to pay attention to the often subtle individual
characteristics that they sought to quash or to perpetuate.* Grounded

---

*The rise in animal breeding was spurred on by the growing industrialization of
England, which brought working people in from the country, where they could
keep a few barnyard animals of their own, to the cities, where they were fed from
ever larger herds bred to maximize profits. More generally, the advent of Dar-

in the specifics of biological variety, Darwin's thought was a mosaic of the particular: Scores of his publications consists of little notes in the *Gardeners' Chronicle and Agricultural Gazette* and the *Journal of Horticulture and Cottage Gardener* asking such questions as, "Has any one who has saved seed Peas grown close to other kinds observed that the succeeding crop came up untrue or crossed?"[13] and, "Is any record kept of the diameter attained by the largest Pansies?"[14]

Darwin's second premise is that all living creatures tend to produce more offspring than the environment can support. It's a cruel world, in which only a fraction of the wolves and turtles and dragonflies that come into existence manage to find sustenance and avoid predators long enough to reproduce. The English economist Thomas Malthus had quantified these harsh facts of life by pointing out that most species reproduce geometrically, while the environment can support no better than a linear increase in their populations.* Darwin read Malthus's *An Essay on the Principle of Population* in London in 1838—"for amusement," he recalled—and the hypothesis of evolution by natural selection began to take form in his mind. "One may say," he wrote, that "there is a force like a hundred thousand wedges trying [to] force every kind of adapted structure into the gaps in the [e]conomy of nature, or rather forming gaps by thrusting out weaker ones."[15] It was in the combination of the boundless fecundity of living things with the limited resources available to support them that Darwin found a natural, global mechanism that worked constantly to extinguish most variations, preserving only those carried by individuals who managed to survive and reproduce.

Which leads to the third premise—that the differences among individuals, combined with the environmental pressures emphasized by Malthus, affect the probability that a given individual will survive long enough to pass along its genetic characteristics. This is the process that Darwin called "natural selection." White moths fare better in snow, where their coloration serves as camouflage and

---

winism itself might be said to have been fostered by a certain distancing of human beings from the creatures they studied; it was only once people stopped cohabiting with animals that they began to entertain the idea that they were the animals' relations.

*Malthus, incidentally, appears to have been inspired in part by reading Darwin's grandfather Erasmus. It's a small world, or was so in Victorian England.

protects them from predator birds, while brown moths do better in snowless autumnal forests, where their color blends in against the brown tree trunks.* It is in this sense that the "fittest" (the phrase is Herbert Spencer's) survive, not because they are in some sense superior to their colleagues, but because they better "fit" their environment. When environmental conditions change, the most exquisitely adapted individuals may suddenly find themselves no longer fit; then it is the freaks and misfits who inherit the future.

Darwin's conclusion was that natural selection leads to the origin of new species. Because the world is constantly in a state of change, nature favors the varied—a community of predominantly white moths is better off if it contains a few dark moths, against a smoggy day—and the geographically dispersed, those who do not keep all their eggs in one basket. As a result, the degree of individual variations found within a given species tends to increase with the passage of time, until some groups have become so different from others that they can no longer mate and produce fertile offspring. At that point, a new species has emerged. As Darwin wrote:

> During the modification of the descendants of any one species, and during the incessant struggle of all species to increase in numbers, the more diversified the descendants become, the better will be their chance of success in the battle for life. Thus the small differences distinguishing varieties of the same species, steadily tend to increase, till they equal the greater differences between species . . .[16]

Darwin noted that in some ways his theory recalled the biblical image of the Tree of Life. But now the tree, instead of being static as in the creationist view, had come alive and was still growing:

> The green and budding twigs may represent existing species; and those produced during former years may represent the

---

*A striking example of adaptive color change occurred among British peppered moths in the vicinity of Manchester. In the eighteenth century, all such moths collected were pallid in color; in 1849 a single black moth was caught in the vicinity, and by the 1880s the black moths were in the majority. Why? Because industrial pollution had blackened tree trunks in the vicinity, robbing the original moths of their camouflage while bestowing its benefits upon the few black moths there. Once pollution-control ordinances came into effect, the soot slowly washed from the tree trunks and the pale peppered moth population rebounded.

long succession of extinct species. . . . The Tree of Life . . .
fills with its dead and broken branches the crust of the earth,
and covers the surface with its ever-branching and beautiful
ramifications.[17]

Critics with a preference for Bible stories complained that natural
selection was cold and mechanical. But in Darwin's eyes it both
animated and illuminated the natural world:

When we no longer look at an organic being as a savage looks
at a ship, as something wholly beyond his comprehension;
when we regard every production of nature as one which has
had a long history; when we contemplate every complex structure and instinct as the summing up of many contrivances,
each useful to the possessor, in the same way as any great
mechanical invention is the summing up of the labor, the experience, the reason, and even the blunders of numerous workmen; when we thus view each organic being, how far more
interesting—I speak from experience—does the study of natural history become![18]

And he added, in what was to become an evolutionists' credo:

There is a grandeur in this view of life, with its several powers,
having been originally breathed by the Creator into a few forms
or into one; and that, whilst this planet has gone cycling on
according to the fixed law of gravity, from so simple a beginning endless forms most beautiful and most wonderful have
been, and are being evolved.[19]

Darwin had formulated the essential elements of his theory by
the time of his marriage in 1839, and by 1844 had outlined it, in
a 230-page essay. Yet he withheld it from publication for the next
fifteen years. While the essay lay in his desk drawer, accompanied
by strict instructions to publish it in the event of his death, Darwin
settled in the country, fathered ten children, corresponded with
Lyell and a hundred other scientists, and wrote books—among
them a journal of the voyage of the *Beagle*, an account of his theory
of coral reefs, a treatise on volcanos and another on the geology of
South America, and a masterful study of barnacles that consumed
seven years of work and left him fuming that "I hate a barnacle as

no man ever did before."[20] In all, Darwin kept his theory of evolution a secret for nearly as long as Copernicus had concealed his heliocentric cosmology. Why the delay?

One explanation, still sometimes put forward, is that Darwin was constantly ill. This will not wash. Ill he certainly was: From about the time of his marriage and probably long before he was subject to intense headaches, vomiting, and heart palpitations. He consulted the best doctors in England in search of a cure, had himself hypnotized, and resorted to hydrotherapy, spending winter days wrapped in a cold, wet sheet. "His life," wrote his son Francis Darwin, "was one long struggle against the weariness and strain of sickness."[21] The ailment was never conclusively diagnosed and has since been attributed to many agencies, from Chagas' disease, brought on by what Darwin called the "attack (for it deserves no better name) of the Benchuca, the great black bug of the pampas" on March 26, 1835, to the psychosomatic affects of internal conflict between this former candidate for the priesthood and the anticlerical implications of his own theory. A more likely if less colorful possibility is that he suffered from severe allergies. But in any case, illness alone cannot explain why Darwin suppressed the theory of natural selection, since during those same years he wrote prolifically on other subjects.

It is much more likely that Darwin feared the storm of opposition he knew his ideas would provoke. He was a gentle, straightforward, almost childishly simple man, habitually respectful of the outlook of others and disinclined toward disputation. His theory, he knew, would draw down fire, not only from the clergy but from many of his fellow scientists as well.

The religious opposition promised to be formidable. Darwin did not have to strain his imagination to foresee what the orthodoxy would make of his assertion that animals and men are kin and that chance mutations drive evolution; to advocate such a thing, he told his friend Joseph Hooker, would be like admitting to a murder. (The murder of Adam, it was to be called.) Nor did he need look beyond England to envision what lay in store for him once word of the theory got out. When William Lawrence, later president of the Royal College of Surgeons, suggested that man evolves through the inheritance of innate rather than acquired traits, the Lord Chancellor declared his book contrary to Scriptures and denied it copyright. The legendary erudition of Benjamin Jowett of Oxford is recalled in a famous Balliol College masque's quatrain:

> First come I; my name is Jowett.
> There's no knowledge but I know it.
> I am the master of this college;
> What I don't know isn't knowledge.

But when Jowett in 1855 published a controversial interpretation of the Epistles of Saint Paul, he was accused of heresy and his salary was frozen. Darwin, puttering happily with wormstones and petunias in gardens that his insight rendered luminescent as Eden, was not eager to see the day when a thousand country parsons would turn his name into a synonym for the antichrist.

The scientific opposition arose in large measure from professional disdain for the very concept of evolution, which had long been an enthusiasm of ecstatics and occultists devoted to seances and tales of fairies flitting across the moors at dawn. To advocate so amateurish a theory was to invite learned ridicule. When in 1844 a theory of evolution was championed in the anonymous and enormously popular *Vestiges of the Natural History of Creation*, the book was pilloried by such authorities as the Cambridge mineralogist William Whewell (of whom it was said that "science was his forte, omniscience his foible"), the astronomer John Herschel, and the geologist Adam Sedgwick, who devoted eighty-five pages of the *Edinburgh Review* to its demolition (and who, indeed, was to subject Darwin's book to comparable scorn once it finally appeared).

Against these forces Darwin, like Copernicus, would have to defend a theory that he knew to be incomplete, for neither he nor anyone else understood the micromechanism of heredity. "The laws governing inheritance," as Darwin admitted, "are quite unknown."[22] Missing was proof of the existence of the fundamental hereditary unit, the biological quantum—in short, the *gene*. Without the stability imparted by genes, innovative mutations would be diluted away like drops of blood in the ocean, before they had time to spread to any significant numbers of individuals. In such a situation natural selection might occur, but it could scarcely account for the origin of species.

The first evidence of the existence of genes did not appear until

1866, eight years after Darwin was obliged to publish *The Origin of Species*, when the Moravian monk Gregor Mendel published the results of his extensive experiments with green peas in the garden of an Augustinian monastery—results that demonstrated the requisite persistence of the quanta of heredity—and Mendel's findings were in any event universally ignored until attention was called to them in 1900, by which time Darwin was dead. Darwin sought to make up the deficiency by proposing a theory of "pangenesis" to account for the transmission of hereditary traits, but he remained sensitive to his vulnerability on this count. As he once remarked, he appreciated the shortcomings of his theory better than did most of its censurers.

It was, then, a reluctant Darwin who at Lyell's urging finally began writing an exhaustive account of the origin of species through natural selection. He intended it to be a massive tome, the completion of which could safely be expected to take years; perhaps, like Copernicus, he would not have to live to read the reviews. But then, on June 3, 1858, when he had written only the first few chapters, everything changed. A letter bearing the postmark of the Malay Archipelago arrived at Darwin's home. It came from the naturalist Alfred Russel Wallace. It contained the draft of an essay by Wallace titled, "On the Tendency of Varieties to Depart Indefinitely from the Original Type." Wallace asked for Darwin's reactions to the paper.

Darwin had a reaction, all right, and it was one of horrified astonishment: The theory outlined in the essay was identical to Darwin's own. "I never saw a more striking coincidence," he wrote to Lyell that afternoon.[23]

Wallace, like Darwin, was an indefatigable collector of plants and insects.* He, too, had been impressed by reading Lyell's book, had long pondered "the question of *how* changes of species could have been brought about," and had hit upon the answer after reading Malthus. He was, he recounted, recovering from malaria when "it suddenly flashed upon me that . . . in every generation the

---

*It had been Wallace's misfortune, however, to lose his specimens in a fire at sea. Watching from an open lifeboat as the blazing ship sank beneath the waves, Wallace recalled, "I began to feel the greatness of my loss. . . . I had not one specimen to illustrate the unknown lands I had trod, or to call back the recollection of the wild scenes I had beheld! But such regrets were vain . . . and I tried to occupy myself with the state of things which actually existed."[24]

inferior would inevitably be killed off and the superior would remain—that is, the *fittest would survive* (Wallace's italics).[25] Wallace drafted the theory in three nights and sent it by the next mail to Darwin, who was known in scientific circles to have some sympathy for the hypothesis of evolution.

Darwin's initial inclination was to take the high road, renouncing his priority and giving all the credit to Wallace. "I should rather burn my whole book, than that he or any other many should think that I had behaved in a paltry spirit," he told Lyell.[26] But Lyell and Hooker prevailed upon Darwin instead to publish a joint announcement of his and Wallace's conclusions, and then to get to work writing a briefer account of his theory for prompt publication in book form. This he did, rushing to complete what he called an "abstract" of his theory within a year. This was *The Origin of Species by Means of Natural Selection.*

More than two hundred thousand words in length, the *Origin* reads less like an abstract than like a steady, not to say relentless, recounting of specifics: The incidence of beetle spoilage in American purple plums; the size of the stem of the Swedish turnip; the exact number of tail feathers sported by the trumpeter pigeon; the tactics employed by male alligators when they fight over female alligators. The book is objective to the point of bloodlessness; here are to be found no ecstatic outbursts comparable to Copernicus's tributes to the sun, no philosophizing on a level with Newton's descriptions of the workings of God, none of the fiery contentiousness of Galileo's dialogues. Instead there is a constant amassing of factual detail, gradual as a silt deposit hardening into sedimentary rock.*

Indeed, the book was *so* detailed and modest that it struck many readers as self-evident. This was a source of strength, in that nothing so persuades a man to accept a novel idea as the sense that he already knew it to be true. ("How extremely stupid of me not to have thought of that," said Thomas Huxley, previously an evolutionary skeptic, upon reading the *Origin.*[27]) Many scientists and

---

*Readers who tire of the details they encounter in the *Origin* may take comfort in considering that until he was interrupted by Wallace's letter, Darwin had intended to include a great many more of them. "To treat this subject properly, a long catalogue of dry facts ought to be given," he wrote, in Chapter Two of the *Origin,* "but these I shall reserve for a future work."[28] He kept this promise in his exhaustive, not to say exhausting, book *The Variation of Animals and Plants Under Domestication.*

scholars soon came around to Darwin's point of view—Hooker at once, the botanist Asa Gray soon thereafter, and Lyell, remarkably for a public figure so prominently established as an antievolutionist, only five years later—though more than a few of them would have agreed with Whitehead, who in a conversation in 1944 declared that "Darwin is truly great, but he is the dullest great man I can think of."[29] Darwin replied to contemporary criticism in this vein with his customary restraint:

> Some of my critics have said, "Oh, he is a good observer, but he has no power of reasoning." I do not think that this can be true, for the *Origin of Species* is one long argument from the beginning to the end, and it has convinced not a few able men. No one could have written it without having some power of reasoning.[30]

But he conceded that, though the study of living things had never lost its fascination for him, the years of drudgery had taken a toll on his nonscientific interests: Neither music nor literature nor even "fine scenery" held much pleasure for him any longer; he wrote in his *Autobiography*: "My mind seems to have become a kind of machine for grinding general laws out of large collections of facts."[31]

The religious reaction was every bit as vehement as Darwin had feared, but much of it was so florid, compared to Darwin's quiet reasonableness, that it flowed around the *Origin* like water around a rock. Bishop Wilberforce of Oxford set the tone for the long burlesque that was to follow. A passionate lecturer, called "Soapy Sam" after his habit of rubbing his hands together as he preached, Wilberforce condemned Darwin's theory as "a dishonoring view of Nature. . . . absolutely incompatible with the word of God." A prisoner of his own passion, he soon overplayed his hand. The scene was a meeting of the British Association for the Advancement of Science, at Oxford on June 30, 1860. Taking part in the discussion was Thomas Huxley, who loved a good argument and styled himself "Darwin's bulldog" for his tireless sallies against the opponents of evolution. With a sarcastic smile, Wilberforce turned to Huxley and asked "was it through his grandfather or his grand*mother* that he [Huxley] claimed his descent from a monkey?"[32] "The Lord hath delivered him into mine hands," whispered Huxley to his friend Benjamin Brodie, seated beside him. Then he rose, savoring the moment, and replied:

> A man has no reason to be ashamed of having an ape for his grandfather. If there were an ancestor whom I should feel shame in recalling it would rather be a *man*—a man of restless and versatile intellect—who, not content with success in his own sphere of activity, plunges into scientific questions with which he has no real acquaintance, only to obscure them by an aimless rhetoric, and distract the attention of his hearers from the real point at issue by eloquent digressions and skilled appeals to religious prejudice.[33]

The audience broke into laughter. In the general excitement that followed, one Lady Bruster fainted and had to be carried from the hall, while Captain Fitz-Roy of the *Beagle* marched up and down the aisles, holding a Bible aloft and chanting, "The Book, the Book!"[34] The drama of Darwinism versus Christian fundamentalism went on to play to packed houses in the Dayton, Tennessee, courthouse where Clarence Darrow defended John Scopes, and road-show productions were still drawing crowds to the so-called "creation science" trials of the 1980s. One such case reached the Supreme Court of the United States, which voted in 1987 that the state of Louisiana did not have the right to require that creationism be taught alongside evolution in the public schools (Chief Justice William Rehnquist dissenting). But science is not rhetoric, and the evolutionary debates, though entertaining, were always more show than substance.

The ascent of Darwin's theory brought new vitality to the question of the age of the earth. Darwinism was a *time* bomb: For species to have evolved to their present-day diversity through the slow workings of random mutation and natural selection required that the duration of the past be much longer than the six thousand or so years suggested by the Bible. Darwin grasped this nettle firmly: "He who . . . does not admit how vast have been the past periods of time may at once close this volume," he wrote in the *Origin*.[35]

But while Darwin's evolution and Lyell's geology implied that the earth was old, they did not prove it. That issue was left to the physicists, who approached the question of the age of the earth by way of thermodynamics, the developing science of the transfer of heat. The earth, as coal miners know, is hotter in its depths than at the surface. Therefore it must be radiating heat into space, rather than receiving all its warmth from the sun. (Were it the other way around, the earth's surface would be hotter than its interior.) If, then, one assumed that the earth began as a molten ball and has been cooling ever since, and if one could determine the rate at which it is cooling, it ought to be possible to calculate its age.

The first significant experiments along these lines had been conducted in the 1770s by Buffon, an early champion of deep time. In a thermally stable basement laboratory, Buffon fashioned little spheres one to five inches in diameter from suitably earthy materials, heated them, determined how long it took them to cool, and extrapolated the results to the much larger sphere of the earth. He made his measurements by sitting in the dark and observing how long it took a white-hot ball to fade to invisibility, or by touching them with his hand until they seemed to have returned to room temperature. The results, though admittedly crude, yielded a geochronology generous by the standards of the day: Buffon calculated that the earth was some 75,000 to 168,000 years old, and he guessed privately that the true figure was probably closer to half a million years. This, however, was still far too little time for Darwinian evolution to have brought life on Earth from a single-celled organism to the present-day world of orchids and adders and chimpanzees. *That* feat would have required *billions* of years.

Thermodynamics had advanced a long way by the time Darwin came on the scene. Thanks in large measure to its important practical applications in the design of steam engines, the study of heat attracted some of the most intrepid intellects of the nineteenth century—men of the stature of Lord Kelvin, Hermann von Helmholtz, Rudolf Clausius, and Ludwig Boltzmann. But when all this brainpower was brought to bear upon the question of geochronology, the verdict was bad news for Darwin and the uniformitarian geologists.

The titans of physics chose to focus less on the earth than on that suitably grander and more luminous body, the sun. Helmholtz was helpful: An able philosopher as well as a scientist, he was amused to read that the late Immanuel Kant (with whom he disagreed over just about everything) had thought that the sun was "a flaming body, and not a mass of molten and glowing matter."[36]

This Helmholtz the physicist knew to be wrong; were the sun simply burning like a giant campfire, it would have run out of fuel in but a thousand years. Casting about for an alternative solar energy-source, Helmholtz hit upon gravitational contraction: The material of the sun, he reasoned, settles in toward the center, releasing gravitational potential energy in the form of heat. This, the most efficient solar energy-production mechanism that could be envisioned by nineteenth-century physics, yielded an age for the sun of some twenty to forty million years—a lot longer than the chronology of Buffon or the Bible, though still not enough to satisfy the Darwinians.

The question of the age of the sun then was taken up by Lord Kelvin, an imposing figure by any intellectual standard. Born in Belfast in 1824, Kelvin (né William Thompson) was admitted to the University of Glasgow at the age of ten, had published his first paper in mathematics before he was seventeen, and was named professor of natural philosophy at Glasgow at age twenty-two. An adept musician and an expert navigator as well as a distinguished mathematician and physicist and inventor, Kelvin was a hard man with whom to differ. Moreover, his forte was heat: The Kelvin scale of absolute temperature is named after him, and he was instrumental in identifying the first law of thermodynamics (that energy is conserved in all interactions, meaning that no machine can produce more energy than it consumes) and the second law (that some energy must always be lost in the process). When Kelvin declared the verdict of thermodynamics as to the question of the age of the sun, few mortals, and fewer biologists, could expect both to differ with him and to prevail.

Kelvin calculated that the sun, releasing heat by virtue of gravitational contraction, could not have been shining for more than five hundred million years. This was a disaster for Darwin. "I take the sun much to heart," he wrote to Lyell in 1868. "I have not as yet been able to digest the fundamental notion of the shortened age of the sun and earth," he wrote to Wallace three years later.[37] Huxley the bulldog dutifully debated Kelvin on geochronology, at a meeting of the Geological Society of London, but Kelvin was no Bishop Wilberforce and Huxley got nowhere. Clearly either Darwin's theory or Kelvin's calculations were wrong. Darwin died not knowing which.

To their credit, both Darwin and Kelvin allowed that something important might be missing from their considerations. As Darwin put it, pleading his case in a late edition of the *Origin*, "We are confessedly ignorant; nor do we know how ignorant we are."[38] Kelvin, for his part, admitted that his assessments of the age of the sun depended upon the accuracy of Helmholtz's hypothesis that solar energy came from the alleged contraction of the sun. He remarked, in one of the most pregnant parenthetical phrases in the history of physics, that "(I do not say there may not be laws which we have not discovered.)"[39]

It was in conceding that their views might be incomplete that both men proved most prophetic. What they lacked was an understanding of two of the fundamental forces of nature, known corporately as nuclear energy. It is the decay of radioactive material —via the weak nuclear force—that has kept the earth warm for nearly five billion years. It is nuclear fusion—which also involves the strong force—that has powered the sun for as long, and that promises to keep it shining for another five billion years. With the discovery of nuclear energy the time-scale debate was resolved in Darwin's favor, the doors to nuclear physics swung open, and the world lost its innocence.

The nuclear age may be said to have dawned on November 8, 1895, in a laboratory at the University of Würzburg, at the hands of the physicist Wilhelm Conrad Röntgen. Röntgen was experimenting with electricity in a semivacuum tube. The laboratory was dark. He noticed that a screen across the room, coated with barium, platinum, and cyanide, glowed in the dark whenever he turned on the power to the tube, as if light from the tube were reaching the screen. But ordinary light could not be responsible: The tube was enclosed in black cardboard and no light could escape it. Puzzled, Röntgen placed his hand between the tube and the screen and was startled to see the bones in his hand exposed, as if the flesh had become translucent. Röntgen had detected "X rays"—high-energy photons generated by electron transitions at the inner shells of atoms.*

---

*Not long before Röntgen's discovery, Frederick Smith at Oxford was informed by an assistant that photographic plates stored near a cathode-ray tube were being fogged; but Smith, rather than pondering the matter, simply ordered that the plates be kept somewhere else.

Among the scores of physicists who took notice of Röntgen's detection of X rays was Henri Becquerel, a third-generation student of phosphorescence who shared with his father and grandfather a fascination with anything that glowed in the dark. Becquerel's discovery, like Röntgen's, was accidental, though both illustrated the validity of Louis Pasteur's dictum that chance favors the prepared mind. Between experiments in his laboratory in Paris, Becquerel stored some photographic plates wrapped in black paper in a drawer. A piece of uranium happened to be sitting on top of them. When Becquerel developed the plates several days later, he found that they had been imprinted, in total darkness, with an image of the lump of uranium. He had detected radioactivity, the emission of subatomic particles by unstable atoms like those of uranium—which, Becquerel noted in announcing his results in 1896, was particularly radioactive. His work helped initiate a path of research that would lead, eventually, to Einstein's realization that every atom is a bundle of energy.

At McGill University in Montreal, the energetic experimentalist Ernest Rutherford, a great bear of a man whose roaring voice sent his assistants and their laboratory glassware trembling, found that radioactive materials can produce surprisingly large amounts of energy. A lump of radium, Rutherford established, generates enough heat to melt its weight in ice every hour, and can continue to do so for a thousand years or more. Other radioactive elements last even longer; some keep ticking away at an almost undiminished rate for billions of years.

This, then, was the answer to Kelvin, and one that spelled deliverance for the late Charles Darwin: The earth stays warm because it is heated by radioactive elements in the rocks and molten core of the globe. As Rutherford wrote:

> The discovery of the radioactive elements, which in their disintegration liberate enormous amounts of energy, thus increases the possible limit of the duration of life on this planet, and allows the time claimed by the geologist and biologist for the process of evolution.[40]

Understandably pleased with this conclusion, the young Rutherford rose to address a meeting of the Royal Institution, only to find himself confronted by the one scientist in the world his paper could most deeply offend:

> I came into the room, which was half dark, and presently spotted Lord Kelvin in the audience and realized that I was in for trouble at the last part of my speech dealing with the age of the earth, where my views conflicted with his. To my relief, Kelvin fell fast asleep, but as I came to the important point, I saw the old bird sit up, open an eye and cock a baleful glance at me! Then a sudden inspiration came, and I said Lord Kelvin had limited the age of the earth, provided no new source [of energy] was discovered. That prophetic utterance refers to what we are now considering tonight, radium! Behold! the old boy beamed upon me.[41]

Radioactive materials not only testified to the antiquity of the earth, but provided a way of measuring it as well. Rutherford's biographer A. S. Eve recounts an exchange that signaled this new insight:

> About this time Rutherford, walking in the Campus with a small black rock in his hand, met the Professor of Geology. "Adams," he said, "how old is the earth supposed to be?" The answer was that various methods lead to an estimate of one hundred million years. "I *know*," said Rutherford quietly, "that this piece of pitchblende is seven hundred million years old."[42]

What Rutherford had done was to determine the rate at which the radioactive radium and uranium in the rock gave off what he called alpha particles, which are the nuclei of helium atoms, and then to measure the amount of helium in the rock. The result, seven hundred million years, constituted a reasonably reliable estimate of how long the radioactive materials had been in there, emitting helium.

Rutherford had taken a first step toward the science of radiometric dating. Every radioactive substance has a characteristic half-life, during which time half of the atoms in any given sample of that element will decay into another, lighter element. By comparing the abundance of the original (or "parent") isotope with that of the decay product (or "daughter"), it is possible to age-date the stone or arrowhead or bone that contains the parent and daughter isotopes.

Carbon-14 is especially useful in this regard, since every living thing on Earth contains carbon. The half-life of carbon-14 is 5,570 years, meaning that after 5,570 years half of the carbon-14 atoms in any given sample will have decayed into atoms of nitrogen-14. If we examine, say, the remains of a Navaho campfire and find that half the carbon-14 in the charred remains of the burnt logs has decayed into nitrogen-14, we can conclude that the fire was built 5,570 years ago. If three quarters of the carbon has turned to nitrogen, then the logs are twice as old—11,140 years—and so forth. After about five half-lives the amount of remaining parent isotope generally has become too scanty to be measured reliably, but geologists have recourse to other, more long-lived radioactive elements. Uranium-238, for one, has a half-life of over 4 billion years, while the half-life of rubidium-87 is a methuselian 47 billion years.

In practice, radiometric dating is a subtle process, fraught with potential error. First one has to ascertain when the clock started. In the case of carbon-14, this is usually when the living tissue that contained it died. Carbon-14 is constantly being produced by the collision of high-energy subatomic particles from space with atoms in the earth's upper atmosphere. Living plants and animals ingest carbon-14, along with other forms of carbon, only so long as they live. The scientist who comes along years later to age-date their remains is, therefore, reading a clock that started when the host died. The reliability of the process depends upon the assumption that the amount of ambient carbon-14 in the enviroment at the time was roughly the same as it is today. If not—if, for instance, a storm of subatomic particles from space happened to increase the amount of carbon-14 around thousands of years ago—then the radiometric date will be less accurate. In the case of inorganic materials, one may be dealing with radioactive atoms older than the earth itself; their clocks may have started with the explosion of a star that died when the sun was but a gleam in a nebular eye. But if such intricacies complicate the process of radiometric age-dating they also hint at the extraordinary range of its potential applications, in fields ranging from geology and geophysics to astrophysics and cosmology.

The process of radiometrically age-dating geological strata got under way only ten years after the discovery of radioactivity itself, when the young British geologist Arthur Holmes, in his book *The Age of the Earth*, correlated the ages of uranium-bearing igneous rocks with those of adjacent fossil-bearing sedimentary strata. By the 1920s it was becoming generally accepted by geologists, physicists, and astronomers that the earth is billions of years old and that radiometric dating presents a reliable way of measuring its age. Since then, ancient rocks in southwestern Greenland have been radiometrically age-dated at 3.7 billion years, meaning that the crust of the earth can be no younger than that. Presumably the planet is older still, having taken time to cool from a molten ball and form a crust. Moon rocks collected by the Apollo astronauts are found to be nearly 4.6 billion years old, about the same age as meteorites—chunks of rock that once were adrift in space and since have been swept up by the earth in its orbit around the sun. It is upon this basis that scientists generally declare the solar system to be some 5 billion years old, a finding that fits well with the conclusions of astrophysicists that the sun is a normal star about halfway through a 10-billion-year lifetime.

When nuclear fission, the production of energy by splitting nuclei, was detailed by the German chemists Otto Hahn and Fritz Strassmann in 1938, and nuclear fusion, which releases energy by combining nuclei, was identified by the American physicist Hans Bethe the following year, humankind could at last behold the mechanism that powers the sun and the other stars. In the general flush of triumph, few paid attention to the dismaying possibility that such overwhelming power might be set loose with violent intent on the little earth. Einstein, for one, assumed that it would be impossible to make a fission bomb; he compared the problem of inducing a chain reaction to trying to shoot birds at night in a place where there are very few birds. He lived to learn that he was wrong. The first fission (or "atomic") bomb was detonated in New Mexico on July 16, 1945, and two more were dropped on the cities of Hiroshima and Nagasaki a few weeks later. The first fusion (or "hydrogen") bomb, so powerful that it employed a fission weapon as but its detonator, was exploded in the Marshall Islands on November 1, 1952.

A few pessimists had been able to peer ahead into the gloom of the nuclear future, though their words went largely unheeded at the time. Pierre Curie had warned of the potential hazards of nuclear weapons as early as 1903. "It is conceivable that radium in criminal hands may become very dangerous," said Curie, accepting the Nobel Prize.* ". . . Explosives of great power have allowed men to do some admirable works. They are also a terrible means of destruction in the hands of the great criminals who lead nations to war."[43] Arthur Stanley Eddington, guessing that the release of nuclear energy was what powered the stars, wrote in 1919 that "it seems to bring a little nearer to fulfillment our dream of controlling this latent power for the well-being of the human race—or for its suicide."[44] These and many later admonitions notwithstanding, the industrialized nations set about building bombs just as rapidly as they could, and by the late 1980s there were over fifty thousand nuclear weapons in a world that had grown older if little wiser. Studies indicated that the detonation of as few as 1 percent of these warheads would reduce the combatant societies to "medieval" levels, and that climatic effects following a not much larger exchange could lead to global famine and the potential extinction of the human species. The studies were widely publicized, but years passed and the strategic arsenals were not reduced.

It was through the efforts of the bomb builders that Darwin's century-old theory of the origin of coral atolls was at last confirmed. Soon after World War II, geologists using tough new drilling bits bored nearly a mile down into the coral of Eniwetok Atoll and came up with volcanic rock, just as Darwin had predicted. The geologists' mission, however, had nothing to do with evolution. Their purpose was to determine the structure and strength of the atoll before destroying it, in a test of the first hydrogen bomb. When the bomb was detonated, its fireball vaporized the island on which it had been placed, tore a crater in the ocean floor two miles deep, and sent a cloud of freshly minted radioactive atoms wafting across the paradisiacal islands downwind. President Truman in his final State of the Union message declared that "the war of the future would be one in which Man could extinguish millions of lives at one blow, wipe out the cultural achievements of the past, and destroy the very structure of civilization.

"Such a war is not a possible policy for rational men," Truman added.[45] Nonetheless, each of the next five presidents who suc-

---

*Curie's wife Marie, winner of two Nobel Prizes, died of the effects of radiation contracted in years of experimental research into radioactive isotopes. Her laboratory apparatus and even her cookbooks at home, inspected fifty years later, were found to be contaminated by lethal radiation.

ceeded him in office found it advisable to threaten the Soviets with the use of nuclear weapons. As the British physicist P.M.S. Blackett observed, "Once a nation pledges its safety to an absolute weapon, it becomes emotionally essential to believe in an absolute enemy."[46]

Einstein, sad-eyed student of human tragedy, closed the circle of evolution, thermodynamics, and nuclear fusion in a single sentence. "Man," he said, "grows cold faster than the planet he inhabits."[47]

14

THE
EVOLUTION
OF ATOMS
AND STARS

At quite uncertain times and places,
  The atoms left their heavenly path,
And by fortuitous embraces,
  Engendered all that being hath.
                    —James Clerk Maxwell

For I have already at times been a boy and a
girl, and a bush and a bird and a mute fish in
the salty waves.

                    —Empedocles

By the dawn of the twentieth century it was becoming evident that some sort of "atomic" energy must be responsible for powering the sun and the other stars. As early as 1898,
only two years after Becquerel's discovery of radioactivity, the
American geologist Thomas Chrowder Chamberlin was speculating
that atoms were "complex organizations and seats of enormous
energies" and that "the extraordinary conditions which reside in
the center of the sun may . . . set free a portion of this energy."[1]
But no one could say what this mechanism might be, or just how

it might operate, until a great deal more was understood about both atoms and stars. The effort to garner such an understanding involved a growing level of collaboration between astronomers and nuclear physicists. Their work was to lead, not only to a resolution of the stellar energy question, but to the discovery of a golden braid of cosmic evolution intertwining atomic and stellar history.

The key to understanding stellar energy was, as Chamberlin foresaw, to discern the structure of the atom. That there *was* an internal structure to the atom could be intimated from several lines of research, among them the study of radioactivity: For atoms to emit particles, as they were found to do in the laboratories of Becquerel and the Curies, and for these emissions to change them from one element to another, as Rutherford and the English chemist Frederick Soddy had established, atoms must be more than the simple, indivisible units that their name (from the Greek for "cannot be cut") implied. But atomic physics still had a long way to go in comprehending that structure. Of the three principal constituents of the atom—the proton, neutron, and electron—only the electron had as yet been identified (by J. J. Thomson, in the waning years of the nineteenth century). Nobody spoke of "nuclear" energy, for the existence of the atomic nucleus itself had not been established, much less that of its constituent particles the proton and the neutron, which were to be identified, respectively, by Thomson in 1913 and James Chadwick in 1932.

Rutherford, Hans Geiger, and Ernest Marsden ranked among the Strabos and Ptolemies of atomic cartography. In Manchester from 1909 through 1911 they probed the atom by launching streams of subatomic "alpha particles"—helium nuclei—at thin foils made of gold, silver, tin, and other metals. Most of the alpha particles flew through the foil, but, to the experimenters' astonishment, a few bounced right back. Rutherford thought long and hard about this strange result; it was, he remarked, as startling as if a bullet were to bounce off a sheet of tissue paper. Finally, at a Sunday dinner at his house in 1911, he announced to a few friends that he had hit on an explanation—that most of the mass of each atom resides in a tiny, massive nucleus. By measuring the back-scattering rates obtained from foils comprised of various elements, Rutherford could calculate both the charge and the maximum diameter of the atomic nuclei in the target. Here, then, was an atomic explanation

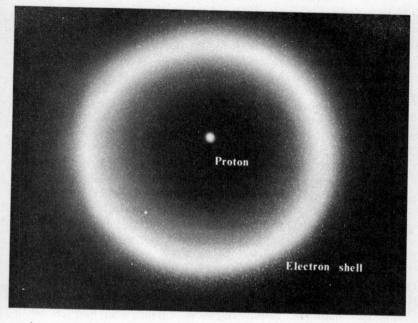

An atom of simple hydrogen consists of a single proton (its nucleus) surrounded by a shell containing one electron. Atoms of heavier elements have more protons, as well as neutrons, in the nucleus, and additional electrons in the shells. (Not to scale: Were the proton the size of a grain of sand, the shell would be larger than a football field.)

of the weights of the elements: Heavy elements are heavier than light elements because the nuclei of their atoms are more massive.

The realm of the electrons was then explored by the Danish physicist Niels Bohr, who established that electrons inhabit discrete orbits, or shells, surrounding the nucleus. (For a time Bohr thought of the atom as a miniature solar system, though this analysis soon proved inadequate; the atom is ruled not by Newtonian but by quantum mechanics.) Among its many other felicities, the Bohr model laid bare the physical basis of spectroscopy: The number of electrons in a given atom is determined by the electrical charge of the nucleus, which in turn is due to the number of protons in the nucleus, which is the key to the atom's chemical identity. When an electron falls from an outer to an inner orbit it emits a photon. The wavelength of that photon is determined by the particular orbits between which the electron has made the transition. And *that* is why a spectrum, which records the wavelengths of photons, reveals the chemical elements that make up the star or other object the spectroscopist is studying. In the words of Max Planck, the founder of quantum physics, the Bohr model of the atom provided "the long-sought key to the entrance gate into the wonderland of spectroscopy, which since the discovery of spectral analysis had obstinately defied all efforts to breach it."[2]

But, marvelous though it might be to realize that the spectra evinced the leaps and tumbles of electrons in their Bohr orbits, nobody could yet read the spectra of stars for significant clues to what made them shine. In the absence of a compelling theory, the field was left to the taxonomists—to those who went on doggedly recording and cataloging the spectra of stars, though they knew not where they were going.

At Harvard College Observatory, a leader in the dull but promising business of stellar taxonomy, photographic plates that revealed the color and spectra of tens of thousands of stars were stacked in front of "computers"—spinsters, most of them, employed as staff members at a university where their sex barred them from attending classes or earning a degree. (Henrietta Leavitt, the pioneer researcher of the Cepheid variable stars that were to prove so useful to Shapley and Hubble, was a Harvard computer.) The computers were charged with examining the plates and entering the data in neat, Victorian script for compilation in tomes like the *Henry Draper Catalog*, named in honor of the astrophotographer and physician who had made the first photograph of the spectrum of a star. Like prisoners marking off the days on their cell walls, they tallied their progress in totals of stars cataloged; Antonia Maury, Draper's niece, reckoned that she had indexed the spectra of over five hundred thousand stars. Theirs was authentically Baconian work, of the sort Newton and Darwin claimed to practice but seldom did, and the ladies took pride in it; as the Harvard computer Annie Jump Cannon affirmed, "Every fact is a valuable factor in the mighty whole."[3]

It was Cannon who, in 1915, first began to discern the shape of that whole, when she found that most stars belonged to one of about a half-dozen distinct spectral classes. Her classification system, now ubiquitous in stellar astronomy, arranges the spectra of the stars by color, from the blue-white O stars through yellow G stars like the sun to the red M stars.* Here was a sign of simplicity beneath the astonishing variety of the stars.

A still deeper order was soon disclosed, when in 1911 the Danish engineer and self-taught astronomer Ejnar Hertzsprung analyzed Cannon's and Maury's data for stars in two clusters, the Hyades and the Pleiades. Clusters like these are intuitively recognizable as genuine assemblies of stars and not merely chance alignments; even an inexperienced observer will start with recognition when sweeping a telescope across the Pleiades, its ice-blue stars tangled in gossamer webs of diamond dust, or the Hyades, whose stars range in color from bone-white to Roman gold. Since all the stars in a given cluster may thus be assumed to lie at about the same distance from the earth, any observed differences in their apparent magnitudes can be ascribed, not to their differing distances, but to actual differences in their absolute magnitudes. Hertzsprung took advantage of this situation, treating the clusters as laboratory samples wherein he could look for a relationship between the colors and intrinsic brightnesses of stars. He found just such a relationship: Most of the stars in each cluster fell along two gently curved lines. This, in sapling form, was the first intimation of a tree of stars that has since been designated the Hertzsprung-Russell diagram.

The applicability of Hertzsprung's method soon broadened to include noncluster stars as well. In 1914, Walter Adams and Arnold Kohlschutter at Mount Wilson found that the relative intensities of lines in stellar spectra suggested their absolute magnitudes. Hereafter, whenever the distance to a single star of a given variety—a class B giant, say, or a class K dwarf—was measured by the parallax method, then the distances of all other stars that displayed comparable spectra could be estimated as well. This meant that Hertzsprung's approach of graphing the absolute magnitude of stars against their colors could be applied to field stars as well as to the relatively few stars found in clusters.

Henry Norris Russell, a Princeton astrophysicist with an en-

---

*After many false starts, Cannon designated the classes by the letters O, B, A, F, G, K, and M. Students ever since, in a largely unconscious tribute to her memory, have learned the sequence via the mnemonic phrase "Oh, Be a Fine Girl, Kiss Me."

cyclopedic command of his field, promptly set to work doing just that. Without even knowing of Hertzsprung's work, Russell plotted the absolute magnitudes of a few hundred field stars against their colors, and found that most lie along a narrow, slanting zone—the trunk of the tree of stars.

The tree has been growing ever since, and today it is embedded in the consciousness of every stellar astronomer in the world. Its trunk is the "main sequence," a gently curving S along which lie 80 to 90 percent of all visible stars. The sun, a typical yellow star, resides on the main sequence a little less than halfway up the trunk. A slender branch departs from the trunk and makes its way upward and to the right, where it blossoms into a bouquet of brighter, redder stars—the red giants. Below and to the left sits a humus pile of dim, blue to white stars—the dwarfs.

The Hertzsprung-Russell diagram provided astronomers with a frozen record of evolution, the astrophysical equivalent of the fossil record geologists study in rock strata. Presumably, stars somehow evolve, spending most of their time on the main sequence (most stars today, in the snapshot of time given us to observe, are found there) but beginning and ending their careers somewhere else, among the branches or in the humus pile. One could not, of course, wait to see this happen; the lifetimes of even short-lived stars are measured in millions of years. To find the answers required understanding the physics of how stars work.

Progress on the physics side, meanwhile, was blocked by a seemingly insurmountable barrier. Literally so: The agency responsible was known as the Coulomb barrier, and for a time it stymied the efforts of theoretical physicists to comprehend how nuclear fusion might produce energy in stars.

The line of reasoning that led to the barrier was impeccable. Stars are made mostly of hydrogen. (This is evident from studying their spectra.) The nucleus of the hydrogen atom consists of but a single proton, and the proton contains nearly all the mass in the atom. (This we know from Rutherford's experiments.) Therefore the proton must also contain nearly all of a hydrogen atom's latent energy. (Recall that mass equals energy: $E = mc^2$). In the heat of a star, the protons are flung about at high velocities—heat *means* that the particles involved are moving fast—and, as there are plenty of protons milling about in close quarters at the dense core of a star, they must collide quite a lot. In short, the energy of the sun

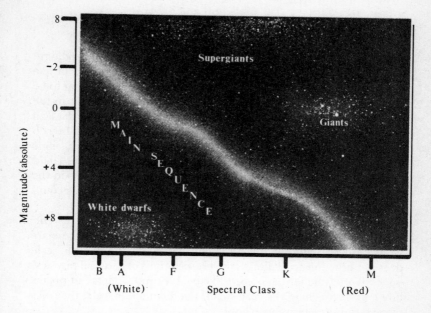

The Hertzsprung-Russell diagram plots the spectral classes (or colors) of stars against their brightnesses. This version of the diagram is thought to represent the general stellar population of our galaxy.

and stars could reasonably be assumed to involve the interactions of protons. This was the basis of Eddington's surmise that the stellar power source could "scarcely be other than the subatomic energy which, it is known, exists abundantly in all matter."[4]

What happens when protons collide? Well, we know that they can stick together—"fuse"—because they are found, stuck together, in the nuclei of all the heavier elements. Might the fusion of protons release energy? A strong hint that this is so lay in the fact that the heavier nuclei weigh a little less that the sum of their parts. There was some confusion about this point, but the basic idea was correct—that energy is released in stars when the nuclei of the light atoms fuse to make those of heavy atoms. Rutherford already had been performing experiments in what he called "the newer alchemy," bombarding nuclei with protons and changing them into the nuclei of different elements, and, as Eddington wryly noted, "what is possible in the Cavendish Laboratory may not be too difficult in the sun."[5]

So far, so good; science was close to identifying thermonuclear fusion as the secret of solar power. But it was here that the Coulomb barrier intervened. Protons are positively charged; particles of like charge repel one another; and this obstacle seemed too strong to be overcome, even at the high velocity of protons flying about in the intense heat at the center of a star. Seldom, according to classical physics, could two protons in a star get going fast enough to breach the walls of their electromagnetic force fields and merge into a single nucleus. The calculations said that the proton collision rate could not possibly suffice to sustain fusion reactions. Yet there stood the sun, its beaming face laughing at the equations that said it could not shine.

There was nothing wrong with the argument, so far as it went: Were classical physics declared the sole law of nature, the stars would indeed wink out. Fortunately, nature on the nuclear scale does not function according to the proscriptions of classical physics, which works fine for big objects like pebbles and planets but breaks down in the realm of the very small. On the nuclear scale, the rules of quantum indeterminacy apply.

In classical mechanics, subatomic particles like protons were viewed as analogous to macroscopic objects like grains of sand or cannonballs. Viewed by these lights, a proton hurled against the Coulomb barrier of another proton had no more chance of penetrating it than a cannonball has of penetrating a ten-foot-thick fortress wall. Introduce quantum indeterminacy, however, and the picture changes dramatically. Quantum mechanics demonstrates that the proton's future can be predicted only in terms of probabilities: Most of the time the proton will, indeed, bounce off the Coulomb barrier, but from time to time it will pass right through it, as if a cannonball were to fly untouched through a fortress wall.*

This is "quantum tunneling," and it licenses the stars to shine. George Gamow, eager to exploit connections between astronomy and the exotic new physics at which he was adept, applied quantum probabilities to the question of nuclear fusion in stars and found that protons could surmount the Coulomb barrier—almost. Quantum tunneling took the calculations from the dismal, classical pre-

---

*There is a quantum chance that a real cannonball will do the same thing, but as this requires nearly all its protons to get lucky at once, the odds of its happening are small. One can calculate, via quantum theory, that it has almost certainly never occurred, anywhere in the universe, even if—unhappy thought—cannonballs and fortress walls are a cosmic commonplace.

diction, which had protons fusing at only one one-thousandth of the rate required to account for the energy released by the sun, up to fully one tenth of the necessary rate. It then took less than a year for the remaining deficit to be accounted for: The solution was completed in 1929, when Robert Atkinson and Fritz Houtermans combined Gamow's findings with what is called the Maxwellian velocity distribution theory. In a Maxwellian distribution there are always a few particles moving much faster than average; Atkinson and Houtermans found that these fleet few were sufficient to make up the difference. Now at last it was clear how the Coulomb barrier could be breached often enough for nuclear fusion to function in stars.

But how, exactly, do the stars do it? Within another decade, two likely fusion processes were identified—the proton-proton chain reaction and the carbon cycle.

The key figure in both developments was Hans Bethe, a refugee from Nazi Germany who had studied with Fermi in Rome and gone on to teach at Cornell. Like his friend Gamow, the young Bethe was an effervescent, nimble thinker, so gifted that he made his work look like play. Though untrained in astronomy, Bethe was a legendarily quick study. In 1938 he helped Gamow's and Edward Teller's student C. L. Critchfield calculate that a reaction beginning with the collision of two protons could indeed generate approximately the energy—some 3.86×10^{33} ergs per second— radiated by the sun.* And so, in the span of less than forty years, humankind had progressed from ignorance of the very existence of atoms to an understanding of the primary thermonuclear fusion process that powers the sun.

The proton-proton reaction was insufficiently energetic, however, to account for the much higher luminosities of stars much larger than the sun—stars like the blue supergiants of the Pleiades, which occupy the higher reaches of the Hertzsprung-Russell diagram. This Bethe was to remedy before the year was out.

In April 1938, Bethe attended a conference organized by Gamow and Teller at the Carnegie Institution in Washington to bring astronomers and physicists together to work on the question of

---

*Carl Friedrich von Weizsäcker, the physicist and later philosopher of science, had shown how the proton-proton reaction would work, but had not calculated its energy output for a solar-mass star.

stellar energy generation. "At this conference the astrophysicists told us physicists what they knew about the internal constitution of the stars," Bethe recalled. "This was quite a lot [although] all their results had been derived without knowledge of the specific source of energy."[6] Back at Cornell, Bethe attacked the problem with such alacrity that Gamow would later joke that he had calculated the answer before the train that carried him home arrived at the Ithaca station. Bethe wasn't *that* quick, but within only a matter of weeks he had succeeded in identifying the carbon cycle, the critical fusion reaction that powers stars more than one and a half times as massive as the sun.

Publication of the paper, however, was delayed. Bethe finished it that summer and sent it to the *Physical Review*, but then was informed by a graduate student, Robert Marshak, that the New York Academy of Sciences offered a five-hundred-dollar prize for the best *unpublished* paper on energy production in stars. Bethe, who had need of the money, coolly asked that the paper be sent back, entered it in the competition, and won. "I used part of the prize to help my mother emigrate," he told the American physicist Jeremy Bernstein. "The Nazis were quite willing to let my mother out, but they wanted two hundred and fifty dollars, in dollars, to release her furniture. Part of the prize money went to liberate my mother's furniture."[7] Only then did Bethe permit publication of the paper that was to win him a Nobel Prize. He had, for a time, been the sole human being who knew why the stars shine.

Curiously stutter-stepped were the fusion reactions Bethe perceived. The proton-proton reaction begins with the collision, deep inside the sun, of two protons that have sufficient velocity and good fortune to penetrate the Coulomb barrier. If the collision succeeds in transforming one of the protons into a neutron—another rather unlikely event, involving a weak-force interaction called beta decay—the result is a nucleus of heavy hydrogen. The interaction releases a neutrino, which flies out of the sun, and an electron, which plows into the surrounding gas and thus helps heat the sun. The average proton at the center of the sun finds it necessary to wait more than thirty million years before chancing to experience this brief fling.

The next step, however, comes quickly. Within a few seconds, the heavy hydrogen nucleus snaps up another proton, transforming itself into helium-3 and releasing a photon that carries off further energy into the surrounding gas. Nuclei of helium-3 are rare, and so most are obliged to wait another few million years before encountering a second helium-3 nucleus. Then the two nuclei can fuse, forming a stable helium nucleus and releasing two protons, which are free to join the dance in their turn. The result has been to release energy: The helium end-product weighs sixth tenths of 1 percent less than did the particles that went into the reaction. This mass has been converted into energy, in the form of quanta that slowly make their way to the surface, blundering into atoms and being absorbed and reemitted as they go, until, centuries later, they at last break into the clear and are released into space as sunlight.

The proton-proton reaction has ramifications that are not completely understood—measurements of the neutrino flux on earth have to date yielded only a third as many neutrinos as the theory says should be released—and the carbon cycle is more complicated still. Nonetheless, enough is now known about solar fusion for us humans to begin to appreciate the elegance of the workings of our mother star. We have learned, for one thing, that the sun is not a bomb, although nuclear fusion is the same mechanism that functions in a thermonuclear weapon. When a chain reaction occurs in one tiny area in the center of the sun, it does not normally touch off other reactions in the surrounding gas; instead, the additional heat expands the gas slightly, lowering its density and so decreasing the probability of further proton-proton collisions for the moment. Owing to the operation of this self-regulating process, as averaged out for countless interactions, the entire star equilibrates, expanding to damp the rate of thermonuclear processes when they can attain a runaway rate, then contracting and heating to increase the rate when the center begins to cool. Although only one five-billionth of the sun's light strikes the earth, that has been sufficient to endow the earth with warmth, and life, and with bipeds clever enough to decipher the particulars of their debt to Sol.

With the basic physics of solar fusion now in hand, it became possible to rework Kelvin's estimates of the age of the sun. The sun's mass can be determined, and very accurately so, from Newton's laws and the orbital velocity of the planets. The result is 1.989×10^{33} grams, the equivalent of three hundred thousand Earths. The sun's composition, at the surface at least, is revealed by the spectrograph to be principally hydrogen and helium. Knowing, then, the mass, volume, and approximate composition of the sun, one can ascertain the conditions that pertain at its center, where the thermonuclear processes take place. One can, for instance, calculate that the temperature at the core is about 15 million degrees, that the density is about twelve times that of lead (though the heat keeps the dense material in a gaseous and not a solid state), and that the fusion reaction rate is such that some 4.5 million tons of hydrogen are fused into helium inside the sun every second. Since the sun contains a finite amount of hydrogen, it must eventually run low on fuel, at which time its nuclear furnaces will falter. The total hydrogen-burning "lifetime" for the sun can thus be calculated. It turns out to be about ten billion years. Since radiometric dating of the asteroids and the earth yields an age for the solar system of a little less than five billion years old, we conclude that the sun now is in its middle age, and has another five billion years of hydrogen-burning ahead of it. And so the investigation of stellar energy sources, which had been driven in part by the demands of the geologists and biologists for a time scale longer than the old ideas permitted, opened up immensities of astronomical history even longer than the Darwinians had required.

The lifetimes of other stars can be calculated similarly. The fusion rate increases by the fourth power of the mass; consequently, dwarf stars last much longer than giants. The least massive stars have about 1 percent of the mass of the sun. (Much less and they would fail to generate sufficient interior heat for fusion to take place, and would instead be planets.) These little dwarfs, residents of the lower tiers of the Hertzsprung-Russell diagram, burn their hydrogen fuel so prudently that they can last for a trillion years or more. At the other end of the scale, toward the top of the diagram, stand giant stars with up to sixty times the mass of the sun. (If much larger, they would blow themselves apart as soon as they got fired up.) These huge stars squander their fuel profligately, and run out of hydrogen almost immediately; a star ten times as massive as the sun lasts less than one hundred million years.

These considerations greatly enriched and enlivened human appreciation of what might be called the ecology of the Milky Way. They revealed that the most spectacular stars in the galaxy, the giant, blue-white O and B stars, are also the stars that have the least time to live: Giants typically burn for only ten million to one hundred million years, and some may last no longer than a million years. This means that the brilliant giants that trace out the spiral arms are, by galactic standards, flowers that bloom for but a day. Indeed, that is *why* they trace out the arms. Stars of various masses condense along the arms, but while more modest stars last long enough to drift off into the surrounding disk, the brilliant superstars die before they ever get far from their birthplaces, which, consequently, they demark.

How do stars die? This, too, depends principally on their mass. When an ordinary star like the sun runs low on fuel it takes on a split personality: Its core contracts, no longer propped up by the radiation of energy from thermonuclear processes at the center, while its outer portion—its "atmosphere," so to speak—expands and cools. The star's color changes from a yellow-white to a deepening red: It has become a "red giant." Ultimately the stellar atmosphere boils away into space, leaving behind the naked core, a massive, dense sphere only about the size of the earth—a "white dwarf" star.

Such a prognosis, plotted on the Hertzsprung-Russell diagram,

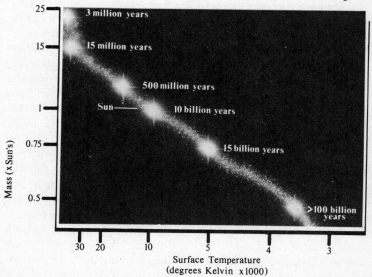

The main-sequence lifetimes of stars are determined principally by their masses: Massive stars exhaust their fuel much more rapidly than do low-mass stars.

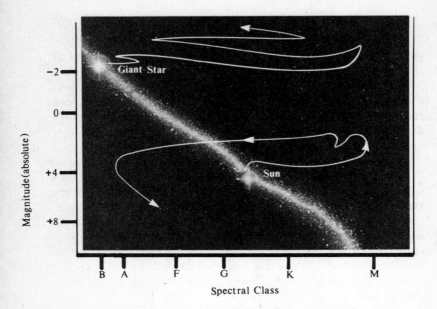

The destinies of stars once they leave the main sequence also differ greatly, according to their masses. When the sun runs low on fuel it will exit the main sequence toward the right, becoming a red giant. After another billion years or so it will eject its outer atmosphere, skidding from right to left across the diagram as it does so, then plunge down into the graveyard of the white dwarfs. A star with five times the mass of the sun remains on the main sequence for less than a tenth as long, then begins oscillating back and forth near the top of the diagram as an unstable giant. For stars of ten solar masses or more, such instabilities may culminate in the explosion of the star as a supernova.

serves to animate the tree of stars. When an average star like the sun exhausts its hydrogen fuel, it leaves the main sequence and moves upward—since the growing size of its outer atmosphere briefly makes it brighter—and to the right, since it is getting redder. Many stars during this phase may become unstable, staggering back and forth from right to left on the diagram. When the star sheds its atmosphere, it drops down the diagram and skids to the left, settling finally into the zone of the white dwarfs. Giant stars follow an approximately similar course, but start higher on the main sequence (since they are brighter) and leave it sooner (since they run out of fuel more rapidly).

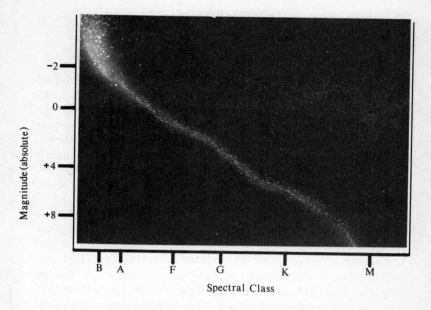

Magnitude (absolute)

-2
0
+4
+8

B A F G K M

Spectral Class

The ages of star clusters may be inferred from their Hertzsprung-Russell diagrams. In a young cluster like the Pleiades, nearly all the visible stars lie on the main sequence: There are few red giants or white dwarfs to be found, because the cluster is not yet old enough for many of its stars to have run out of hydrogen fuel and departed from the main sequence.

The Hertzsprung-Russell diagram for any given population of stars—a star cluster, say—therefore provides evidence of its age. When the cluster is in its infancy, virtually all its stars lie on the main sequence, contentedly burning hydrogen. Soon the giant stars—those at the upper-left extremity of the main sequence— run out of fuel and balloon into red giants; each, as it does so, leaves the main sequence and moves to the right. As more time goes by, the same fate afflicts stars of ever less mass. The result, on the diagram, is a "cutoff point," a place along the main sequence where the tree branches off to the right. The diagram is only a snapshot of a moment amid billions of years of stellar history, but the location of the cutoff point tells us how long the cluster has been there: The farther down the trunk the cutoff point falls, the older the tree.

The Hertzsprung-Russell diagram of the Pleiades cluster, for example, shows almost entirely main sequence stars. This tells us that the Pleiades is a young cluster, in which not enough time has passed for even the giant stars to burn down to the red giant stage. (The stars of the Pleiades are estimated to be less than one hundred million years old.) The diagram of the globular cluster M3, however, looks dramatically different. Here the great majority of stars are either in the red giant phase or are on their way to becoming dwarfs. (We don't see the dwarfs themselves because they are too dim; M3 is an ample thirty thousand light-years away.) The cutoff branch points like the hand of a clock at the age of the cluster: For M3, the age reads out to some fourteen billion years, making it one of the oldest ever dated.

To envision the pace of stellar evolution more directly, imagine that the sun was a star in a young star cluster and that we were present on the earth right from the outset, when our planet had just cooled sufficiently for its crust to have solidified. Imagine, further, that we could speed up the passage of time, so that ten billion years would pass in a single night. As the sun sets, at time zero, we find the sky studded with main-sequence stars. There are as yet no red giants and no dwarfs. A few bright giants stand out, as well as a number of stars about as luminous as the sun, but the great majority of stars are dimmer and less bright than the sun.

Almost immediately, the giant stars exhaust their fuel, become unstable and explode as supernovae, flooding the landscape with scalding white light. On our compressed time scale, where each hour equals a billion years, all these spectacular stars die within the first few minutes. Conceivably their explosions may shock any remaining gas in the cluster into collapsing to form new stars, but any giant stars produced in this fashion will also consume themselves quickly, so that the fireworks are over by the time we've settled down to watch the show.

In the hours that follow, successively less massive stars in turn leave the main sequence; we watch them swell into red giants, shed shells of multicolored gas, and reduce themselves to dim dwarf stars. These events are rare enough to hold our attention, however, because relatively few stars in the cluster are more massive than the sun. By dawn some ten billion years have gone by. Now it is the sun's turn to die. There is a sudden, shuddering contraction of the sun's core, and the solar atmosphere balloons into an aethereal red cloud that expands and swallows up the planets Mercury and Venus, and then Earth. Backing away to a prudent distance, we

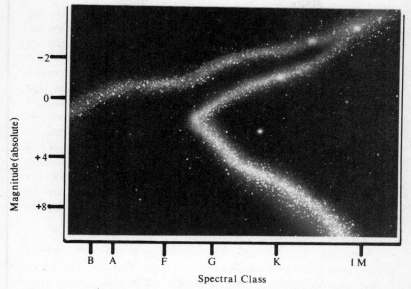

Magnitude (absolute)

Spectral Class

An old star cluster like the globular cluster M3 displays a strikingly different Hertzsprung-Russell diagram. Here, the more massive stars have had ample time to burn up their fuel and become red giants, moving up and toward the right on the diagram, and then to slide down and to the left as some evolve into dwarfs. The result is a dramatic "cutoff point" at which the main sequence is interrupted. All else being equal, the lower the cutoff point, the older the cluster.

watch the cloud disperse and see the naked, helium-rich core of the sun exposed as a dim, dense dwarf.

The night is over, but the story has hardly begun. Most of the cluster stars, less massive than the sun, continue to burn steadily, with an unexceptional, candle-yellow glow. These members of the silent majority have long lives ahead of them on the main sequence; they will still be shining aeons after the evacuated atmosphere of the sun has been gathered up to make new stars and planets. The study of stellar evolution teaches us that the meek shall inherit the galaxy.

Once it had been established that stars shine by means of nuclear fusion, it became apparent that they must also be in the business of building light elements into heavy elements. They could hardly do otherwise, inasmuch as nuclear fusion involves the fusing of the nuclei of light atoms to make the nuclei of heavier atoms. Through a variety of fusion processes, stars build hydrogen into helium; helium into carbon; carbon into oxygen and magnesium, and so forth. Indeed, given that the energy released amounts to but a tiny fraction of the mass being shuffled about, we could say that element-making is the primary business of stars, and that their light and heat, though subjectively important to creatures like ourselves who owe their lives to it, is but a by-product of that process, as incidental as the heat ventilated out of the smokestack of a tool and die works. If, as the textbooks like to say, atoms are the building blocks of matter, stars are the place where the building blocks are built. As Eddington wrote presciently in 1920, "The stars are the crucibles in which the lighter atoms which abound in the nebulae are compounded into more complex elements."[8]

Two essential questions remained.

One was just *how* stars make the heavy elements. Bethe's proton-proton reaction yields nothing heavier than helium, which is the second *lightest* element. If stars build heavier atoms, they must do so by means of other fusion processes. The carbon cycle won't do the trick; it employs carbon, nitrogen, and oxygen merely as catalysts, leaving no new elements behind. Clearly, it would take some fancy nuclear physics to better reconstruct the full complexity of stellar fusion.

The other question, closely related to the first, was whether stars are the sole, or even the primary, source of the elements. There was a competing hypothesis. It held that most of the elements were fused, not in stars, but in the big bang.

For fusion to have taken place in the big bang, the universe at the very onset of its expansion would have had to be hot. The hypothesis that this was the case came in part from the basic laws of thermodynamics, which show that any given volume of material will become hotter if it is compressed. Suppose, for instance, that the Milky Way galaxy were to be enclosed in a gigantic hydraulic press, like the ones used to crush the hulks of old cars into cubes of scrap metal, and were squeezed down into a volume of, say, only one cubic foot. (This is thought to have been its state when the universe was but a fraction of a second old.) While the compression process was taking place, the stars and planets would be melded together, then the molecules would break down, and finally, when the temperature exceeded that of a stellar interior, even the nuclear structures of the matter in the galaxy would begin to decompose, reducing everything to a hot, dense gas made of subatomic particles—what physicists call a plasma. Release the press, and the plasma would expand and cool, recombining into atoms and molecules in the process. This, then, is a small-scale model of what is thought to have happened in the big bang, with the universe evolving from a high-density plasma into the structures—nuclear, atomic, molecular, stellar, and planetary—that we see around us today.

If astronomers at first regarded the hot big bang idea with reservation, the nuclear physicists were more open to it. They were growing accustomed to envisioning conditions of high temperatures and high densities, if only because of their work on chain reactions in nuclear bombs. Gamow in particular was interested in the question of whether the chemical elements that compose the universe today could have been forged in the fires of the big bang. It was a reasonable supposition—the heavier the element, the more energy was required to build it, and where was there more energy than in the big bang?—and Gamow went to work painting in the details with the broad brush and vivid colors that characterized his approach to physics.

Alas, he was soon in trouble. He and his collaborators were able to determine how hydrogen nuclei could fuse to make nuclei of helium (Von Weizsäcker and others had suggested earlier that helium originated in the big bang) but thereafter their calculations stalled. As the physicists Enrico Fermi and Anthony Turkevich learned, there was no way for nuclei heavier than those of helium to be built in any quantity in the rapidly expanding fireball. The conditions just were not right; by the time helium had been synthesized, the primordial material ("ylem," Gamow called it, after an old Greek word for the substance of the cosmos prior to the evolution of form) would have thinned out too much for further fusion reactions to take place. The Hungarian-American physicist Eugene Wigner tried to find a way to negotiate what Gamow called the "mass five crevasse" that divides helium from the next stable nucleus, that of lithium. Gamow, who liked to illustrate his own books, published a sketch of Wigner, in mountaineer's garb, leaping the gap while crying, "Please!"[9] Wigner never made it across. Nevertheless, many big bang enthusiasts held out hope that Gamow was right in thinking of the big bang as the birthplace of the elements, and imagined that the difficulties he was encountering would, like the Coulomb barrier problem in astrophysics, eventually be overcome.

This was not the view, however, taken by researchers skeptical about the big bang theory, the most formidable among whom was the British astrophysicist Fred Hoyle. A born outsider who had by sheer intellectual energy made his way from the gray textile valleys of the north of England to the high table at Cambridge, Hoyle was individualistic to the point of iconoclasm, and as combative as if he had earned his knighthood on horseback. He lectured charismatically, in a working-class accent that seemed if anything to deepen as his scholarly credentials accumulated, and he was equally effective with the written word, publishing incisive technical papers, engrossing popularizations of science, and sprightly science fiction yarns with a seemingly effortless facility. Fearful was his scorn, and withering were his critiques of the big bang theory.

Hoyle damned the theory as epistemologically sterile, in that it seemed to place an inviolable, temporal limitation on scientific inquiry: The big bang was a wall of fire, past which science at the time did not know how to probe. Hoyle found it "highly objectionable that the laws of physics should lead us to a situation in which we are forbidden to calculate what happened before a certain moment in time."[10] He poked fun at the theory's creationist overtones: Had it not been proposed by a priest, Lemaître, and had not Pope Pius XII, at the opening of a meeting of the Pontifical Academy of Sciences on November 22, 1951, declared that it accorded with the Catholic concept of creation (an endorsement that, Gamow joked, demonstrated its "unquestionable truth")?[11] Empirically, Hoyle was unsparing in calling attention to the big bang theory's most telling liability, the time-scale problem. Owing to a number of errors, chief among them an inadequate understanding of the absolute magnitude of the Cepheid variable stars employed as intergalactic distance indicators, Hubble and Humason had severely underestimated the dimensions of the expanding universe—and, therefore, its age as well. Hubble's original statement of the expansion law had been that H_0, the expansion parameter, equaled 550 kilometers per second per megaparsec—meaning that for every megaparsec (or 3.26 million light-years) that one looks out into space, one finds galaxies moving apart at an additional 550 kps.

The trouble was that this value for H_0 yielded an elapsed time since the big bang of only about two billion years. This was smaller than the age of the sun and the earth. Since the universe cannot be younger than the stars and planets it contains, obviously something was wrong.

In Hoyle's view, what was wrong was the big bang concept itself. As an alternative, he and two colleagues, Herman Bondi and Thomas Gold, promulgated in 1948 what they called the steady state model. According to their theory, the universe was infinitely old and generally unchanging: There had been no creation event, no high-density infancy from which the universe had evolved.* The steady state theory was not destined to prosper; it lost its raison d'être once the errors in Hubble's distance figures were repaired, and it predicted that some galaxies ought to be very much older than others, of which no evidence has ever been found. But it had the salubrious effect of concentrating its advocates' attention on the question of where the heavier elements had come from. The steady staters could scarcely imagine, as Gamow did, that the elements had been synthesized in the big bang, since they denied that there had ever *been* a big bang. Consequently they were obliged to find another furnace in which to cook up such wonderfully complex atoms as those of iron, aluminum, and tin. The obvious candidate was the stars.

Hoyle, who possessed a command of nuclear physics unsurpassed among the astronomers of his generation, had begun working on the question of stellar fusion reactions in the mid-1940s. He had published little, however, owing to a running battle with "referees," anonymous colleagues who read papers and vet them for accuracy, whose adversity to Hoyle's more innovative notions prompted him to stop submitting his work to the journals. Hoyle paid a price for his rebelliousness, though, when in 1951, while he stood stubbornly in the wings, Ernst Öpik and Edwin Salpeter worked out the syn-

*To explain why the galaxies are not infinitely far apart in an infinitely old, expanding universe, the theory proposed that hydrogen atoms materialize spontaneously, out of empty space, and thence condense into new stars and galaxies. This hypothesis, though much ridiculed at the time, is not so implausible as it might at first appear. Owing to quantum indeterminacy, "virtual" particles do materialize out of space all the time, though their lifetimes normally are short. In some versions of the big bang theory, notably the "inflationary universe" model, *all* matter is said to have appeared out of a vacuum—though all at once, and long ago.

thesis in stars of atoms up through beryllium to carbon. Rankling
at the missed opportunity, Hoyle then broke his silence, and in a
1954 paper demonstrated how red giant stars could build carbon
into oxygen-16.

Ahead still lay the seemingly insurmountable obstacle of iron.
Iron is the most stable of all the elements; to fuse iron nuclei into
the nuclei of a heavier element consumes energy rather than releasing it; how, then, could stars fuse iron and still shine? Hoyle
thought that supernovae might do the job—that the extraordinary
heat of an exploding star might serve to forge the elements heavier
than iron, if that of an ordinary star could not. But this he could
not yet prove.

Then, in 1956, fresh impetus was lent to the question of stellar
element production when the American astronomer Paul Merrill
identified the telltale lines of technetium-99 in the spectra of S stars.
Technetium-99 is heavier than iron. It is also an unstable element,
with a half-life of only two hundred thousand years. Had the tech-
netium atoms that Merrill detected originated billions of years ago
in the big bang, they would since have decayed and there would
be too few of them left to show up today in S stars or anywhere
else. Yet there they were. Clearly the stars knew how to build
elements beyond iron, even if the astrophysicists didn't.

Spurred on by Merrill's discovery, Hoyle renewed his investigations into stellar nucleogenesis. It was a task he took very
seriously; as a boy, hiding atop a stone wall while playing hide-
and-seek one night, he had looked up at the stars and resolved to
find out what they were, and the adult astrophysicist never forgot
his childhood pledge. Visiting the California Institute of Technology, Hoyle found himself in the company of Willy Fowler, a res-
ident faculty member with an encyclopedic knowledge of nuclear
physics, and Geoffrey and Margaret Burbidge, a talented husband
and wife team who, like Hoyle, were English big bang skeptics.

A break came when Geoffrey Burbidge, scrutinizing recently
declassified data from a Bikini Atoll bomb test, noticed that the
half-life of one of the radioactive elements produced by the explosion, californium-254, was fifty-five days. This rang a bell: Fifty-
five days was just the period that a supernova that Walter Baade
studied had taken to fade away. Californium is one of the heaviest
of all elements; if it were created in the intense heat of exploding
stars, then surely the elements between iron and californium—

which comprise, after all, most of the periodic table—could have formed there, too. But how?

Happily, nature had provided a Rosetta stone against which Hoyle and his collaborators could test their ideas, in the form of the cosmic abundance curve. This was a plot of the weight of the various atoms—some twelve hundred species of nuclei, when the known isotopes were taken into account—against their relative profusion in the universe, as determined by studying the rocks of the earth, meteorites that have fallen to earth from space, and the spectra of the sun and stars. Physicists working on the Manhattan Project and the hydrogen bomb tests that followed had grown accustomed to deciphering the chain reactions involved by studying the relative abundances of various isotopes found in the debris left behind by the explosion. The cosmic element abundance curve was, in a sense, just another such table writ large; Gamow called it "the oldest document pertaining to the history of our universe."[12] But where for Gamow that history was principally the story of the big bang, for Hoyle and his colleagues the important thing was what had gone on since, inside a billion trillion stars. "The problem of element synthesis," they would write, "is closely allied to the problem of stellar evolution."[13]

The differences in abundances are great—there are, for instance, two million atoms of nickel for every four atoms of silver and fifty of tungsten in the Milky Way galaxy—and the abundance curve consequently traced out a series of jagged peaks more rugged than the ridgeline of the Andes. The highest peaks were claimed by hydrogen and helium, the atoms created in the big bang—more than 96 percent of the visible matter in the universe is composed of hydrogen or helium—and there were smaller but still distinct peaks for carbon, oxygen, iron, and lead. The pronounced definition of the curve put welcome constraints on any theory of element synthesis in stars: All one had to do (though this was quite a lot) was to identify the processes by which stars had come preferentially to make some elements in far greater quantities than others. Here the genealogy of the atoms was inscribed, as in some as yet untranslated hieroglyph: "The history of matter," wrote Hoyle, Fowler, and the Burbidges, ". . . is hidden in the abundance distribution of the elements."[14]

Their work culminated in 1957 in an epochal paper, 103 pages long, that showed how fusion processes operating in addition to

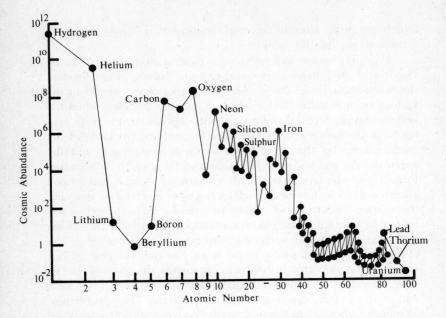

The cosmic element abundance curve depicts the relative numbers of various sorts of atoms found in the universe at large. It serves as a constraint on theories of how the elements formed. (After Taube, 1982.)

Bethe's proton-proton reaction and carbon cycle could build the atoms of the heavy elements—the "metals," which in astrophysical parlance means everything heavier than helium. The tentpole of the paper was the arrow of time: The evolution of atoms, it revealed, is bound up in the evolution of stars, and the mix of elements found in the universe today is largely the result of what stars did in the past. At first, a star is powered by "hydrogen burning," the fusing of hydrogen nuclei to build helium. This is the proton-proton reaction discerned by Bethe, and it can go on for a long time, from about a million years for a furiously burning giant star to ten billion years or so for a more tepid star like the sun. "But," as Hoyle, Fowler, and the Burbidges noted, "no nuclear fuel can last indefinitely."[15] Eventually the supply of hydrogen runs low and the star's core contracts. The contraction heats the core, and in the hotter environment helium burning can begin. The fusion of helium nuclei forms atoms of carbon, oxygen, and neon—but not lithium, beryllium, or boron, which explains why the former elements show up as peaks on the cosmic element abundance curve and the latter as valleys. When this process falters, the core contracts and heats further, fusing helium nuclei with those of neon to build magnesium, silicon, sulphur, and calcium. Now the old picture of a split-personality star could be refined into multiple personalities: A highly evolved star sorts itself into layers, like an onion, its gaseous iron core surrounded by concentric shells where silicon, oxygen, neon, carbon, helium, and, in the outermost shell, hydrogen are being burned. And so it goes, through previously undiscerned displays of the virtuosity of stellar alchemy.

Iron spells death, and death deliverance. The iron core grows like a cancer in the heart of the star, damping nuclear reactions in all that it touches, until the star becomes fatally imbalanced and falls victim to a general collapse. If the mass of the core is a tenth to two or three times that of the sun—here we draw on research by Gamow, Baade, Robert Oppenheimer, Fritz Zwicky, and others—the core rapidly crystallizes into a steely sphere, a "neutron star." Smooth as a ball bearing and smaller than a city but as massive as the sun, a neutron star spins rapidly on its axis and emits pulses of radio energy as it spins, creating a beacon of the sort that betrayed the locations of Tycho's and Kepler's supernovae. It resembles nothing so much as a giant atomic nucleus—as if the real business of the star, the conjuring of nuclei, was now at last monumentalized as a colossal nuclear tombstone.

The bright side, literally so, is that the explosion of the star generates sufficient energy to synthesize an enormous variety of atoms heavier than iron. When the iron core collapses it emits a single great clang, and this final ringing of the gong sends a sound wave climbing upward through the inrushing gas from the envelope of starstuff left behind. As the sonic wave rushing outward meets the waves of gas falling in, the result is a shock stronger than any other in the known universe. In a moment, tons of gold and silver, mercury, iron and lead, iodine and tin and copper are forged in the fiery collision zone. The detonation blows the outer layers of the star into interstellar space, and the cloud with its freight of valuable cargo expands, marching out over the course of aeons to become entangled with the surrounding interstellar clouds. When latter-day stars condense from these clouds, their planets inherit the star-forged elements. The earth was one such planet, and such is the ancestry of the bronze shields and steel swords with which men have fought, and the gold and silver they fought over, and the iron nails that Captain Cook's men traded for the affections of the Tahitians.

Lesser stars contribute less dramatically to the chemical evolution of the universe, but they too play their part, wafting the nuclei of heavy elements into space through stellar winds, shedding their outer atmospheres as planetary nebulae, or blowing them into space in the less disruptive but still imposing explosions called novae. One can see their handiwork in the chemical gradients that show up across the faces of galaxies: Metals are scarce in the spectra of stars near the galactic center, where few stars have formed since the early days, while stars in the spiral arms, where star formation continues apace, are rich in these heavier elements. We see and touch—indeed, *are*—the products of the evolution of atoms and stars.

Much remains to be learned about dying stars and their chemical legacies. We conclude this unfinished story with a coda by the Berkeley astronomer Frank Shu, drawing on research by the Soviet scientists Yakob Zel'dovich and Igor Novikov. Writes Shu:

> Stars begin their lives as a mixture mostly of hydrogen nuclei and their stripped electrons. During a massive star's luminous phase, the protons are combined by a variety of complicated reactions into heavier and heavier elements. The nuclear binding energy released this way ultimately provides entertainment and employment for astronomers. In the end, however, the supernova process serves to undo most of this nuclear evolution. In the end, the core forms a mass of neutrons. Now, the final state, neutrons, contains *less* nuclear binding energy than the initial state, protons, and electrons. So where did all the energy come from when the star was shining all those millions of years? Where did the energy come from to produce the sound and the fury which is a supernova explosion? Energy is conserved: who paid the debts at the end? Answer: Gravity! The gravitational potential of the final neutron star is much greater (negatively: that's the debt) than the gravitational potential energy of the corresponding main-sequence star. So, despite all the intervening interesting nuclear physics, ultimately Kelvin and Helmholtz were right after all! The ultimate energy source in the stars which produce the greatest amount of energy is gravity power.[16]

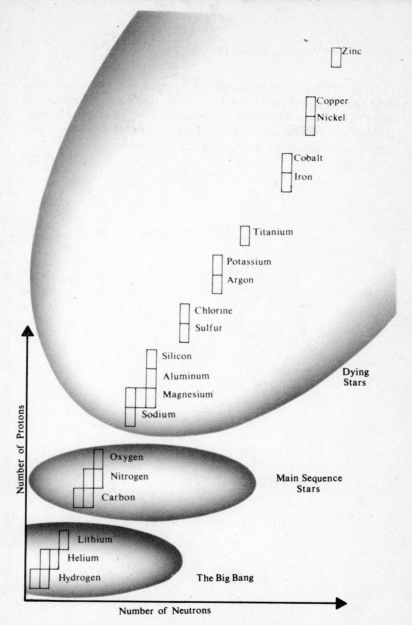

Cosmic evolution of elements involves the building of simple atomic nuclei in the big bang, and the subsequent fusion of these light nuclei into heavier and more complex nuclei inside stars. (After Reeves, 1984.)

Let that be the human image we call to mind as we watch the gold dust and diamonds bid farewell to the exhausted star as they parade away to be woven into future worlds and minds: The face of Kelvin, the old boy who cowed Rutherford one day when the century was young, shaking off his sleep to scowl, then smile.

PART THREE

CREATION

O landless void, O skyless void,
O nebulous, purposeless space,
Eternal and timeless,
Become the world, extend!
                    —Tahitian creation tale

What really interests me is whether God had
any choice in the creation of the world.
                    —Einstein

15

THE
QUANTUM
AND ITS
DISCONTENTS

What is the path? There is no path.
> —Niels Bohr, quoting Goethe

Progress in physics has always moved from
the intuitive toward the abstract.
> —Max Born

The act of exploration alters the perspective of the
explorer; Odysseus and Marco Polo and Columbus returned home
as changed men. So it has been with scientific investigation of the
extremities of scale, from the grand sweep of cosmological space
down to the cramped and frantic world of the subatomic particles:
These journeys changed us, challenging many of the scientific and
philosophical conceptions we had most cherished. Some had to be
discarded, like baggage left behind on a trek across a desert. Others
were altered and repaired almost beyond recognition, like the veteran mountaineer's hand-hammered pitons or the old seaman's knife
with its twine-encrusted handle and bone-thin blade. Exploration
of the realm of the galaxies extended the reach of human vision by
a factor of some 10^{26} larger than the human scale, and brought about
the revolution we identify with relativity, which revealed that the

Newtonian world view was but a parochialism in a wider universe where space is curved and time becomes pliant. Exploration of the subatomic realm carried us far into the realm of the small, to some 10^{-15} of the human scale, and it, too, wrought a revolution. This was *quantum* physics, and all that it touched it transformed.

Quantum theory was born in 1900, when Max Planck realized that he could account for what was called the black-body curve— the spectrum of energy generated by a perfectly radiating object —only if he abandoned the classical assumption that energy is emitted continuously and replaced it with the unprecedented hypothesis that energy comes in discrete units. Planck called these units *quanta*, after the Greco-Latin word for "how much" (as in *quantity*), and he defined them in terms of the quantum of action, symbolized by the letter *h*. Planck was no revolutionary—at age forty-two he was an old man by the standards of mathematical science, and a pillar of nineteenth-century German high culture to boot—but he readily appreciated that the quantum principle would shatter much of the classical physics to which he had devoted his career. "The greater [its] difficulties," he wrote, ". . . the more significant it finally will show itself to be for the broadening and deepening of our whole knowledge in physics."[1] His words proved prophetic: Constantly changing and developing, altering its coloration as unpredictably as a reflection in a soap bubble, quantum physics soon expanded into virtually every area of physics, and Planck's *h* came to be regarded as a fundamental constant of nature, on a par with Einstein's *c*, the velocity of light.

The quantum principle was very strange—it was, as Gamow remarked, as if one could drink a pint of beer or no beer at all, but were barred by a law of nature from drinking any quantity of beer between zero and one pint—and it got stranger as it evolved. The decisive break with classical physics came in 1927, when the young German physicist Werner Heisenberg arrived at the indeterminacy principle. Heisenberg found that one can learn either the exact position of a given particle or its exact trajectory, *but not both*. If, for instance, we watch a proton fly through a cloud chamber, we can by recording its track discern the direction in which it is moving, but in the process of plowing through the water vapor in the chamber the proton will have slowed down, robbing us of information about just where it was at any given instant. Alternately, we can irradiate the proton—take a flash photograph of it, so to speak—

The Scale of the Known Universe

Radius (meters)	Characteristic Objects
10^{26}	Observable universe
10^{24}	Superclusters of galaxies
10^{23}	Clusters of galaxies
10^{22}	Groups of galaxies (e.g., the Local Group)
10^{21}	Milky Way galaxy
10^{18}	Giant nebulae, molecular clouds
10^{12}	Solar system
10^{11}	Outer atmospheres of red giant stars
10^{9}	Sun
10^{8}	Giant planets (e.g., Jupiter)
10^{7}	Dwarf stars, Earthlike planets
10^{5}	Asteroids, comet nuclei
10^{4}	Neutron stars
1	Human beings
10^{-2}	DNA molecules (long axis)
10^{-5}	Living cells
10^{-9}	DNA molecules (short axis)
10^{-10}	Atoms
10^{-14}	Nuclei of heavy atoms
10^{-15}	Protons, neutrons
10^{-35}	Planck length: Quantum of space; radius of "dimensionless" particles in string theory

and thus determine its exact location at a given instant, but the light or other radiation we employ to take the photograph will knock the proton off its appointed rounds, depriving us of precise knowledge of where it would have gone had we left it alone. We are, therefore, limited in our knowledge of the subatomic world: We can extract only partial answers, the nature of which are decided to some extent by the questions we choose to ask. When Heisenberg calculated the inescapable minimum amount of uncertainty that limits our understanding of events on the small scale, he found that it is defined by nothing other than h, Planck's quantum of action.

Quantum indeterminacy does not depend upon the design of

the experimental apparatus employed to investigate the subatomic world. It is, so far as anyone can tell, an absolute limitation, one that the most wizardly scholars of an advanced extraterrestrial civilization would share with the humblest string-and-sealing-wax physicist on Earth. In classical atomic physics it had been assumed that one could, in principle, measure the precise locations and trajectories of billions of particles—protons, say—and from the resulting data make exact predictions about where the protons would be at some time in the future. Heisenberg showed that this assumption was false—that we can *never* know everything about the behavior of even *one* particle, much less myriads of them, and, therefore, can never make predictions about the future that will be completely accurate in every detail. This marked a fundamental change in the world view of physics. It revealed that not only matter and energy but knowledge itself is quantized.

The more closely physicists examined the subatomic world, the larger indeterminacy loomed. When a photon strikes an atom, boosting an electron into a higher orbit, the electron moves from the lower to the upper orbit instantaneously, *without having traversed the intervening space*. The orbital radii themselves are quantized, and the electron simply ceases to exist at one point, simultaneously appearing at another. This is the famously confounding "quantum leap," and it is no mere philosophical poser; unless it is taken seriously, the behavior of atoms cannot be predicted accurately. Similarly, we saw earlier, it is by virtue of quantum indeterminacy that protons can leap the Coulomb barrier, permitting nuclear fusion to occur at a sufficiently robust rate to keep the stars shining.

Those who find such considerations nonsensical are in good company; as Niels Bohr remarked, when one of his students at Copenhagen complained that quantum mechanics made him giddy, "If anybody says he can think about quantum problems *without* getting giddy, that only shows he has not understood the first thing about them."[2] The reason, however, is simply that we human beings, having grown up in the macroscopic world, tend to think of things in terms of macroscopic similes—subatomic particles are like buckshot, light waves are like waves in the ocean, atoms are like little solar systems, and so forth—and these similes break down on the microscopic scale.

Our mental pictures are drawn from our visual perceptions of the world around us. But the world as perceived by the eye is itself exposed as an illusion when scrutinized on the microscopic scale. A bar of gold, though it looks solid, is composed almost entirely of empty space: The nucleus of each of its atoms is so small that if one atom were enlarged a million billion times, until its outer electron shell was as big as greater Los Angeles, its nucleus would still be only about the size of a compact car parked downtown. (The electron shells would be zones of insubstantial heat lightning, each a mile or so thick, separated by many miles of space.) Nor, to return to the old classical metaphor, does a cue ball strike a billiard ball. Rather, the negatively charged fields of the two balls repel each other; on the subatomic scale, the billiard balls are as spacious as galaxies, and were it not for their like electrical charges they could, like galaxies, pass right through each other unscathed.

The quantum revolution has been painful, but we can thank it for having delivered us from several of the illusions that afflicted the classical world view.

One such was the delusion of apartness—the assumption that man is separate from nature and that acts of observation can, there- fore, be conducted with complete objectivity. Traditionally, sci- entists were free to think of themselves as passive observers, sealed off by a pane of laboratory glass or a telescope's lens from the outer world they examined. But on the microscopic level, every act of observation is disruptive—countless photons of starlight die upon the eye, protons smash into accelerator targets—and the manner in which we choose to make the observation (to "collapse the wave function," as the physicists say) influences the results of the inter- action. Subatomic particles sometimes resemble particles, some- times waves, depending upon how we examine them. They are not "really" one or the other—and, in any event, the two images are mathematically equivalent. Rather, they are participants in an act of observation, the nature of which influences the qualities they present to us. Quantum physics obliges us to take seriously what had previously been a more purely philosophical consideration: That we do not see things in themselves, but only aspects of things. What we see in an electron path in a bubble chamber is not an electron, and what we see in the sky are not stars, any more than a recording of Caruso's voice *is* Caruso. By revealing that the ob- server plays a role in the observed, quantum physics did for physics what Darwin had done in the life sciences: It tore down walls, reuniting mind with the wider universe.

Likewise with the dilemma of strict causation. Classical physics was deterministic: If A, then B; the bullet fired at the window shatters the glass. On the quantum scale this is only probably true: Most of the particles in the bullet encounter those of the glass, but some go elsewhere, and the trajectory of any one of them can be predicted only by invoking the statistics of probabilities. Einstein was deeply troubled by this aspect of the new physics. "God does not play dice," he said, and he argued that the indeterminacy principle, though useful in practice, does not represent the fundamental relationship between mind and nature. As he wrote to his friend and colleague Max Born:

> I find the idea quite intolerable that an electron exposed to radiation should choose *of its own free will*, not only its moment to jump off, but also its direction. In that case, I would rather be a cobbler, or even an employee in a gaming-house, than a physicist.[3]

Einstein presented Bohr with a series of thought experiments aimed at disproving the theory of quantum indeterminacy. He was then near the peak of his powers, and his ideas were often startling in their originality and ingenuity, but Bohr and his students found flaws in them all. Nothing in nature, then or now, indicates that the universe is built upon a strictly deterministic underpinning, and no philosopher has been able to prove that we need to believe in hidden, deterministic mechanisms—"hidden variables"—that produce no observable results.

Defeated in battle if unbowed in the greater campaign, Einstein took refuge in the long view: "Quantum mechanics is certainly imposing," he told Born, "but an inner voice tells me that it is not yet the real thing. The theory says a lot, but does not really bring us any closer to the secret of the 'old one.' I, at any rate, am convinced that *He* is not playing at dice."[4] Ultimately, Einstein insisted, he would be proved right: "I am quite convinced that someone will eventually come up with a theory whose objects, connected by laws, are not probabilities but considered facts."[5]

That might conceivably be so, but it is not clear why we should *wish* it to be so. Strict causation, for all its classical pedigree, was ultimately a monstrous doctrine. Consider its stark formulation by the French mathematician Pierre-Simon de Laplace:

An intelligence knowing, at a given instance of time, all forces acting in nature, as well as the momentary position of all things of which the universe consists, would be able to comprehend the motions of the largest bodies of the world and those of the lightest atoms in one single formula, provided his intellect were sufficiently powerful to subject all data to analysis; to him nothing would be uncertain, both past and future would be present in his eyes.[6]

What was there here worth clinging to, against the hard evidence of quantum physics? The invocation of an all-knowing intelligence, which could only be that of God? The depiction of men as machines, deprived of free will? The pretense that every occurrence, from the radioactive decay of a barium atom to the Battle of Hastings, was fated to occur just when and how it did, in a universe devoid of originality and surprise? We are free (or fated) to answer, none of the above, and to recoil from Laplace's deterministic vision with a revulsion just as deep as Einstein's over Bohr's interpretation of Heisenberg. Quantum indeterminacy may have nothing to do with human will, but as a matter of philosophical taste there are good reasons to celebrate the return of chance to the fundamental affairs of the world.

And, of course, the test of a scientific theory has to do less with whether one finds it philosophically palatable than with whether it works. Quantum physics works. It depicts the world as an assembly of animated fields, and the field equations are often too abstract to seem familiar, but they tell the story of the subatomic world more accurately than do the homier metaphors to which the prior intellectual history of our species had accustomed us.

Not that the quantum precept escaped without its share of growing pains. Far from it: In charting the unfamiliar terrain of the small, its practitioners embroiled themselves in misconceptions and perplexities that made the astronomers' earlier bewilderments over spiral nebulae and the age of the sun look inviting by comparison. Quantum numbers and airy abstractions were hurled at tough problems with both hands, until microphysics in its darkest days was justly and scathingly compared to Ptolemaic cosmology, with its wheels within wheels and its abandonment of all but the most abstract claim to model the real world. There were too many particles, so many that physicists eventually were obliged to consult a booklet, the *Particle Properties Data Handbook*, just to keep track of them. "If I could remember the names of all these particles I would have been a botanist,"[7] fumed Enrico Fermi, and the physicist Martinus Veltman later mused that "as the number of particles increases all we are doing is increasing our ignorance."[8] There were, for some decades, too many theories as well, and many inconsistencies among them. A few physicists became so frustrated that they quit science altogether.

Yet it is turmoil and confusion and not calm assurance that mark the growth of the mind, and when the dust began to clear, quantum physics emerged as not only a vital and rapidly developing field of science, but as one of the greatest intellectual achievements in the history of human thought. Though by no means complete, it was now able to make accurate predictions about an imposing array of phenomena, from optics and computer design to the shining of the stars, and to do so in terms of theoretical structures that could already be seen to possess a beauty and scope worthy of the universe they sought to describe.

The patchwork of theories that came to constitute quantum physics by the final quarter of the century was known collectively as the standard model. Viewed by its lights, the world is composed of two general categories of particles—those of fractional spin (½), called *fermions*, after Enrico Fermi, and those of integer spin (0, 1 or 2), called *bosons*, after Satyendra Nath Bose, who, with Einstein, developed the statistical laws that govern their behavior.*

Fermions comprise matter. They obey what is called the Pauli exclusion principle, enunciated by the Austrian physicist Wolfgang Pauli in 1925, which establishes that no two fermions can occupy a given quantum state at the same time. It is owing to this characteristic of fermions that only a limited number of electrons can occupy each shell in an atom, and that there is an upper limit to the number of protons and neutrons that can be assembled to form a stable atomic nucleus. Protons, neutrons, and electrons are all fermions.

Bosons convey force. To hazard a hyperbolic image, one might think of the fermions as akin to ice skaters who are busy tossing medicine balls back and forth; the medicine balls are bosons, and

---

*The "spin" referred to here is a familiar, mechanical spin, though it is quantized, and is measured in terms of *b*, the quantum of action.

the change in the trajectory of each skater that occurs when they throw or catch the balls betrays, in Newtonian language, the presence of a force.* Bosons do not obey the exclusion principle, and consequently several different forces can operate in the same place at the same time: The atoms in this book, for instance, are simultaneously subject both to the electrical attraction among their protons and electrons and to the gravitational force of the earth.

There are four known fundamental forces (or classes of *interactions*, in quantum terminology)—gravitation, electromagnetism, and the strong and weak nuclear forces. Each plays a distinct role. Gravitation, the universal attraction of all particles of matter for one another, holds each star and planet together, and retains planets in their orbits around stars and stars in their orbits in galaxies. Electromagnetism, the attraction of particles with opposite electrical or magnetic charge for one another, produces light and all other forms of electromagnetic radiation, including the long-wavelength radiation called radio waves and the short-wavelength radiation called X rays and gamma rays. Electromagnetism also bundles atoms together as molecules, making it responsible for the structure of matter as we know it. The strong nuclear force binds protons and neutrons (known as nucleons) together in the nuclei of atoms, and binds the elementary particles called quarks together to form each nucleon. The weak nuclear force mediates the process of radioactive decay, the source of energy emitted by the chunks of radium studied by Rutherford and the Curies.

The differing behavior of the forces is reflected in the nature of the bosons that convey them. Gravitation and electromagnetism are infinite in range—which is why our galaxy "feels" the gravitational pull of the Virgo Cluster of galaxies, and why we can see starlight coming from billions of light-years away—because the bosons that carry these two forces, known respectively as gravitons and photons, have zero mass. The weak nuclear force has a very short range because the particles that convey it, called weak bosons, are massive. The strong force is carried by particles called gluons;

---

*The medicine balls are purely repulsive, as in the interaction between two electrons or other fermions of like charge. For attractive forces (as between a proton and an electron), imagine that the bosons are elastic bands that stretch when the skaters move apart, drawing them together. For the exclusion principle, let each skater wear a hoopskirt that forbids their colliding. . . . And that is quite enough of that.

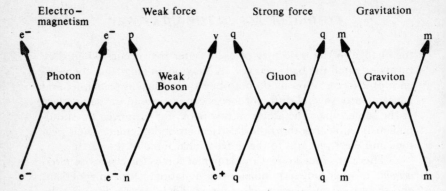

The four fundamental forces known to operate in nature today are here depicted in terms of characteristic interactions. In a typical electromagnetic interaction, a pair of electrons (symbolized $e-$) exchange a photon. In the weak force interaction portrayed here, a neutron (n) decays into a proton (p) via the exchange of a weak boson; the event also converts a positron ($e+$) into a neutrino (v). In a strong interaction, quarks (q) exchange a gluon. Gravitation involves the exchange of a graviton between any two massive particles (m).

they are massless, but have the curious and quite beautiful property of increasing, not decreasing, in strength when the bosons between which they are exchanged (the quarks) move apart: A quark that starts to stray from its two companions soon finds itself hauled back by a gluon lattice, like a finger trapped in a woven Chinese finger-cuff. Consequently quarks in the contemporary universe remain bound up inside their protons and neutrons; no free quark has yet been observed, though they have been searched for in everything from accelerator collisions to moon dust to oysters (which filter seawater and so might catch stray quarks).

The fermions that constitute matter, though notoriously numerous and varied, can all be classified as either *quarks*, which respond to the strong force, or *leptons*, which do not. Leptons are light particles; their ranks include the electrons that orbit atomic nuclei. Quarks are the building blocks of protons and neutrons: Three quarks make a nucleon.* There are thought to be six varieties

*The name "quark" was conferred by Murray Gell-Mann, the Caltech physicist who came up with the idea. It comes from a line in James Joyce's *Finnegans Wake*, "Three quarks for Muster Mark!" George Zweig, a physicist at Caltech who arrived at the same idea independently, called the entities "aces," a term that lost out to Gell-Mann's, perhaps because there are four aces, not three, in a deck of cards.

The Fundamental Interactions

Force	Range (meters)	Relative strength	Role
Strong nuclear	$< 3 \times 10^{-15}$	10^{41}	Binds protons, neutrons together in atomic nuclei
Weak nuclear	$< 10^{-15}$	10^{28}	Involved in radioactive decay
Electromagnetism	Infinite	10^{39}	Binds atoms together to form molecules; propagates light, radio waves, other forms of electromagnetic energy
Gravitation	Infinite	1	Holds planets and stars together; holds stars in galaxies

each of leptons and quarks. Neither quarks nor leptons show any sign of having an internal structure, though their anatomy has been probed on scales down to some 10^{-18} meter. This is to say that if a single atom were enlarged to the dimensions of the earth, any subcomponents of quarks and leptons would have to be smaller

The Building Blocks of Matter

Particle	Description	Examples
Leptons	"Dimensionless" (i.e., radius $< 10^{-35}$ meter); do not participate in the strong force.	Electron Muon Neutrino
Quarks	Small ($< 10^{-18}$ meter) but finite in size; do participate in the strong force.	Hadrons (three quarks) Mesons (two quarks)

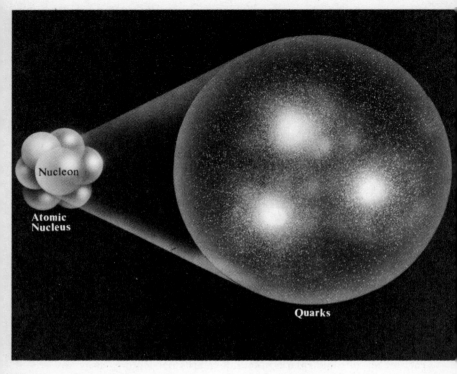

Nucleon

Atomic
Nucleus

Quarks

Trios of quarks are thought to compose the nucleons—protons and neutrons—that in turn constitute the nuclei of atoms. According to this model, a proton consists of two "up" quarks, each of which carries an electrical charge of $+\frac{2}{3}$, and one "down" quark, which has a charge of $-\frac{1}{3}$; the total charge of the proton therefore is $\frac{4}{3} - \frac{1}{3} = +1$. A neutron consists of two down quarks and one up quark; consequently its charge equals 0.

than a grapefruit to have escaped detection. So quarks and leptons are the bedrock particles of matter, so far as we know.

Every fundamental—meaning simple—event in the universe can in principle be interpreted by means of the standard model. When a child looks at a star, photons of starlight strike electrons in the outer atoms of the receptors of the child's retina, setting off further electron interactions that convey the image to the brain; all this is the work of electromagnetism. The nuclear processes that produced the starlight are generated by the strong and weak nuclear forces at work inside the star. And gravitation is the force that holds

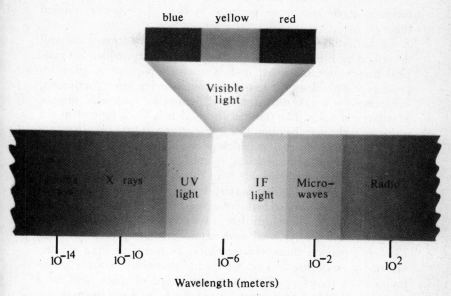

Wavelength (meters)

Electromagnetic energy is generated by natural processes across a wide range of wavelengths, including gamma rays and X rays from gas falling into black holes, light from stars, microwaves from the cosmic background radiation, and radio from interstellar clouds.

the star together and keeps the child's feet (if only intermittently) on the ground.

The scientific accounts of how the various particles of matter behave under the influence of three of the four forces are known as relativistic quantum field theories. They are so called because they incorporate both the quantum precept and the special theory of relativity, in order to take into account such effects as increases in the mass of particles traveling at close to the velocity of light. Electromagnetism is described, with exquisite accuracy, by the theory of quantum electrodynamics, or QED. The strong force is described by quantum chromodynamics, or QCD. (The "chromo" comes from a quantum number, whimsically called "color," that plays a role for quarks comparable to that played by electrical charge in the affairs of electrons.) The weak force, as we will see, has recently come under the purview of the "electroweak" unified theory.

Gravitation remains the odd man out. Its workings are still described by Einstein's general theory of relativity, which is a classical theory, meaning that it does not incorporate the quantum principle. This does not cause problems under most conditions, but relativity breaks down when it comes to extremely intense gravitational fields, like those at the edge of a black hole or in the universe at the very beginning of its expansion. There the curvature of space goes to infinity, at which point the theory tips its hat and makes a graceful exit. There was, by the late 1980s, still no quantum theory of gravitation with which to supplement general relativity. One reason for this is that gravity is weak. Individual subatomic particles normally are so little influenced by the gravitational force exerted by their colleagues that gravity can be ignored. Another reason is that gravitational interactions are interpreted, through Einstein's general theory of relativity, as resulting from the geometry of space itself. The "gravitons" thought to convey gravitation must therefore dictate the very shape of space, and for a theory to elucidate how they manage *that* is no simple matter.

Particle physics today is a house divided, and though the standard model gets results, few imagine that it represents the last word on the subject. The model is a crazy quilt, not a mandala. To fire it up on all cylinders requires inputting some seventeen separate parameters, numbers the values of which have been determined experimentally but whose fundamental significance is not yet understood. We know, for instance, that the electrical charge carried by an electron is equal to $1.6021892 \times 10^{-19}$ coulomb, and that the mass of the proton is 938.3 MeV, equal to 0.9986 the mass of the neutron, but nobody knows why these numbers are as they are and not otherwise. The roots of discontent with the standard model were described this way by Leon Lederman, the director of the Fermilab particle accelerator in Illinois:

> The trouble we're in now is that the standard model is very elegant, it's very powerful, it explains so much—but it's not complete. It has some flaws, and one of its greatest flaws is aesthetic. It's too complicated. It has too many arbitrary parameters. We don't really see the creator twiddling seventeen

knobs to set seventeen parameters to create the universe as we know it. The picture is not beautiful, and that drive for beauty and simplicity and symmetry has been an unfailing guidepost to how to go in physics.[9]

So it was that physicists late in the century were still searching for a simpler and more efficient account of the fundamental interactions. The object of their quest went by the name "unified" theory, by which they usually meant a single theory that would account for two or more of the forces currently handled by separate theories. They were guided, to be sure, by experimental data and by the challenges immediately at hand—the theorist resembles, as Einstein said, an "unscrupulous opportunist," more often trying to find a specific solution to an immediate problem than to write a grand explication of everything. But they were guided as well, as Lederman mentions, by the hope that their accounts of nature could more nearly approach the elegant simplicity and superlative creativity of nature herself.

RUMORS OF
PERFECTION

Spirit of BEAUTY, that dost consecrate
With thine own hues all thou dost shine
                                    upon
Of human thought or form, where art
                            thou gone?
Why dost thou pass away, and leave our
                                    state,
This dim vast vale of tears, vacant and
                                desolate?
—Shelley, "Hymn to Intellectual
                            Beauty"

The Universe is built on a plan the profound
symmetry of which is somehow present in the
inner structure of our intellect.

—Paul Valéry

Theoretical physicists, like artists (one is tempted
to say, like *other* artists) are guided in their work by aesthetic as
well as rational concerns. "To make any science, something else
than pure logic is necessary," wrote Poincaré, who identified this
additional element as intuition, involving "the feeling of mathe-

matical beauty, of the harmony of numbers and forms and of geometric elegance."[1] Heisenberg spoke of "the simplicity and beauty of the mathematical schemes which nature presents us. You must have felt this too," he told Einstein, "the almost frightening simplicity and wholeness of the relationship which nature suddenly spreads out before us."[2] Paul Dirac, the English theoretical physicist whose relativistic, quantum mechanical description of the electron ranks with the masterpieces of Einstein and Bohr, went so far as to maintain that "it is more important to have beauty in one's equations than to have them fit experiment."[3]*

Aesthetics are notoriously subjective, and the statement that physicists seek beauty in their theories is meaningful only if we can define beauty. Fortunately this can be done, to some extent, for scientific aesthetics are illuminated by the central sun of *symmetry*.

Symmetry is a venerable and all but bottomless concept, with many implications in both science and art; long after the Chinese-American physicist Chen Ning Yang had won a Nobel Prize for his work in developing a symmetry theory of fields, he was still pointing out that "we do not yet comprehend the *full scope* of the concept of symmetry" (Yang's italics).[4] In Greek, the word means "the same measure" (*sym*, meaning "together," as in *sym*phony, a bringing together of sounds, and *metron*, for "measurement"); its etymology thus informs us that symmetry involves the repetition of a measurable quantity. But by symmetry the Greeks also meant "due proportion," suggesting that the repetition involved ought to be harmonious and pleasing; this suggests that a symmetrical relationship is to be judged by a higher aesthetic criterion, an idea to which I will return at the end of this chapter. In twentieth-century science, however, the former aspect of the old definition is emphasized; symmetry is said to exist when a measurable quantity remains *invariant* (meaning unchanging) under a *transformation* (meaning an alteration). Because this definition is the most relevant

---

*Dirac meant, of course, not that one should ignore the empirical results altogether, but that a beautiful theory need not be abandoned just because it fails an initial test. He had in mind Erwin Schrödinger's reluctance to publish his estimable equations of wave mechanics merely because they conflicted with experimental data. "It is most important to have a *beautiful* theory," Dirac told the science writer Horace Freeland Judson. "And if the observations don't support it, don't be too distressed, but wait a bit and see if some error in the observations doesn't show up."[5]

to the subject at hand, I will employ it in discussing all aspects of symmetry, including those that were in general use before there was such a thing as science.

Most of us were first introduced to symmetry through its visual manifestations in geometry and art. When we say, for instance, that a sphere is rotationally symmetrical, we indicate that it pos-

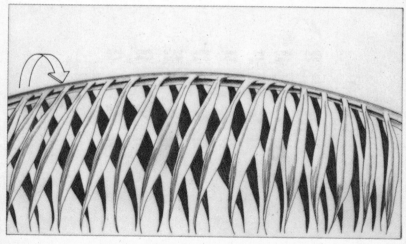

sesses a characteristic—in this case, its circular silhouette—that
remains invariant throughout the transformation introduced by rotating it. The sphere can be rotated on any axis and to any degree
without changing its silhouette, which makes it more symmetrical
than, say, a cylinder, which enjoys a similar symmetry only when
rotated on its long axis; if rotated on its short axis, the cylinder
shrinks to a circle. Translational symmetries, like those found in
palm fronds and building facades like that of the Doge's Palace in
Venice, occur when a shape remains invariant when moved ("translated") a given distance along one axis. (See previous page.)

Symmetries are commonplace in sculpture, beginning with the
human nude, which is (approximately) bilaterally symmetrical when
viewed from the front or back, and in architecture, as in the cross-
shaped floor plans of medieval cathedrals, and they turn up elsewhere in everything from weaving to square dancing. There are
many symmetries in music. Bach, in the following passage from
the *Toccata and Fugue in E Minor*, moves little tentlike trios of notes
up and down the staff. Except for the occasional difference in a
note here and there, the construct is translationally symmetrical:
If we were to peel off any one trio and lay it over another, it would
fit perfectly:

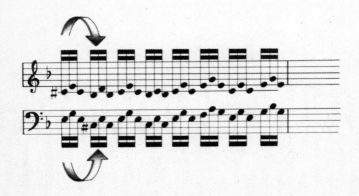

The first two bars of Claude Debussy's *Deux Arabesques* are bilaterally symmetrical both within themselves and relative to each other: The sheet music can be folded vertically at the bar, or midway within each bar, and the notes will still fit atop one another:

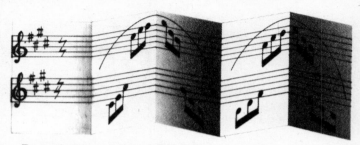

Beneath these visible and audible manifestations of symmetry lie deeper mathematical invariances. The spiral patterns found inside the chambered nautilus and on the faces of sunflowers, for instance, are approximated by the Fibonacci series, an arithmetic operation in which each succeeding unit is equal to the total of the preceding two (1, 1, 2, 3, 5, 8 . . .). The ratio created by dividing any number in such a series by the number that follows it approaches the value 0.618* (see page 306). This, not incidentally, is the formula of the "golden section," a geometrical proportion that shows up in the Parthenon, the *Mona Lisa*, and Botticelli's *The Birth of Venus*, and is the basis of the octave employed in Western music since the time of Bach. All the fecund diversity of this particular symmetry, expressed in myriad ways from seashells and pine cones to the *Well-Tempered Clavier*, therefore derives from a single invariance, that of the Fibonacci series. The realization that one abstract symmetry could have such diverse and fruitful manifestations occasioned delight among Renaissance scholars, who cited it as evidence of the efficacy of mathematics and of the subtlety of God's design. Yet it was only the beginning. Many other abstract

---

*The ratio is approximate because the numbers generated by the Fibonacci series are "irrational"—i.e., the ratio upon which they converge cannot be expressed exactly in terms of a fraction. The Pythagoreans discovered irrational numbers, and are said to have been so unsettled by them that they prescribed the death penalty to any of their sect who revealed their existence to the untutored multitudes. Hippasus was banished for defying the ban. He drowned at sea, a fate that the Pythagoreans ascribed to divine retribution.

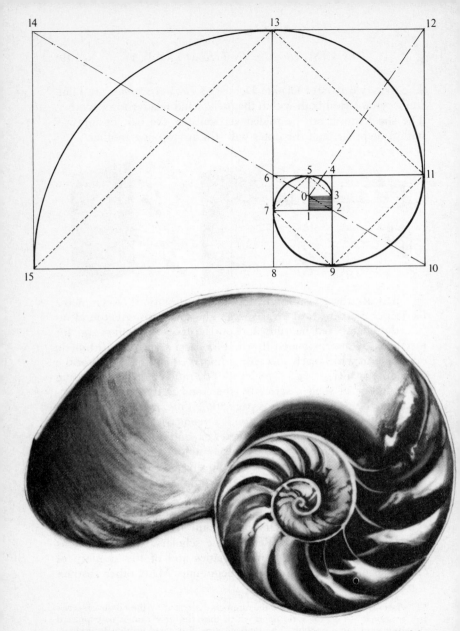

The Fibonacci series, represented in the abstract *(above)* is embodied in the architecture of the chambered nautilus *(below)*.

symmetries have since been identified in nature—some intact and some "broken," or flawed—and their effects appear to extend to the very bedrock foundations of matter and energy.

Which brings us back to science. When mathematicians in the early twentieth century began to look more closely at the concept of symmetry, they realized that the laws that science finds in nature are expressions of invariances—and may, therefore, be based on symmetries. This first became evident with regard to the conservation laws: The laws of thermodynamics, for instance, identify a quantity (energy) that remains invariant under a transformation (work). The German mathematician Emmy Noether demonstrated in 1918 that every conservation law implies the existence of a symmetry. The same evidently is true of the other laws as well. As the Hungarian physicist Eugene Wigner put it, "Laws of nature could not exist without principles of invariance,"[6] and invariance, keep in mind, is the signature of symmetry.

If natural laws express symmetries, then one ought to be able to search for previously unknown laws by looking for symmetrical relationships (invariances) in nature. Einstein absorbed this lesson in his bones, and employed symmetry as a lamp to guide his way in the creation of new theories. In special relativity (which, we recall, he originally called "invariance theory"), he employed the Lorentz transformations to maintain the invariance of Maxwell's field equations for observers in motion; in the general theory of relativity he did much the same thing for observers in strong gravitational fields. As his friend Wigner remarked in 1949, speaking at a Princeton celebration in Einstein's honor, "It is now natural for us to try to derive the laws of nature and to test their validity by means of the laws of invariance, rather than to derive the laws of invariance from what we believe to be the laws of nature."[7]

If the symmetry concept was powerful in relativity, it proved to be even more so when applied to the quantum physics of particles and fields. We can appreciate why this would be the case if we consider that subatomic particles of a given variety are indistinguishable from one another. All protons are identical. So are all neutrons and electrons. The same is true of fields; if we have two electromagnetic or gravitational fields with the same quantum numbers, and a trickster switches them while our backs are turned, we can never tell that we have been tricked, for the fields are identical. Inasmuch as identity is a form of invariance, we can, therefore, choose to regard the individual electron or photon as representative of a symmetry group that embraces all its fellow particles of the same species. Moreover, we can look for larger symmetries that might link the various groups involved—revealing, say, a previously undiscerned invariance linking photons and electrons.

This, much simplified, is the basic concept behind unified field theory, and it is no mere intellectual exercise: Wigner, for one, found that by applying relativistic invariances to quantum mechanics he could organize all known subatomic particles into symmetry groups, classifying them according to their rest mass and their spin. An even more dramatic example of the pathfinding power of symmetry came in 1928, when Dirac derived the relativistic quantum equation of the electron, preserving the symmetries of both special relativity and quantum mechanics, and found that his equation mandated the existence of a positively charged electron. This was the first intimation that there might be such a thing as *antimatter*, particles with mass and spin identical to those of ordinary matter, but with opposite electrical charge.

Dirac was mathematical to the marrow, an epitome of Karl Friedrich Gauss's dictum that whenever possible one should count; when out on a stroll with a colleague who remarked that there were fourteen ducks on the lake, Dirac replied, "Fifteen. I saw one going under the water."[8] He was also empirical to a fault; when a newcomer to Cambridge High Table ventured to say, as a conversational icebreaker, "It is very windy, Professor," Dirac got up, went to the door, opened it, looked out, returned, took his seat, thought for a moment, then replied, "Yes."[9] Yet the concept of antimatter seemed so outlandish that even Dirac initially denied the verdict of his own equations. At first he tried to portray his new particles as but the familiar protons, but this avenue of escape was soon closed off by the German mathematician Herman Weyl (himself the author of a classic treatise on symmetry), who demonstrated that unless Dirac's theory of the electron was nonsense, there had to be such a thing as antimatter. The question was resolved in symmetry's favor in 1932, when the antielectrons (called "positrons") predicted by the Dirac equation were discovered, by Carl Anderson, in a cloud chamber at Caltech. Before the decade was out Dirac and Anderson had won the Nobel Prize.

Just as symmetry can be employed to discern the identity of previously unknown particles, it can guide the search for unknown fields. The foremost application of this realization came with the development of "gauge" field theory, a revolution that has been classed, along with relativity and quantum physics, as the third great theoretical advance in twentieth-century physics.

The first gauge field theory beyond electromagnetism (the term "gauge" has a complex history, and is in any case a misnomer) was invented by Yang and his colleague Robert Mills at Brookhaven National Laboratory in 1954. Yang, a mathematician's son, grew up in China amid the destitution of the war; he studied statistical and quantum mechanics in K'un-ming, taking notes in unheated lecture halls and subsisting on a bare minimum of food, and once had to dig his schoolbooks out of the ruins of the house his family had been renting when it was demolished by a Japanese bomb. (His family, ensheltered, survived.) While still a student, Yang became fascinated with what is called the gauge invariance of the electromagnetic field—the implicit symmetry from which Einstein had deduced that the velocity of light is the same for all observers.

Yang hoped to identify a similar invariance for the strong nuclear force. A clue that such a thing was possible could be found in the fact that the strong force treats protons and neutrons identically, even though the proton has an electrical charge and the neutron does not—which is to say that the strong force is invariant under transformations of electrical charge. This symmetry, first noted in the 1930s, had been encoded as a quantum number called isospin. It was in this guise that the two particles came to be thought of as but varieties of a single kind of particle, the *nucleon*, their differences attributable to their differing isospin.

Yang tried for years to generalize gauge invariance by writing a suitably symmetrical equation for the strong force, and each time he failed. Yet the idea would not let him alone. His line of thought, though rather technical, can be depicted in terms of the relationship between a *global* symmetry, meaning a symmetry that applies everywhere, with a *local* symmetry, one that applies to a given system at a given place and time. Yang's question was this: How does the local system "know" about the global symmetry? How, in other words, is a global symmetry communicated to a local system?

To draw a simile from Yang's childhood, we could say in economic terms that the poverty imposed upon Yang's family and their neighbors by the war was a local invariance, embedded in the global invariance of the overall poverty of wartime China. Here the means of communication was the medium of exchange—money, or barter—that propagated the global conditions locally: Yang's father's life savings were wiped out by national wartime inflation because they were in a medium (currency in a bank account) that conveyed the general devaluation of the Chinese yuan. What, wondered Yang, might be the means of exchange in physics, the agency that connects local invariances with the wider invariances that form the skeleton of universal natural law?

The answer, Yang ultimately realized, was that the medium that communicates between local and global invariances is nothing other than force itself. This was a wholly new idea. Prior to Yang and Mills, force had been viewed as a given. Yang-Mills gauge theory gave force a raison d'être. It proposed that *symmetry* is the overweening principle, and that force is but nature's way of expressing global symmetries in local situations—that force, to speak teleologically, exists in order to maintain the invariances by virtue of which there is such a thing as natural law.

The paper that Yang and Mills originally wrote was limited and flawed and incomplete, and it did not yet fit the experimental results. But in time its problems were cleared up, and its potential beauty and power began to be recognized. Yang-Mills gauge theory offered a new approach to the practice of theoretical physics: What one could now do was to first identify an invariance, the signal of a symmetry; then construct, mathematically, a gauge field capable of maintaining that invariance locally; then derive the characteristics of the particles that would convey such a field; then go and see (or urge the experimenters to go see) whether any such particles actually exist in nature. Viewed from this perspective, Einstein's photons, the carriers of the electromagnetic force, are gauge particles, messengers of symmetry. So are the gravitons thought to be the carriers of gravitation. But what were the gauge particles of the strong and weak forces?

That question was taken up by one of the first to appreciate the beauty of the Yang-Mills approach, the American physicist Murray Gell-Mann. Some smart scientists (Dirac, Bohr, and the elder Einstein) are modest in demeanor. Others are brash. (Wolfgang Pauli disrupted Yang's first explication of gauge invariance so persistently that J. Robert Oppenheimer finally had to tell him to shut up and sit down.) Gell-Mann was very smart—he spoke more languages than his friends could keep count of, displayed an expert knowledge of everything from botany to Caucasian carpet-weaving, and was said, with forgivable exaggeration, to rank as a great physicist not because he had any particular aptitude for physics but simply because he deigned to include physics among his many interests—and very brash. From the popular assertion that he was the smartest man in the world Gell-Mann was not predisposed to demur; when he won the Nobel Prize he remarked, echoing Newton's comment that if he had see farther than others it was because he stood on the shoulders of giants, that if he, Gell-Mann, could see farther than others it was because he was surrounded by dwarfs. An intellectual wrestler who could stoop to bullying, he corrected strangers on the spellings and pronunciations of their own names, while himself pronouncing foreign terms with such an impeccable accent that he sometimes could not make himself understood.* If such habits tended to build a moat around Gell-Mann, perhaps, like Newton, he needed a moat.

Of Gell-Mann's scientific acumen and his love for nature there was no doubt, and when he put gauge field theory to work on the strong force the result was a symphony. Combining the Yang-Mills concept with group theory—a group is an ensemble of mathematical entities linked by a symmetry—Gell-Mann found a symmetrical arrangement of hadrons (particles that respond to the strong force) that he called "the eightfold way." (The Israeli physicist Yuval Ne'eman independently reached the same conclusion.) The eightfold way achieved experimental verification when a previously undetected baryon the existence of which it predicted, the omega minus, was subsequently identified in a bubble chamber experiment at Brookhaven.

The symmetry group involved was designated "SU(3)"—"SU" meaning "special unitary" group, one of a set of symmetry groups identified by the French mathematician Elie-Joseph Cartan, and "(3)" meaning that the symmetry operates in three-dimensional in-

*Richard Feynman, Gell-Mann's chief competitor for the title of World's Smartest Man but a stranger to pretension, once encountered Gell-Mann in the hall outside their offices at Caltech and asked him where he had been on a recent trip; "Moon-TRAY-*ALGH!*" Gell-Mann responded, in a French accent so thick that he sounded as if he were strangling. Feynman—who, like Gell-Mann, was born in New York City—had no idea what he was talking about. "Don't you think," he asked Gell-Mann, when at length he had ascertained that Gell-Mann was saying "Montreal," "that the purpose of language is communication?"

ternal space.* Investigating SU(3) further, Gell-Mann arrived at
the idea that protons and neutrons are each composed of triplets of
still smaller particles: Thus did quark theory spring from symmetry's forehead.

The Yang-Mills equations indicated that gluons, the gauge
particles that convey the strong force and so bind the quarks together inside nucleons, ought to be massless, as are photons and
gravitons. Why, then, does the strong force make itself felt only
over a short range, when light and gravitation are infinite in range?
The answer, according to quantum chromodynamics, the new theory of the strong force, is that the strong force increases in strength
when the quarks it imprisons try to move apart, rather than growing
weaker as do electromagnetism and gravity. This was the origin of
the concept of quark confinement and the gluon lattice that we
touched on in the last chapter. Quantum chromodynamics illuminated the workings of the weak force as well: The previously
mysterious phenomenon of radioactive beta decay could now be
interpreted as the conversion of a "down" quark to an "up" quark,
changing the neutron, which is made of two down quarks and one
up quark, into a proton, which consists of two up and one down
quark.

Symmetry, as we shall see, was to play an undiminished role
in the further development of quantum field theory, even pointing
the way toward a unified, "supersymmetric" theory that might
gather all particles and fields under the umbrella of a single set of
equations. "Nature," as Yang wrote,

> seems to take advantage of the simple mathematical represen-
> tation of the symmetry laws. The intrinsic elegance and beau-
> tiful perfection of the mathematical reasoning involved and the
> complexity and depth of the physical consequences are great
> sources of encouragement to physicists. One learns to hope

---

*Quantum interactions customarily are depicted as taking place not in the con-
ventional space that makes up the theater for macroscopic events, but in an as-
sessment complex space described in part by the quantum wave functions. Quarks,
for instance, are for convenience depicted as existing in a three-dimensional "color"
space described by quantum chromodynamics—color is a quantum number that
plays a role in the strong force analogous to that of the charge in
electrodynamics—while electrons normally occupy a one-dimensional space the
two directions of which represent positive and negative electrical charge.

that nature possesses an order that one may aspire to comprehend.[10]

But by no means are all nature's symmetries manifest. We live in an imperfect world, in which many of the symmetries that show up in the equations are found to be broken. Yang himself, working with Tsung Dao Lee, identified a discrete asymmetry in the weak force called parity violation. In 1956, Yang and Lee predicted, on theoretical grounds, that the spin of particles emerging from beta decay events would show a slight preference for one direction over another—i.e., that the weak force does not function symmetrically with regard to spin. Experiments conducted by Chien-Shiung Wu and others promptly confirmed their prediction, bringing the Nobel Prize the following year to Lee and Yang (though not, for some reason, to Wu) and turning renewed attention to the question of why nature is symmetric in some ways but asymmetric in others.

It was by investigating asymmetries that Steven Weinberg, Sheldon Glashow, and Abdus Salam formulated the unified electroweak theory that revealed a kinship between the weak and electromagnetic forces. Weinberg was intrigued by the fact that nature is replete with broken symmetries—asymmetrical relationships that have arisen from the functioning of symmetrical natural laws. The question, Weinberg observed, was how "symmetrical problems can have asymmetrical solutions."[11] Suppose that you take a handful of sharpened pencils, gather them into a perfectly cylindrical bundle, balance them on their points, and let go. For a moment, the arrangement remains rotationally symmetrical: Looking down from above you can walk around it, and all you will see is a circle made of the pencil erasers. But you'd better look quickly, for the symmetry is unstable: In an instant the pencils will fall, and the result will be an asymmetrical tangle like that encountered at the outset of a game of pickup sticks. In this simile, the jumble of fallen pencils is the universe today, and the original bundle is the symmetric state in which the universe is thought to have begun. The physicist's task is to identify the deeper symmetry hidden beneath the extant broken symmetry. This, indeed, could be the key to writing unified theories of ample scope. "Nothing in physics," Weinberg wrote in 1977, "seems so hopeful to me as the idea that it is possible for a theory to have a very high degree of symmetry which is hidden from us in ordinary life."[12]

The tangled world lines that led Weinberg, Glashow, and Salam to the triumph of the electroweak unified theory were themselves redolent with the tensions and broken symmetries that animate human affairs. Born in the Bronx in 1933, Weinberg attended the Bronx High School of Science, where his close friend was Shelly Glashow. The two went on together to Cornell, then parted when Weinberg went to Princeton and Glashow to Harvard. Aside from their common fascination with science and science fiction they were a study in oppositions, and the differences in their personalities were only magnified once they entered the adult world of theoretical physics. Weinberg was intensely curious, rigorously studious, and compulsively hardworking. He set himself to learning whole branches of physics, less because he saw in them any immediate application to the questions that most concerned him than because he felt that a physicist *ought* to know these things: Though primarily a particle physicist rather than a relativist, he once wrote a textbook on relativity, in part, he said, to help bridge the gap between general relativity and the theory of elementary particles. His self-discipline extended beyond physics: When he joined the faculty at the University of Texas and found that the furnished house he had rented in Austin came with a study full of books on the American Civil War, he simply read his way through them, emerging as something of an expert on the Civil War. Though painfully individualistic, he cultivated the art of communication, becoming an eloquent public speaker and the author of a best-selling popular science book, *The First Three Minutes.*

Glashow, on the other hand, was naturally gregarious, easygoing to the point of indolence, and a stranger to the rigors of study. If Weinberg excelled at Cornell, missing Phi Beta Kappa only because he failed physical education, Glashow barely scraped by; in accepting the Nobel Prize, he thanked "my high school friends Gary Feinberg and Steven Weinberg for making me learn too much too soon of what I might otherwise have never learned at all."[13] He spoke indistinctly, in fragmentary sentences built on an unimposing vocabulary, and smiled perpetually, as if contemplating a private joke. Physics seemed to come to him as naturally and effortlessly as a dream.

Glashow studied at Harvard under the elegant and venturesome Julian Schwinger, called "the Mozart of physics" both for his brilliance and for the uncaring way he wore it. A child prodigy, Schwinger as an adult remained impatient with the fragmented state of quantum physics, and he implored his students and colleagues never to rest until they had arrived at unified theories capable of describing a far wider scope of phenomena through fewer precepts. Even in the 1950s, when the quantum electrodynamics he had helped to create was the rising sun of quantum field theory, Schwinger was writing that

> a full understanding . . . can exist only when the theory of elementary particles has come to a stage of perfection that is presently unimaginable. . . . No final solution can be anticipated until physical science has met the heroic challenge to comprehend the structure of the sub-microscopic world.[14]

Glashow absorbed from Schwinger the conviction that the weak and electromagnetic interactions ought to be explicable by means of a single, unified gauge theory.* "A fully acceptable theory" of the two forces, Glashow wrote in his graduate thesis, echoing Schwinger, ". . . may only be achieved if they are treated together."[15]

His thesis completed, Glashow went to Copenhagen to study with Niels Bohr. There he pieced together a unified Yang-Mills theory of the weak and electromagnetic forces. The glaring problem with this theory, as would be the case for Weinberg and others later, was that its equations produced nonsensical infinities. Glashow tried to solve this problem by "renormalizing" his equations. Renormalization is a mathematical procedure that involves canceling the unwanted infinities by introducing other infinities; it smacks of mathematical trickery, but when adroitly manipulated can produce the desired, finite results. Among other credentials, renormalization had played an essential role in the perfection of quantum electrodynamics—which had made some of the most precise predictions ever confirmed by experiment, and had become a model

---

*Connections between the weak and electromagnetic interactions had been noted before; Fermi in 1933 formulated the first model of the weak force by analogy with electromagnetism. But a great many such threads weave their way through the history of physics, and this book is not the place to attempt to trace more than a few of them.

of what a quantum field theory ought to be.* By late 1958 Glashow was satisfied that he had renormalized his unified theory, and he presented a paper saying so the following spring, in London.

In the audience was the Pakistani physicist Abdus Salam, seven years Glashow's senior but seemingly older, a dignified, composed man in whom strong intellectual currents flowed beneath an exterior of oceanic calm. Born in 1926, the son of a high school English teacher who had prayed nightly to Allah for a son of intellectual brilliance, Salam at age fourteen scored the highest marks in the history of the Punjab University matriculation examination, a feat that brought cheering throngs out to greet him when he bicycled home to the little town of Jhang in what is now Pakistan. While working for his Ph.D., Salam managed to prove the renormalizability of quantum electrodynamics as applied to mesons, an accomplishment that garnered him a reputation as an expert on renormalization. Since then, he and a colleague, John Ward, had devoted considerable effort to the renormalization of a unified theory of the electromagnetic and weak interactions, without success. So when Glashow claimed that he had solved the problem, he got Salam's attention.

"My God! This young boy was claiming that this theory was renormalizable!" Salam recalled, in a 1984 interview with Robert Crease and Charles Mann.

> It cut me to the quick! Both of us considered ourselves *the* experts on renormalizability, wrestling for months with the problem—and here was this slip of a boy who claimed he had renormalized the whole thing! Naturally, I wanted to show he was wrong—which he was. He was completely wrong. As a consequence, I never read anything else by Glashow, which of course turned out to be a mistake.[16]

Glashow, however, was not easily discouraged, and despite any embarrassment he may have felt at having mistakenly claimed to have solved the renormalization problem, he persisted in searching for links between electromagnetism and the weak force. In this

---

*Compare, for instance, the value of g, the gyromagnetic ratio of the electron, as predicted by the theory of quantum electrodynamics and as tested experimentally:
Theory: $g = 1.00115965241$
Experiment: $g = 1.00115965238, \pm 0.00000000026$

effort he was encouraged by Gell-Mann if by few others. ("What you're doing is good," Gell-Mann recalled having told Glashow, over a seafood lunch in Paris, "but people will be very stupid about it.")[17] In 1961, Glashow produced a paper, "Partial-Symmetries of Weak Interactions,"[18] that called attention to "remarkable parallels" between electromagnetism and the weak force, depicted them as linked by a broken symmetry, and predicted the existence of the W and Z force-carrying particles—later known as the W^+, W^-, and Z^0. These hitherto undetected particles were to play an important role in experimental tests of the unified electroweak theory, but Glashow was unable to predict their masses, which left the experimenters with nothing to go on. Glashow and Gell-Mann then wrote a paper demonstrating that all the symmetries evinced in what are known as noncommutative or Cartan groups correspond to Yang-Mills gauge fields. Their efforts to identify a gauge symmetry group that would embrace both the strong force and Glashow's protounified electroweak forces, however, came to naught. Glashow, discouraged, set aside his work on electroweak unified theory.

Meanwhile, in 1959, Salam and Ward had, like Glashow, arrived at insights about links between the weak and electromagnetic forces, but had, like Glashow, met with an indifferent response from the scientific community, and likewise grew discouraged. "A broken symmetry breaks your heart," said Salam.[19]

The situation then brightened, thanks to new insights into the mechanism of spontaneous symmetry-breaking first presented by Yoichiro Nambu, Jeffrey Goldstone, and others and culminating in work published by Peter Higgs in 1964 and 1966. This research demonstrated that symmetry-breaking events could create new kinds of force-carrying particles, some of them massive. (The particles envisioned by Yang-Mills gauge theory had been massless.) If the particles that carry the weak and electromagnetic forces were related by a broken symmetry, these new tools might make it possible to estimate the masses of the W and Z particles characteristic of the unified, more symmetrical force from which the two forces were thought to have arisen.

Weinberg in particular was captivated by the concept of spontaneous symmetry breaking. "I fell in love with this idea," he said in his Nobel Prize address in 1979, "but as often happens with love affairs, at first I was rather confused about its implications."[20] Initially he tried to apply the new symmetry-breaking tools to the strong force. This worked well insofar as global symmetries were concerned—specifically, Weinberg found that he was able to make successful predictions of the scattering of pi mesons—but when he sought to extend the technique to local symmetries, the results were disappointing. "The theory as it was working out was making nonsensical predictions that didn't look like the strong interactions at all," Weinberg recalled in a 1985 interview. "I could fiddle with it and make it come out right, but then it looked too ugly to bear."[21] The worst problem was that the particle masses predicted by the breaking of the symmetry group Weinberg was contemplating did not match those of the particles involved in the strong interactions.

But then, in Weinberg's recollection, "at some point in the fall of 1967, I think while driving to my office at MIT, it occurred to me that I had been applying the right ideas to the wrong problem."[22] The particle descriptions that kept bobbing up out of his equations—one set massive, the other massless—resembled nothing in the strong force, but fit perfectly with the particles that carry the weak and electromagnetic forces. The massless particle was the photon, carrier of electromagnetism; the massive particles were the Ws and Zs. Moreover, Weinberg found, he could calculate the approximate masses of the Ws and Zs. Here, finally, was an electroweak theory that made a verifiable prediction. Salam independently reached a similar conclusion the following year—testimony, Weinberg said, to "the naturalness of the whole theory."[23]

With that, the work that would win the 1979 Nobel Prize in physics was complete. Yet little heed was paid to it at first. Weinberg's paper, the first complete statement of the electroweak theory, was cited not once in the scientific literature for four full years after it appeared. The main reason was that the theory had not yet been shown to be renormalizable. Once that was accomplished—in 1971, when its dolorous infinities were scotched in a heroic effort by the Dutch physicist Gerard 't Hooft—interest in the electroweak theory intensified, and the focus of attention turned to the question of testing the theory through experiment. This called upon those embodiments of big science, the particle accelerators.

Accelerators are to particle physics what telescopes and spectrographs are to astrophysics—both an exploratory tool for finding new things and a supreme court for testing existing theories. Their operating principle is based on Einstein's $E = mc^2$. One accelerates charged particles to nearly the speed of light by propelling them along an electromagnetic wavefront created by pulsing electromagnets, then smashes them into a target, creating tiny explosions of intense power. New particles condense from the tiny fireball, like raindrops precipitating in a storm cloud, and are recorded by surrounding detectors as they come reeling out. The original detectors were photographic plates; later these were replaced by electronic sensors coupled to computers.

Engaged in the race to test the electroweak theory were researchers at two of the world's most powerful accelerators—CERN, the European center for nuclear research near Geneva, and Fermilab, named after the physicist Enrico Fermi, on the Illinois plains west of Chicago. Both are proton accelerators.* The protons come from a little bottle of hydrogen gas, small enough for a backpacker to carry, that contains a year's supply of atoms. Computer-controlled valves release the gas in tiny puffs, each scantier than a baby's sigh but each containing more protons than there are stars in the Milky Way galaxy. The gas enters the electrically charged cavity of what is called a Cockroft-Walton generator.† The field strips the electrons away from the hydrogen atoms and sends the protons speeding down a tunnel and into a pipe the size of a garden hose that describes an enormous circle—three miles in circumference in the case of Fermilab. The protons are accelerated around the ring by pulses sent through surrounding electromagnets, while focusing magnets gather them to a beam thinner than a pencil lead. When they reach a velocity approaching that of light—at which point, thanks to special relativity effects, their mass has increased by some three hundred times—they are diverted from the ring and slammed into a stationary target inside a detector. Their tracks, subjected to yet another magnetic field in the detector, betray their charge and mass and thus their identity.

Though similar in design, the CERN and Fermilab accelerators exemplified rather different styles of doing big science. Fer-

---

*Electrons, since they also carry an electrical charge, can also be employed; the resulting explosions are cleaner and therefore easier to study, but as electrons are less massive than protons they collide less violently, and so electron accelerators yield weaker collisions relative to their energy consumption.
†After Ernest Walton, an Irish physicist, and John Cockroft, the English physicist who on one fine day in 1932 could be seen stopping strangers on the streets of Cambridge and exclaiming, "We have split the atom! We have split the atom!"

milab, built under the direction of the American physicist and sculptor Robert Wilson, was conceived and executed as a work of art, an embodiment of the aesthetics of science. A Wilson sculpture, a looming set of steel arches titled "Broken Symmetry," was erected at the main entrance. The accelerator tunnel, buried underground, was delineated, for purely aesthetic purposes, by an earthwork berm. Within the ring buffalo grazed; swans swam in the waters employed to cool the electromagnets. The administration building, a sweeping, convex tower, was set against the berm like a diamond on an engagement ring; Wilson modeled it on the proportions of Beauvais Cathedral in France. As he recalled his reasons for this decision:

> I found a striking similarity between the tight community of cathedral builders and the community of accelerator builders. Both of them were daring innovators, both were fiercely competitive on national lines, but yet both were basically internationalists. . . . They recognized themselves as technically oriented; one of their slogans was *Ars sine scientia nihil est!*— art without science is nothing.[24]

Wilson defended the aesthetics of his creation—which, it should be added, was completed under budget—by drawing further parallels between art and science:

> The way that science describes nature is based on aesthetic decisions. Physics is very close to art in the sense that when you examine nature on a small scale, you see a diversity in nature, you see symmetries in nature, you see forms in nature that are just utterly delightful. Eventually, in the way that one looks at sculpture or art, people will also begin to look at those great simple facts.[25]

CERN, for its part, looked about as aesthetically unified as a Bolshevik boiler factory. *Its* administration building, slapped together from prefabricated plastic panels and aluminum alloy window frames that bled pepper-gray corrosives in the rain, called to mind less Beauvais Cathedral than the public housing projects of suburban Gorky. Its laboratories were scattered across the landscape, as haphazardly as the debris from a trucking accident, on a plot of land that straddled the French-Swiss border outside Geneva.

The prevailing style was late Tower of Babel, with scientists switching from French to German to English in mid-sentence while lunching at the laboratory cafeterias, one of which accepted only French currency and the other only Swiss. Yet for all its air of disorder, CERN worked every bit as well as Fermilab, and by the early 1970s was beginning to surpass it.

It was in this fevered context that the two laboratories raced to test the predictions of the electroweak theory. The new force-carrying particles postulated by the electroweak theory, the W^+ W^-, and Z^0, were massive, meaning that it would take a lot of energy to bring them into existence in an accelerator collision. In 1971, no accelerator could yet summon up sufficient energy to create W and Z particles, if they existed. In the meantime, however, the experimentalists could hope to discern the existence of the Z indirectly, by identifying the effects, in accelerator collisions, of "neutral currents." This consisted of searching through thousands of accelerator events for evidence of the few neutral current interactions in which the Z^0 would have played a role. Encouraged by Weinberg's estimation that such events "are just on the edge of observability," a team working under the experimental physicist Paul Musset at the CERN accelerator began staying up nights, examining thousands of photographs of particle interactions. After a year's work they were finally rewarded when the myopic Musset, who scrutinized the particle tracks with his nose almost touching the print, discerned a kink in the recorded path of a particle that gave away its identity as a pion rather than a muon, indicating that it had emerged from a neutral current reaction. Salam learned of the result shortly after arriving at Aix-en-Provence, where he was to attend a physics conference. He was lugging his suitcase to a student hostel near the train station when a car stopped next to him. Musset looked out and said, "Are you Salam?" Salam said yes. "Get into the car," Musset said. "I have news for you. We have found neutral currents."[26]

This was welcome news to Salam, Glashow, and Weinberg, but it nevertheless fell short of fully vindicating the electroweak theory, for other theories also predicted the existence of neutral currents. The Weinberg-Salam theory surpassed its predecessors in predicting the mass of the carriers of the electroweak force—about 80 GeV for the Ws and 90 GeV for the Z. (A GeV is one billion electron volts; in this context, it is convenient to express

mass in terms of energy.) The Ws and Zs were known collectively
as intermediate vector bosons. To produce enough intermediate
vector bosons to make their detection likely would require a particle
accelerator with a minimum energy of some 500 to 1,000 GeV.

Neither accelerator could reach this level, but both were hur-
riedly being souped up to approach it, by means of a daring new
technique involving the collision of protons, not with a stationary
target, but with an oncoming stream of antiprotons. The universe,
so far as we can tell, contains only trace amounts of antimatter—
this in itself is one of nature's more intriguing broken symmetries
—but antimatter can be created in accelerator collisions, and by
the 1970s accelerator engineers were beginning to talk of collecting
the antiprotons they created and then colliding them with protons
coming the other way. Since matter and antimatter particles annihilate each other when they meet, the result would be to greatly
boost the effective power of the accelerator.

Fermilab approached the problem methodically. They would
first install new magnets to increase the power of the accelerator to
1,000 GeV (equal to one teraelectron volt, or one TeV), and only
thereafter take on the more hazardous business of trying to make
and store antimatter. CERN proceeded in a more intrepid fashion,
going for a matter-antimatter collider right away. Wilson, with his
customary gentility, wished them well: "May they reach meaningful luminosity and may they find the elusive intermediate boson,"
he wrote. "We will exult with them if they do."[27] CERN officials,
with equal courtliness, described the Fermilab plan as "a project
of great vision being attacked with courage and enthusiasm."[28] But
behind the pleasantries raged a fierce competition between rival
teams of the world's brightest and most egocentric scientists and
engineers.

Of these, few were brighter—and none more egocentric—than
Carlo Rubbia, the driving force behind the CERN effort. Born in
northern Italy in 1934 of Austrian parents, Rubbia was by nature
an internationalist ("I have an accent in every language," he said)
who felt at home cajoling and browbeating the scores of scientists
who made up his enormous research teams, among them Italian,
French, English, German, and Chinese researchers, a Finn, a
Welshman, and a Sicilian. A driven man, Rubbia traveled ceaselessly, flying from CERN to Harvard to Berkeley to Fermilab to
Rome so incessantly that friends who monitored his progress cal-

culated that he had a lifetime average velocity of over forty miles per hour. ("Ah," he said, settling into his seat one morning, "my first flight of the day!") Massive and energetic and constantly in motion, he resembled nothing so much as a human proton: Like Rutherford, who told his tailor, "Every year I grow in girth, *and* in mentality," Rubbia ballooned in size until, by 1984, he was boasting that his form now approached the perfection of a Platonic sphere.

Rubbia's hopes of winning a Nobel Prize rested on a conception concocted by an austere CERN engineer named Simon van der Meer. Van der Meer was convinced that one could make antiprotons (albeit at a rate of only one hundred-billionth of a gram of them per day) and keep them in storage until enough had accumulated to collide them with protons in significant numbers. Storing anti-

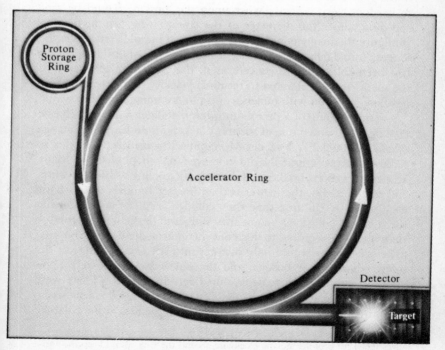

An accelerator speeds subatomic particles—in this case, protons—around a ring, then diverts them to a target inside a detector.

matter would be a tricky business, akin to the old conundrum of how to bottle a universal solvent: If an antiproton made contact with a particle of ordinary matter, both would instantly annihilate. Van der Meer proposed to handle the problem by constructing an antiproton accumulator, a small ring in which the antiprotons could be kept circling for days, suspended in a vacuum in an electromagnetic field. To keep the antiprotons concentrated in tight, secure bundles, Van der Meer proposed a technique called stochastic cooling—stochastic meaning statistical, and cooling meaning reducing random motions among the particles. As little clumps of antiprotons whirled around the storage ring, sensors would detect the drift of those that strayed, and computers would then send a correcting message across the ring to adjust the magnets on the opposite side to correct for the drift. Since the antiprotons were moving at close to the speed of light, the computation would have to be done very quickly, whereupon the message would be sent speeding across the diameter of the storage ring just in time to configure the magnets before the antiproton bundle arrived via the longer, roundabout route. Once a sufficient supply of antiprotons had been collected and concentrated, they could be released into the main ring, accelerated to terminal velocity, and steered into a headlong collision with bunches of protons coming the other way.*

   The building of a proton-antiproton collider fed via stochastic cooling represented one of the most audacious endeavors in an age of high technology. Van der Meer himself considered the idea so radical that he originally did not even try to publish it. Many accelerator experts predicted that stochastic cooling would not work, and that if it did, the matter and antimatter bundles would blow each other up the first time they collided, rather than producing the repeated collisions—some fifty thousand of them per second— that would be required to flush out the intermediate vector bosons. (The accelerator would only just get into the energy range of the intermediate vector bosons, and the physicists had to rely upon quantum probabilities to deliver up a detection event.) There were snickers in the audience when Rubbia first proposed constructing an antimatter collider; when he brought up the idea at Fermilab he

---

*Since antiprotons have opposite electrical charge, the same sequence of magnetic pulses that kept protons moving clockwise around the ring would keep the antiprotons moving counterclockwise. A somewhat more exotic way of looking at the situation, proposed by Feynman years earlier, was to say that antimatter particles move in reverse time.

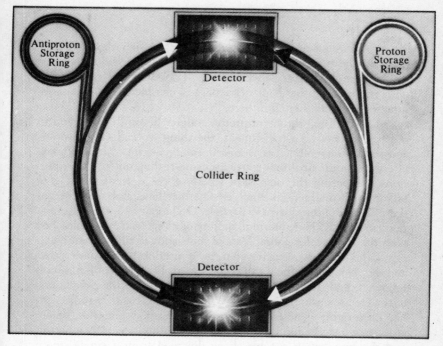

A collider sends particles of matter speeding in one direction and particles of antimatter in the opposite direction, smashing them into one another at detector sites located where the beams intersect.

was invited to leave; and when he and two colleagues submitted a paper on it, the editors of the *Physical Review Letters*, a leading journal, refused to publish it. But Rubbia kept pushing, despite the high stakes—one hundred million dollars to build the antimatter accumulator and to modify the accelerator, plus another thirty million dollars to build the detectors—and he made it a habit to project an air of robust assurance, keeping his reservations to himself. "Let's be serious," he said later. "If we had spelled out these doubts before the project was launched, nobody would have given us the money for it. . . . I was scared stiff the beam wouldn't work."[29]

In the end, CERN took the gamble. For three years, while the antiproton ring was being constructed, Rubbia busied himself building the detector, an instrument with the bulk and weight of a Wall Street bank vault, ten meters long by five meters wide and weighing two thousand tons, buried underground and straddling the accelerator tunnel. He worked himself to new depths of exhaustion and twice was nearly electrocuted, but he seldom faltered and he kept learning as he went along. "Look at this place," he said with pride, once the giant detector was completed. "I know the function of every switch in here."[30]

Tests of the proton-antiproton collider began in 1982, and to nearly everybody's astonishment, the thing worked. The protons and antiprotons collided as promised, producing tiny, intense bursts of energy, and subatomic particles came reeling out of the explosions, peppering the onionskin layers of the detector. Out of a billion such interactions emerged five that held clear evidence of the existence of the elusive W particle. On January 20, 1983, Rubbia stood in the CERN auditorium, in front of a long blackboard bleached with the technicolor palimpsests of thousands of rubbed-out equations, and told his colleagues that the W particle had been detected and the electroweak theory thus confirmed. Detection of the Z soon followed, and the masses of both bosons matched the predictions of the electroweak unified theory. Weinberg, Glashow, and Salam had been right; we live in a universe of broken symmetries, where at least two of the fundamental forces of nature, electromagnetism and the weak nuclear force, have diverged from a single, more symmetrical parent.

The battle of the big accelerators continued in the years that followed. Enormous boring machines toiled in Rembrandtesque gloom beneath the French countryside, digging a tunnel seventeen miles in circumference for a CERN accelerator that would collide electrons with their antimatter opposite numbers, the positrons. Proton accelerators continued to grow as well. The original CERN proton-antiproton machine had achieved an energy of 640 GeV; in America, Fermilab's proton-antiproton collider, which went into operation in 1985, soon was climbing toward a luminosity of over 1 TeV. Two years thereafter, the United States began planning a "superconducting super collider" that would attain energies of 20 TeV, flushing out particles forty times more massive than any previously detectable. With a ring fifty miles or more in circumference, the super collider would be the largest machine ever constructed.

The theorists, meanwhile, kept sifting through the particle zoo in search of further hidden symmetries. A number of grandly titled "grand unified" theories (GUTs for short) were written that purported to identify the electroweak and strong nuclear forces as partners in a single, broken gauge symmetry group. The GUTs made a curious prediction: They implied that the proton, always assumed to be stable, instead decays. Its half-life was estimated at some 10^{32} years. That's a long time—a thousand billion times the age of the universe—but the prediction could be tested by keeping watch on 10^{32} protons, one of which ought then to decay each year on the average. To test the grand unified theories, protons accordingly were gathered together, in the form of thousands of tons of filtered water in a tank in a salt mine near Cleveland and in a lead mine in Kamioka, Japan, a thirty-five-ton block of concrete in an iron mine in Minnesota, sheets of iron in a gold mine in India, and stacks of steel bars adjacent to a highway tunnel under Mont Blanc. (The experiments were conducted deep underground to minimize contamination by cosmic rays.) Light-sensing devices were attached to computers programmed to record the telltale flash of light that would be produced by a spontaneously disintegrating proton.

It was a hard life, waiting for years on end in lead mines and salt mines. ("That's what they get for choosing to be *experimental* physicists," joked one hard-hearted theorist.) The results, moreover, were null, and as years went by and no proton was observed to decay, it became increasingly evident that the GUT theorists had picked the wrong broken symmetries. Meanwhile, looking for something to do while they waited, the experimentalists put their instruments to work detecting neutrinos, a few of which betrayed their presence by smashing into atoms in the vats of water and stacks of concrete and metal that had been assembled to look for proton decay. This came in handy in 1987, when a supernova blazed forth in the Large Magellanic Cloud and a wave of neutrinos was promptly detected at the Kamioka and Lake Erie proton-decay installations. The observation confirmed a theory (authored in part by Bethe, indefatigable student of stars) that supernovae generate enormous quantities of neutrinos, and gave birth to the new science of observational neutrino astronomy.

The waning of the grand unified theories went widely unmourned. The GUTs had lacked the sweeping simplicity that unified theories are supposed to be all about; like the standard model they were full of arbitrary parameters, and, of course, they left out gravity. What the theorists really wanted was a "superunified" theory that would identify symmetrical family relationships among all four forces.

Elements of just such a theory began to appear, first in the Soviet Union and then independently in the West, in the 1970s. Collectively called "supersymmetry," these new theories identified a symmetry linking bosons, the carriers of force, with fermions, the stuff of matter. Gravitation was drawn under the umbrella of the theory in 1976, a development that generated widespread excitement. And yet, by the early 1980s, supersymmetry had begun to stall. In itself it could not generate all the known quarks, leptons, and gauge particles, and it introduced even more unexplained terms than had grand unified theories and the standard model. Something was missing.

That something, a few young theorists proposed, was strings. Traditionally, elementary particles like the electron had been regarded as dimensionless points. In string theory, the particles are instead portrayed as extended objects, longer than they are wide —in short, as strings. They can be mistaken for infinitesimal objects because they are very small—only about one Planck length long, which is just about as small as anything can be. The prospect that particles are strings rather than points made an enormous difference, however, in the way their behavior was interpreted. Strings can vibrate, and the rate at which they vibrate, it turned out, can generate the properties of all known particles—and of an infinite variety of other particles as well. The bewildering diversity of the myriad particles was suddenly, if only potentially, unified, by a stroke as simple as a chord struck on Pythagoras's lyre: All, said the theory, are but differing harmonies of strings.

String theory proffered potential answers to some of the most troubling questions that had been confronting theorists concerned with unification. Why did prior versions of quantum field theory so often generate infinities that had to be "renormalized" away? Because they regarded the elementary particles as having zero dimension: This meant they could draw infinitesimally close together, in which case the energy level of the force being exchanged between them could rise to infinity. Since strings have length, the problem of infinities did not arise in string theory. Why are gravitons spin two and the other force-carrying particles spin one? Because, said the theory, a string can either be open, meaning it has two ends,

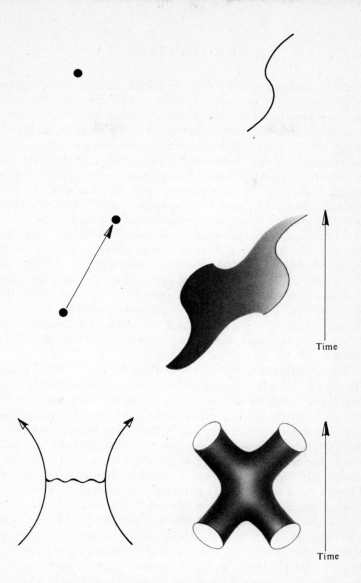

Subatomic particles, traditionally envisioned as points, are depicted in string theory as extended objects *(top)*. Particles in motion trace out world lines; strings, world sheets *(middle)*. A "Feynman diagram" of pointlike particle interactions consists of lines; for closed (i.e., looplike) strings, the Feynman diagram is tubular *(bottom)*.

or closed, meaning that the ends are joined, forming a loop: Open strings can be spin one, closed strings can be spin two. Why has the Yang-Mills gauge field concept enjoyed such broad applicability in understanding the forces? Because a string when in its lowest energy state—straight and nonrotating—acts like a massless, spin-one particle, and that is the description of the gauge particles that convey the Yang-Mills fields. String theory even opened a door toward understanding the conceptual gulf between relativity and quantum mechanics. Indeed, string theory could not work *without* including gravity. It was an inherently unifying conception.

The string concept originally was invoked in the 1960s, by theorists who had in mind larger strings whose harmonies might explain the behavior of the rapidly spinning hadrons. At this task it did not fare well, and most physicists soon dropped the idea. One of the few to appreciate its potential was (once again) the perspicacious Murray Gell-Mann, who encouraged the American physicist John Schwarz that even if string theory appeared sterile at present, "somehow, sometime, somewhere, it would still be useful."[31]

A breakthrough came in 1974, when Schwarz and the young French physicist Joël Scherk realized that an unwelcome particle that kept turning up in their string equations—its mass zero, its spin two—might be none other than the graviton, the boson that carries gravitation. Schwarz and Scherk then began thinking of strings as being only 10^{-35} meter long, the "Planck length" at which gravitation becomes as strong as the other forces and, therefore, presumably begins to function in an obviously quantized manner. Though these ideas initially garnered little enthusiasm in the scientific community, Schwarz stubbornly kept returning to the string concept, working on it in collaboration with Michael Green, who was visiting Caltech from the University of London. The concept was so unfashionable that Schwarz and Green apparently were the only two people in the world conducting research into strings at that time. But their efforts finally began to bear fruit, and in the summer of 1984 they were able to demonstrate that anomalies that had troubled other unified field theories canceled out in string theory. This captured attention, and by 1987 strings were the hottest topic in theoretical particle physics.

Writing a unified theory is something of an ad hoc affair, like

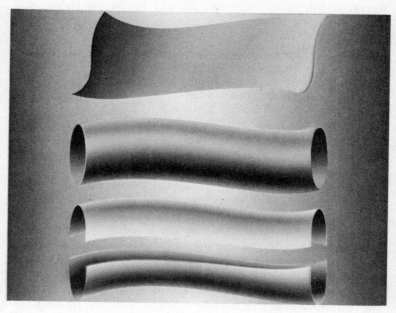

Unification of quantum mechanics and general relativity, long a conundrum, appears to be inherent to string theory. It implies that gravitation, explicated in relativity, is produced by open strings *(top)*, while the other, quantum forces are produced by closed strings *(middle)*. Cutting a closed string produces two open strings *(bottom)*, suggesting a natural affinity between the two classes of force.

putting up a tent in a high wind; while one sets the pegs, the tent flaps free. Einstein's relativity required abandoning classical conceptions of space and time; quantum mechanics required abridging classical causality. The odd thing about string theory was very odd indeed: It required that the universe have at least ten dimensions. As we live in a universe of only four dimensions (three of space plus one of time), the theory postulated that the other dimensions were "compactified," meaning that they had collapsed into structures so tiny that we do not notice them. Weinberg stumped for this idea, and was kidded about it when Howard Georgi, known for his work in grand unified theory, introduced a 1984 Weinberg lecture at Harvard by writing a limerick on the blackboard that read:

> Steve Weinberg, returning from Texas
> Brings dimensions galore to perplex us.
> But the extra ones all
> Are rolled up in a ball
> So tiny it never affects us.[32]

Hyperdimensionality had first been introduced into unified theory by Theodor Kaluza in Germany in 1919. Kaluza wrote to Einstein, proposing that Einstein's dream of finding a unified theory of gravitation and electromagnetism might be realized if he worked his equations in five-dimensional space-time. Einstein at first scoffed at the idea, but later reconsidered and helped Kaluza get his paper published. A few years after that, the Swedish physicist Oskar Klein published a quantum version of Kaluza's work. The resulting Kaluza-Klein theory seemed interesting, but nobody knew what to do with it until the 1970s, when it turned out to be salutary in working on supersymmetry. Soon Kaluza-Klein was on everyone's lips (with Gell-Mann, in his role as linguistic sentry, chiding colleagues who failed to pronounce it "Ka-*woo*-sah-Klein").

Though both string theory in particular and supersymmetry in general invoked higher dimensions, strings had a way of selecting their requisite dimensionality: String theory, it soon became apparent, would work only in two, ten, or twenty-six dimensions, and invoked only two possible symmetry groups—either $SO(32)$ or $E_8 \times E_8$. When a theory points a finger that decisively, scientists pay attention, and by the late 1980s scores were at work on strings. A great deal of toil lay ahead, but the prospects were bright. "The coming decades," wrote Schwarz and his superstring co-workers Green and Edward Witten, "are likely to be an exceptional period of intellectual adventure."[33]

Such optimism may, of course, prove to have been misplaced. The history of twentieth-century physics is strewn with the bleached bones of theories that were once thought to approach an ultimate answer. Einstein devoted much of the later half of his career to trying to find a unified field theory of gravitation and electro-magnetism, with popular expectations running so high that equations from his work in progress were posted in windows along New

York's Fifth Avenue, where they were scrutinized by curious if uncomprehending multitudes. Yet nothing came of it. (Einstein had ignored the quantum principle.) Wolfgang Pauli collaborated with Werner Heisenberg on a unified theory for a while, then was alarmed to hear Heisenberg claim on a radio broadcast that a unified Pauli-Heisenberg theory was close to completion, with only a few technical details remaining to be worked out. Put out by what he regarded as Heisenberg's hyperbole, Pauli sent George Gamow and other colleagues a page on which he had drawn a blank box. He captioned the drawing with the words, "This is to show the world that I can paint like Titian. Only technical details are missing."[34]

Critics of the superstring concept pointed out that claims for its power were based almost entirely upon its internal beauty. The theory had not yet so much as duplicated the achievements of the standard model, nor had it made a single prediction that could be tested by experiment. Supersymmetry did mandate that the universe ought to contain whole families of new particles, among them "selectrons" (supersymmetric counterparts of the electron) and "photinos" (counterparts of the photon), but it did not postulate the hypothetical particles' masses. The absence of evidence adduced in preliminary searches for supersymmetric particles, like those conducted at the PEP accelerator at Stanford and at PETRA in Hamburg, therefore proved nothing; one could always imagine that the particles were too massive to be produced in these machines, or indeed in any newer and more powerful machines that might be built. The prospects of conducting experiments to test string theory were even more remote: The putative strings themselves had a theoretical mass of more than 10^{21} times that accessible to existing accelerators, meaning that their detection, using existing technology, would require building an accelerator larger than the solar system. Supersymmetry and string theory were elegant, but if the theorists working on them had to proceed indefinitely without the benefit of what Weinberg called "that wonderful fertilization that we normally get from experiment,"[35] they seemed in danger of drifting away into the ionospheric reaches of pure, abstract thought. If that happened, argued Glashow and his Harvard colleague Paul Ginsparg, their tongues only slightly in cheek, "contemplation of superstrings may evolve into an activity as remote from conventional particle physics as particle physics is from chemistry, to be conducted at schools of divinity by future equivalents of medieval theologians." They added sardonically that "for the first time since the Dark Ages, we can see how our noble search may end, with faith replacing science once again."[36]

Nonetheless, hope continued to run high that there is a fundamentally beautiful, symmetrical principle to nature that has generated the particles and forces, and that it can perhaps be glimpsed by the human mind. "Maybe it isn't true," Weinberg allowed. "Maybe nature is fundamentally ugly, chaotic and complicated. But if it's like that, then I want out."[37]

Which brings us back to the other Greek definition of symmetry—"due proportion." To the Greeks, symmetry consisted, not simply of invariance, but of an aesthetically pleasing *kind* of invariance. This implies that there is a higher order of perfection, a more perfect world, that we glimpse through the windows proffered by symmetry and by which the elegance of any symmetry theory can be gauged. Supersymmetry portrays this ultimate perfection as a hyperdimensional universe, of which our poor imperfect universe is but a paltry shadow. It implies that physicists—in identifying, say, the weak and electromagnetic forces as having arisen from the breaking of the more symmetrical electroweak force, or in finding concealed symmetries cowering in the cramped nuclear precincts where the strong force does its work—are in effect piecing together the shattered potsherds of that perfect world. Indeed, the theory indicates that there may be countless more such debris, in the form of "shadow matter," supersymmetric particles that have as yet remained undetected because they interact only weakly or not at all with the particles we are made of and have come to know.

Where, then, is the hyperdimensional universe of perfect symmetry to be found? Certainly not here and now; the world we live in is fraught with broken symmetries, and knows but four dimensions. The answer comes from cosmology, which tells us that the supersymmetric universe, if it existed, belonged to the past. The implication is that the universe began in a state of symmetrical perfection, from which it evolved into the less symmetrical universe we live in. If so, the search for perfect symmetry amounts to a search for the secret of the origin of the universe, and the attention of its acolytes may with good reason turn, like the faces of flowers at dawn, toward the white light of cosmic genesis.

17

THE AXIS OF HISTORY

Every present state of a simple substance is
naturally a consequence of its preceding state,
in such a way that its present is big with its
future.

—Leibniz

He who has seen present things has seen all,
both everything which has taken place from
all eternity and everything which will be for
time without end; for all things are of one kin
and of one form.

—Marcus Aurelius

The late twentieth century may be remembered in
the history of science as the time when particle physics, the study
of the smallest structures in nature, joined forces with cosmology,
the study of the universe as a whole. Together these two disciplines
were to sketch the outlines of cosmic history, investigating the
ancestry of natural structures across an enormous range of scale,
from the nuclei of atoms to clusters of galaxies.

It was a shotgun wedding between two very different disciplines. Cosmologists tend to be loners, their gaze fixed on the far

horizons of space and time and their data tenderly garnered from trickles of ancient starlight; none will ever touch a star. Particle physicists, in contrast, are relatively gregarious—they have to be; not even an Einstein knows enough physics to do it all by himself—and *physical:* They are by tradition hands-on students of the here and now, inclined to bend things and blow up things and take things apart.* Physicists work hard and fast, haunted by the legend that they are unlikely to have many useful new ideas after the age of forty, while cosmologists are more often end-game players, devotees of the long view, who can expect to still be doing productive research when their hair turns white. If physicists are the foxes that Archilochus said know many things, cosmologists are more akin to the hedgehogs, who know one big thing.

Yet by the late 1970s, particle physicists were venturing to cosmology seminars to bone up on galaxies and quasars, while cosmologists were hiring on at CERN and Fermilab to do high-energy physics at underground installations blind to the stars. By 1985, Murray Gell-Mann could declare that "elementary particle physics and the study of the very early universe, the two most fundamental branches of natural science, have, essentially, merged."[1]

Their meeting ground was the big bang. As we saw in the previous chapter, the physicists identified symmetries in nature that today are broken but which would have been intact in a high-energy environment. From the cosmologists came word that the universe was once embroiled in just such a high-energy state, during the initial stages of the big bang. Put the two together, and a picture emerges of a more or less perfectly symmetrical universe that fractured its symmetries as it expanded and cooled, creating the particles of matter and energy that we find around us today and stamping them with evidence of their genealogy. Steven Weinberg, a champion of the new alliance, described the electroweak unified theory in terms of its connection with the early universe:

> The thing that's so special about the electroweak theory is that
> the [force-carrying] particles form a tightly knit family, with

*There are exceptions, of course, notably those mathematicians who come into physics with little or no grounding in experimental science. But generally speaking, the best theoretical physicists are willing, if only during their student days, to get their hands dirty in the laboratory. Recall that the young Einstein nearly *lost* a hand this way.

four members: There's the W^+, the oppositely charged W^-, the neutral Z, and the fourth member is our old friend the photon, the carrier of electromagnetism. These are siblings of each other, tightly related by a principle of symmetry that says that they're really all the same thing—but that the symmetry is broken. The symmetry is there, in the underlying equations of the theory, but it's not evident in the particles themselves. That's why the W and the Z are so much heavier than the photon.

But there was a time, in the very early universe, when the temperature was above a few hundred times the mass of the proton, when the symmetry hadn't yet been broken, and the weak and electromagnetic forces were all not only mathematically the same, but *actually* the same. A physicist living then, which is hard to imagine, would have seen no real distinction between the forces produced by the exchange of these four particles—the Ws, the Z, and the photon.[2]

Similarly, if less distinctly, the emerging supersymmetry theories suggested that all *four* forces may have been linked, by a symmetry that evidenced itself in the even higher energy levels that characterized the universe even earlier in the big bang.

The introduction of an axis of historical time into cosmology and particle physics benefited both camps. The physicists provided the cosmologists with a wide range of tools useful in trying to piece together how the early universe developed: Evidently the big bang was not the impenetrable wall of fire that Hoyle had scoffed at, but an arena of high-energy events that might very possibly be comprehensible in terms of relativistic quantum field theory. Cosmology, for its part, lent a tincture of historical reality to the unified theories. Though no conceivable accelerator could attain the titanic energies invoked by the grand unified and supersymmetry theories, these exotic ideas still might be tested, by investigating whether the particle constituency of the present-day universe accords with the sort of early history the theories imply. As Gell-Mann put it, "The elementary particles apparently provide the key to some of the fundamental mysteries in early cosmology. . . . and cosmology, it turns out, provides a sort of testing ground for some of the ideas of elementary particle physics."[3]

Viewed from this new, historical perspective, the proliferation of particle types that had been so discouraging to the physicists (prompting Fermi to muse that he should have been a botanist) began to look less like a burden than a boon. Once it became clear that every particle has arisen from a process of cosmic evolution, about which it can testify, one could regard the variety of particles as evidence of the richness of cosmic history. Physicists no longer needed to feel unhappy about the diversity of the particle world, any more than archaeologists would be disappointed if, say, while excavating the ruins of ancient Herculaneum they unearthed the foundations of an even older city beneath it. Instead, they could consider that nature is complicated and imperfect because it has a past—that, as the American physicist Thomas Gold remarked, things are as they are because they were as they were.

Indeed, one could discern signs of a direct relationship linking the size, binding energy, and age of nature's fundamental structures. A molecule is larger and easier to break apart than an atom; the same is true of an atom relative to an atomic nucleus, and of a nucleus relative to the quarks that comprise it. Cosmology suggests that this relationship results from the course of cosmic history— that the quarks were bound together first, in the extremely high energy of the early big bang, and that as the universe expanded and cooled the protons and neutrons made of quarks adhered to one another to form the nuclei of atoms, which thereafter attracted electrons to set up shop as complete atoms, which in turn linked up to form molecules.

If so, the more closely we examine nature the further we are peering back in time. Look at something familiar—the back of your hand, let us say—and imagine that you can turn up the magnification to any desired power. At a relatively low magnification you will discern individual cells in the skin, each looming as large and complex as a city, its boundaries delineated by the cell wall. Increase the magnification and you will see, within the cell, a tangle of meandering ribosomes and undulating mitochondria, spherical lysosomes and starburst centrioles—whole neighborhoods full of complex apparatus devoted to the respiratory, sanitary, and energy-producing functions that maintain the cell. Here, already, we encounter ample evidence of history: Though this particular cell is only a few years old, its architecture dates back more than a billion years, to the time when eucaryotic cells like this one first evolved on Earth.

To determine where the cell obtained the blueprint that told

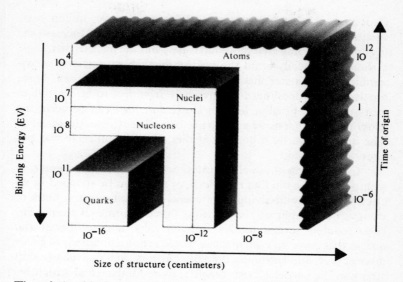

Size of structure (centimeters)

The relationship between the sizes of basic natural structures and their binding energies (i.e., the forces needed to tear them apart) is thought to reflect their origins at differing stages of cosmic history. Quarks, for example, are said to be smaller than nucleons (i.e., protons and neutrons), and to have higher binding energies, because they were formed earlier in cosmic time, when the universe itself was small and relatively energetic.

it how to form, move into the nucleus and behold the lanky contours of the DNA macromolecules secreted within its genes. Each holds a wealth of genetic information accumulated over the course of some four billion years of evolution. Stored in a nucleotide alphabet of four "letters"—made of sugar and phosphate molecules and replete with punctuation marks, reiterations to guard against error, and superfluities accumulated in blind alleys of evolutionary history—its message spells out just how to make a human being, from skin and bones to brain cells.

Turn up the magnification some more and you can see that the DNA molecule is composed of many atoms, their outer electron shells intertwined and festooned in a miraculous variety of shapes, from hourglasses to ascending coils like lanky springs to ellipses fat as shields and threads thin as cheroots. Some of these electrons are new arrivals, recently snatched away from neighboring atoms; others joined up with their atomic nuclei more than five billion years ago, in the nebula from which the earth was formed. Increase the magnification a hundred thousand times, and the nucleus of a single carbon atom swells to fill the field of view. Such nuclei were assembled inside a star that exploded long before the sun was born; the age of this one might be anywhere from five to fifteen billion years or more. Finally, looking closer still, one can perceive the trios of quarks that make up each proton and neutron in the nucleus. The quarks have been bound together since the universe was but a few seconds old.

In venturing to smaller scales, we have also been entering realms of higher binding energies. An atom can be stripped of its electron shell by applying only a few thousand electron volts of energy, but to split up the nucleons that constitute an atomic nucleus requires several million electron volts, and to liberate the quarks that make up each nucleon would require hundreds of times more energy still. Introduce the axis of history, and this relationship attests to the particles' past: Smaller, more fundamental structures are bound by higher levels of energy because the structures themselves were forged in the heat of the big bang.

This implies that accelerators, like telescopes, function as time machines. A telescope looks into the past by virtue of the time it takes light to travel between the stars; an accelerator re-creates, however fleetingly, conditions that pertained in the early universe. The 200 KeV accelerator devised in the 1920s by Cockroft and Walton replicated some of the events that transpired at about one day after the beginning of the big bang. Accelerators built in the 1940s and 1950s hovered at around the one-second mark. The Fermilab Tevatron pushed back the boundary to less than a billionth of a second after the beginning. The proposed superconducting super collider would provide a glimpse of the cosmic environment when the universe was less than one thousand billionth of a second old.

That's pretty early: One ten thousand billionth of a second takes a smaller slice out of a second than a snap of the fingers takes out of all recorded human history. And yet, oddly enough, research into the evolution of the newborn universe indicates that a great deal happened even earlier, during that first tiny fraction of a second. The theorists, accordingly, endeavored to piece together a coherent account of the first moments in cosmic history. Their ideas were of course sketchy and incomplete, and many of their conjectures will doubtless turn out to have been distorted or simply wrong, but they constituted a far more enlightening chronicle of the early universe than was available only a decade or so earlier, and hinted at the extraordinary beauty and explanatory power that could be expected from a more advanced theory once one could be worked out.

To review the story of cosmic history as depicted by the early-universe theories, imagine a staircase leading into the past—a stairway to heaven, if you will. We are standing at its base, in the present, at a time when the universe is some ten to twenty billion years old. (Most of the observational evidence suggests that the age of the universe is between fifteen and eighteen billion years.) The first step upward will take us back to when the universe was only one billion years old, and each step higher will turn back the clock to a tenth of its previous reading—to only a hundred million years after the beginning, then ten million years, then one million, and so on.

Suppose that we ascend this staircase. One step, and the date is one billion years after the beginning of time (or ABT for short). The universe looks quite different. The nucleus of the young Milky Way galaxy burns brilliantly, casting the shadows of galactic thunderheads out across the murky disk; at its core shines a bright, blue-white quasar. The disk, still in the process of formation, is jumbled and thick with dust and gas; it bisects a spherical halo that will be dim in our day, but currently wreathes the galaxy in a glittering chandelier of hot, first-generation stars. Our neighboring galaxies in the Virgo Supercluster float relatively nearby; the expansion of the universe has not yet had time to carry them away to the distances, typically tens of millions of light-years, at which we will encounter them in our own era. The universe is highly radioactive: Torrents of cosmic rays rain through us every millisecond, and if anything lives at this time, it probably mutates rapidly. Indeed, the pace of most events is hectic, an urban bustle compared to the relative placidity of our more mature epoch.

With the second step, we are plunged into darkness. We have reached a time, one hundred million years ABT, before any but the most precocious stars have yet had time to form. Except for

The history of the universe, depicted in terms of a stairway leading exponentially backward in time, displays the evolution of natural structures from quarks to atomic nuclei to atoms and galaxies of stars.

their scarce and smoky beacons, the universe is a dark soup of hydrogen and helium gas, whirlpooling here and there into protogalaxies.

In two more steps, the darkness is replaced by blinding white light. The time is one million years ABT, and the technical term for what has happened is photon decoupling. The ubiquitous cosmic gas has recently thinned sufficiently to permit light particles—photons—to travel for significant distances without colliding with particles of matter and being reabsorbed. (There are plenty of photons on hand, because the universe is rich in electrically charged particles, which generate electromagnetic energy, the quantum of which is the photon.) It is this great gush of light, much redshifted and thinned out by the subsequent expansion of the universe, that human beings billions of years hence will detect with radiotelescopes and will call the cosmic microwave background radiation.

This, the epoch of "let there be light," has a significant effect on the structure of matter. Electrons, relieved from constant harassment by the photons, are now free to settle into orbit around nuclei, forming hydrogen and helium atoms. With atoms on hand, chemistry can proceed, to lead, eons hence, to the formation of alcohol and formaldehyde in interstellar clouds and the building of biotic molecules in the oceans of the early earth.

The ambient temperature of the universe rises rapidly as we continue up the stairway. It was less than 3 degrees above absolute zero on the bottom step, reached room temperature by the third step, and by the sixth step has risen to 10,000 degrees Kelvin—hotter than the surface of the sun. By the eleventh step, at which point the universe is a little under one month old, the temperature everywhere surpasses that of the *center* of the sun, and at the fifteenth step (five minutes ABT) it is fully a billion degrees Kelvin.

Energetic as this may be, the universe at the age of five minutes has already become cool enough for nucleons to stick together to make permanent atomic nuclei. We watch as protons and neutrons adhere to make nuclei of deuterium (a form of hydrogen) and deuterium nuclei pair off to form the nuclei of helium (two protons and two neutrons). In this fashion, one quarter of all the matter in the universe is rapidly combined into helium nuclei—along with traces of deuterium, helium-3 (two protons, one neutron), and lithium. The whole process is over in three minutes twenty seconds.

Above this point—prior to about one minute forty seconds

ABT—there are no stable atomic nuclei. The ambient energy level exceeds the nuclear binding energy. Consequently, any nuclei that form are quickly torn apart again.

Between the seventeenth and eighteenth steps, at about one second ABT, we encounter the epoch of neutrino decoupling. Though the universe at this time is denser than rock (and as hot as the explosion of a hydrogen bomb) it has already begun to look vacuous to the neutrinos. Since neutrinos react only to the weak force, which is extremely short in range, they now find that they can escape its clutches and fly along indefinitely without experiencing any significant further interaction. Thus emancipated, they are free hereafter to roam the universe in their aloof way, flying through most matter as if it weren't there. (Ten million trillion neutrinos will speed harmlessly through your brain and body in the time it takes to read this sentence. By the time you have read *this* sentence, they will be farther away than the moon.) The flood of neutrinos released at one second ABT therefore persists even after, forming a cosmic *neutrino* background radiation comparable to the microwave background radiation produced by the decoupling of the photons. If these "cosmic" neutrinos (as they are called, to differentiate them from neutrinos released later on by supernovae) could be observed by a neutrino telescope of some sort, they would provide a direct view of the universe when it was only one second old.

As we climb on, the universe continues to become hotter and denser, and the level of structure that can exist becomes ever more rudimentary. There are of course no molecules or atoms or atomic nuclei at this early time, and by about the twenty-second step, some 10^{-6} (0.000001) second ABT, there are no protons or neutrons, either. The universe is an ocean of free quarks and other elementary particles.

If we take the trouble to count we will find that for every billion antiquarks there are a billion and one quarks. This asymmetry is important: The few excess quarks destined to survive the general quark-antiquark annihilation will form all the atoms of matter in the latter-day universe. The origin of the inequity is unknown; presumably it involved the breaking of a matter-antimatter symmetry at some earlier stage.

We are approaching a time when the basic structures of natural law, and not only those of the particles and fields whose behavior they dictate, were altered as the universe evolved. The first such transition comes at the twenty-seventh step, 10^{-11} second ABT, when the functions of the weak and electromagnetic forces are found to be handled by a single force, the electroweak. There is now enough ambient energy available to support the creation and maintenance of large numbers of W and Z bosons. These particles—the same kind the conjuring up of which in the CERN accelerator verified the electroweak theory—mediate electromagnetic and weak force interactions interchangeably, making the two forces indistinguishable. Prior to the twenty-seventh step the universe is ruled by only three forces—gravity, and the strong nuclear and electroweak interactions.

The next two dozen or so steps of our ascent are clouded in mystery. Some say that they traverse a "desert," a bleak stretch of time in which little of importance occurred. But it remains to be seen, given further accelerator experiments and the development of more sophisticated theories, whether the desert will prove to have bloomed.

According to the "inflationary universe" theory (about which more in the next chapter) there may here have been a brief period, upward of the fortieth step, during which the universe expanded much more rapidly than it did thereafter. During this inflationary epoch the universe would have been empty, all its latent matter and energy swallowed up by the rapidly expanding vacuum. There would be nothing to write home about (no material structure at all!) other than the vacuum itself, its unfolding fields pregnant with potential but devoid of tangible objects.

Prior to the start of the inflationary epoch—at about the fifty-first step, only 10^{-35} second ABT—we enter a realm in which cosmic conditions are even less well understood. If the grand unified theories are correct, there here occurred a symmetry-breaking event in which the unified electronuclear force split into the electroweak and strong forces. If supersymmetry theory is correct, the transition may have come earlier, and would have involved gravitation. Writing a fully unified theory amounts to trying to understand what went on at this early time, when the perfect symmetry thought originally to have characterized the universe shattered into the broken symmetries we find around us today.

But until we have such a theory, we cannot expect to understand what went on in the infant universe. We approach the limits of our present conjecture at the sixtieth step, when the age of the universe is but 10^{-43} second. Here we encounter a locked door. On the other side lies the Planck epoch, a time when the gravitational attraction exerted by each particle was comparable in strength to the strong nuclear force.* The theoretical key that could open the door would be a unified theory that includes gravitation. The person who arrives at that theory will gaze deepest into the dawn of time. What will he or she see?

One possibility, of course, is that there will be more doors. This prospect has been raised by several researchers, among them Michael Turner, an American cosmologist working on early-universe theory at Fermilab and the University of Chicago. "I suspect that we may always find ourselves in this position—that to go the next tiny fraction of a second we will need some further knowledge that we won't yet have," Turner suggested in a 1985 interview. "If so, it may be a very long time, if ever, before we can answer the question that everyone would like to know—the question of what caused creation."[4] Another possibility is that we will find the answer, behind the Planck door or the one after that. The conviction that such an outcome is possible was expressed this way, in 1985, by the American physicist John Archibald Wheeler:

> To my mind there must be, at the bottom of it all, not an equation, but an utterly simple idea. And to me that idea, when we finally discover it, will be so compelling, so inevitable, that we will say to one another, "Oh, how beautiful. How could it have been otherwise?"[5]

Suppose that a unified theory is written—a year from now or a century from now—that succeeds in delivering up just such a transcendental vision of perfection. How could we be sure that we could trust it? As Kepler realized after wasting years on his spherical universe of Platonic solids, one needs not only elegance from a theory but the verdict of experimental or observational test as well.

---

*At this point gravitons, the carriers of gravitational force, would have decoupled from the other particles, producing a gravitational background radiation much like those generated later by the decoupling of neutrinos and photons. The present-day temperature of the cosmic gravitational background radiation, however, is only 1 degree Kelvin, placing it far below the sensitivity of any conceivable gravitational detector. Still, it is there, and if we could find a way to observe it we could see all the way back to the Planck epoch.

A fully unified theory would in all likelihood purport to describe the universe as it was at less than 10^{-43} second ABT, when the ambient energy level was more than 10^{19} GeV. To re-create such conditions would require an accelerator a million trillion times more powerful than the proposed superconducting super collider—far beyond the reach of any foreseeable technology. Experimental verification of such a theory might remain forever out of reach.

The big bang itself, however, can be regarded as one gigantic accelerator experiment, and the universe we live in as its result. Viewed in this way, our microwave radiotelescopes are like Carlo Rubbia's detectors at CERN, inasmuch as the particles they intercept were hurled off by the first (and still the greatest) experimental run of all time. A proper unified theory ought to specify just how that run turned out, by predicting the existence of all the particles in the present-day universe. Some of these, presumably, would not yet have been detected: One could then test the theory by searching for such "relic" particles in the here and now. Supersymmetry theory, as we saw in the previous chapter, predicts the existence of enormous numbers of as yet undetected particles left over from the early universe—the so-called "shadow matter."* If the theory ripened to the point that it could specify the masses of these particles, it might be possible to test it by looking for them.

A ghostly clue that there may be such undetected material in the universe today is proffered by what astronomers call the "dark matter" problem. The masses of galaxies and their clusters can be deduced by measuring the velocity at which stars orbit the centers of the galaxies to which they belong, and at which galaxies orbit the centers of clusters of galaxies.† In case after case, this turns out

---

*String theory postulates that there exists and has existed only a single variety of particle, but that this particle has an infinite number of manifestations—as in the innumerable tunes that may be composed on a single string of Pythagoras's lyre. Thus a single supersymmetric variety of particle shows up in various harmonics as gravitons and gravitini, quarks and squarks, photons and photinos, and so forth. Since, as Gell-Mann noted, "these infinitely many particles all obey a single very beautiful master equation," the theory suggests how maximum complexity could have arisen from maximum simplicity.

†Recall that, as Newton found, the gravitational force of any object may be regarded as emanating from a point at its center. Each star in a galaxy responds to the total gravitation of the mass of the galaxy that lies within its orbit, as if the gravity were coming from a point source at the galactic center. The orbital velocity of a star lying near the edge of a galaxy therefore constitutes an index of the total mass of the galaxy.

to add up to something like five or ten times the mass of all the visible stars and nebulae. The startling implication is that everything we see and photograph in the sky amounts to only a fraction of the gravitationally interacting matter in our quarter of the universe. The unseen matter might, of course, consist of relatively large objects, such as brown dwarf stars or small black holes. But it might also consist of subatomic particles, many of them left over from the high-energy days of the early universe, in which case the identity of the particles would provide an observational test of supersymmetry or of any comparable unified theory of the early universe.

While awaiting the wished-for apotheosis of supersymmetry theory, we may care to reflect on the role played by symmetry in cosmic history. In doing so we soon confront the realization that perfect symmetry, though beautiful in the abstract, is also sterile. If, for instance, the matter-antimatter symmetry thought to have existed at the outset of cosmic evolution had been preserved, the particles of matter and those of antimatter would have mutually annihilated in the big bang, and no matter of either kind would have survived from which to make stars and planets and people. Had the putative primal force not scattered into the four forces, the universe today would be very different, perhaps uninhabitably so. It just may be, then, that we owe our existence, and that of the stars in the sky, to *im*perfections born of broken symmetry. To investigate the riddle of creation would then involve envisioning a perfectly symmetrical but unlivable universe, then trying to determine how it devolved from that sterile, pristine state toward becoming the less perfect but more variegated and hospitable universe in which we find ourselves today.

18

THE ORIGIN
OF THE
UNIVERSE

Where wast thou when I laid the foundations
of the earth? Declare if thou hast understanding!

— Yahweh, to Job

Who really knows?

— Rig-Veda

Speculation about the origin of the universe is an
old and notorious human activity. Old, I suppose, because there
is no birth certificate for the human species: We are obliged to
investigate our origins on our own, and in doing so have found it
necessary to ponder as well the derivation of the wider world of
which we are a part. Notorious, because the cosmogonic speculations that resulted told us more about ourselves than about the
universe they claimed to describe: All, to some extent, were psychological projections, patterns cast outward from the mind onto
the sky, like dancing shadows from a jack-o'-lantern.

Prescientific creation myths depended for their survival less
on their accordance with the data of observation (of which there
was in any event very little) than on the extent to which they were
satisfying or reassuring or poetically resonant. Cherished insofar as
they were our own, these tales emphasized what mattered most to

the societies that preserved them. The Sumerians, living at a con-
fluence of rivers, envisioned creation as having resulted from what
amounted to a mud-wrestling match among the gods. (From a clod
thrown off, the earth congealed.) The Mayans, obsessed with ball
playing, conjectured that their creator was transformed into a solar
kickball each time the planet Venus disappeared behind the sun.
Tahitian fisherman told of an angler god who tugged their islands
from the ocean floor; the Japanese sword-wielders formed *their* islands from drops of blood dripping from a cosmic blade. To the
logic-loving Greeks, creation was elemental: For Thales of Miletus,
the universe originally was water; for Anaximenes (also of Miletus),
air; for Heraclitus, fire. In the fecund Hawaiian Islands, genesis
was managed by a team of spirits skilled in embryology and child
development. African bushmen huddled around a fire watched the
sparks fly upward into the night sky and recited these words:

> The girl arose; she put her hands into the wood ashes; she
> threw up the wood ashes into the sky. She said, "The wood
> ashes must become the Milky Way. They must lie white along
> in the sky, that the stars may stand outside of the Milky Way,
> and the Milky Way be the Milky Way, while it used to be
> wood ashes."[1]

The advent of science and technology has brought about an
improvement in the sophistication of cosmogonic theorizing—relative at least to what preceded it, if not to the bald reality (if there
be such) of the great yawning cosmos (if it be a cosmos). But science
has by no means freed the creation question from its old entangle-
ment in human presuppositions and desires. The question of how
the universe began is at best elusive, and when we hunt after it,
our quivers bristling with quarks and leptons and curved space
tensors and quantum probabilities, we have an only marginally
better justification for our audacity than was enjoyed by Tahitian
visionaries who imagined that God might cast his fishing line and
catch not a fish but an emerald isle. Many scientists understood
this very well, and many, consequently, would have nothing to do
with cosmogony, the study of the origin of the universe. Some left
the matter alone simply because they could see no practical way of
approaching it. Others, adhering to the doctrine of causation, ban-

ished the issue of a first cause to exile in realms beyond science. As the astronomer Allan Sandage said:

> If there was a creation event, it had to have had a cause. This was Aquinas's whole question, one of the five ways he established the existence of God. If you can find the first effect, you have at least come close to the first cause, and if you find the first cause, that to him was God. What do astronomers say? As astronomers you can't say anything except that here is a miracle, what seems almost supernatural, an event which has come across the horizon into science, through the big bang. Can you go the other way, back outside the barrier and finally find the answer to the question of why is there something rather than nothing? No, you cannot, not within science. But it still remains an incredible mystery: Why is there something instead of nothing?[2]

Such reservations notwithstanding, a few scientists did attempt to investigate the question of how the universe might have originated, while admitting that their efforts were probably "premature," as Weinberg mildly put it. At its best, if viewed with an encouraging squint, their work appeared to shine a lamp into the anterooms of genesis. What they illuminated there was very strange, but this was, if anything, encouraging: We should hardly expect to find the familiar at the wellsprings of creation.

Two of their hypotheses—one called vacuum genesis, the other quantum genesis—seemed best to hint at what the near future might promise for human knowledge of the origin of the universe.

First, vacuum genesis. The central problem of cosmogony is to explain how something came from nothing. By "something" we mean the totality of matter and energy, space and time—the universe that we inhabit. The question of what is meant by "nothing," however, is more subtle. In classical science, "nothing" was a vacuum, the empty space that intervenes between particles of matter. But this conception always posed problems, as witness the long inquiry into whether space was filled with an aether, and in any event it did not long survive the coming of quantum physics.

The quantum vacuum is never really empty, but instead roils with "virtual" particles. Virtual particles may be thought of as representing the possibility, delineated by the Heisenberg indeterminacy principle, that a "real" particle will arrive at a given time and place: Like the pop-up silhouettes on a police firing range, they represent not only what is but what *might* be. As quantum physics sees it, every "real" particle is surrounded by a corona of virtual particles and antiparticles that bubble up out of the vacuum, interact with one another, and then vanish, having lived on borrowed, Heisenberg time. ("Created and annihilated, created and annihilated—what a waste of time," mused Richard Feynman.)[3] A free proton, say, is not alone in its travels, but is surrounded by a corona of virtual protons, the existence of which influences its behavior in ways that are not only observable but are, indeed, fundamental to the interactions of the proton as we know it. One example of the reality of virtual particles resides in the fact that the stars shine: To revisit the Coulomb barrier one last time, it is the structure of the virtual particle clouds surrounding protons that makes it possible for protons at the centers of stars to tunnel through one another's electrical fields often enough for nuclear fusion to be maintained.

The quantum vacuum, then, is a seething ocean, out of which virtual particles are constantly emerging and into which they constantly subside. And this is not merely an abstraction but a practical reality; as the American physicist Charles Misner notes:

> There is a billion dollar industry—the TV industry—which does nothing except produce in empty space potentialities for electrons, were they to be inserted there, to perform some motion. A vacuum so rich in marketable potentialities cannot properly be called a void; it is really an ether.[4]

The rules governing the brief existence of the virtual particles are set by the uncertainty principle and by the law of conservation of matter and energy. They state that the probable frequency with which virtual particles of a given mass can be produced, and the amount of time each can cavort before falling back into nonexistence, is determined by the energy potential of the vacuum. In a low-energy environment, massive particles like the W and Z bosons cannot borrow enough energy to exist in any quantity for any discernible interval: That is why we do not normally encounter these bosons in nature today, and why it was necessary to spend millions of dollars souping up the CERN accelerator until it could inject enough energy into the vacuum to make a few Ws appear and survive long enough to trigger Carlo Rubbia's detector. In the early universe, however, there would have been adequate ambient energy in the vacuum for the W and Z bosons to pop up all the time; this is the historical basis for the assertion of the electroweak theory that these bosons gamboled about in great numbers when the universe was young, managing the affairs of the unified electroweak force.

What has this to do with the origin of the universe? Perhaps little or nothing. Or perhaps, according to the vacuum genesis hypothesis, everything.

The protocols governing virtual particle production are tantalizingly open-ended, in that they place no absolute upper limit on the masses or lifetimes of the particles that can be created out of the vacuum. The known laws of science permit us to deduce the energy potential of the vacuum by observing the rate of particle production, but they set no ceiling on the energy that a given vacuum might contain. A vacuum that had looked quite unprepossessing might suddenly give birth to a particle as massive as a planet: Such an event is highly unlikely, but it is not impossible. Genesis, of course, can be quite unlikely—it need have happened but once—and it is through this keyhole that the vacuum genesis hypothesis entered the halls of science. Its thesis is that the entire universe originated as a single, extraordinarily massive virtual particle, one that sprang unbidden from a vacuum billions of years ago.

The first physicist to think of vacuum genesis was Edward Tryon. A modest messenger for so startling a hypothesis, Tryon had graduated Phi Beta Kappa from Cornell and had won his Ph.D. under Weinberg at Berkeley, but he was only an assistant professor at Columbia University, and his countenance did not seem destined to adorn any scientific Mount Rushmore. One afternoon during the fall 1969 academic semester, assistant professor Tryon was in the audience at a seminar being conducted by the luminary English cosmologist Dennis Sciama. As happens to everyone at times, Tryon drifted off into a reverie at one point during the talk. His thoughts wandered to the boiling quantum vacuum and the virtual particles that appear out of it. Suddenly he was seized by an idea, and was startled to hear himself interrupting Sciama's talk. "Maybe," he blurted out, "the *universe* is a vacuum fluctuation."

Tryon's colleagues laughed. They thought it was a joke. "It just cracked them up," Tryon recalled more than a decade later, still looking pained at the memory. "I was deeply embarrassed. . . . I never told them I'd not been joking."[5]

Humbled, Tryon put the idea out of his mind, but it came back to him in full force three years later, one evening while he was sitting quietly at home. "I had a revelation," he recalled, blushing. "I visualized the universe erupting out of nothing as a quantum fluctuation and I realized that it was possible and that it explained the critical density of the universe. I understood all those things in an instant, and a chill ran through my body."[6]

At the magnetic north of Tryon's speculation stood the realization that the overall energy content of the universe might well be zero. True, when one adds up the energy released by the big bang and by starlight, plus the frozen energy that we call matter that is bound up in the stars and planets, the total is an enormous positive sum. But there is also gravitation, which, since it is purely attractive, belongs on the minus side of the ledger. (Gravity was Tryon's specialty.) Interestingly, the gravitational potential of the earth or of any other object turns out to be approximately equal to its total energy content as calculated via $E = mc^2$. If this were true for the universe as a whole, then the universe would have *no* net positive energy, and could have emerged from a vacuum without violating the law of conservation of energy.

But is it true that the universe has zero net energy? The answer, Tryon realized, could be found in the rate at which cosmic expansion is slowing down. The universe is continuing to expand, owing to impetus generated by the big bang. The rate of expansion, however, is decreasing with time, owing to the mutual gravitational attraction exerted by the galaxies upon one another. The rate of slowing therefore reveals the overall mass density of the universe, a quantity the cosmologists symbolize by the Greek letter omega. If omega is equal to or less than 1, the mass density is insufficient to stop expansion, and the universe will go on expanding forever. Geometrically, such a universe is described as "open," meaning that the overall curvature of space is hyperbolic. If omega is more than 1, the expansion is destined eventually to stop, after which the universe presumably will collapse into another fireball. If omega is exactly 1, then expansion will continue forever, forever slowing but never quite coming to a halt.

Tryon's speculation required that omega be equal to or less than 1. Strangely, omega appears be exactly (or almost exactly) equal to 1. Indeed, the reason that observational cosmologists like Sandage and Tammann had been unable to determine conclusively whether the universe is open or closed was precisely because it is balanced at or close to an omega of value unity. Cosmic space, in other words, is neither dramatically open nor dramatically closed, but is perfectly—or almost perfectly—flat.

That it should be so is nothing short of astonishing. The gross features of the present-day universe are highly dependent upon tiny variations in the early universe—just as, say, a variation of millimeters in the angle at which a bat strikes a baseball can produce variations of hundreds of feet in where the ball lands in the outfield. In the standard big bang model, for the universe to be flat today it must have been incredibly flat at the beginning: At one second ABT, the cosmic matter density would have to have fallen within one trillionth of 1 percent of the critical value. At 10^{-35} second the permitted deviation would have been even smaller—less than one part in 10^{49}. If this happened by pure chance, it was very lucky indeed; the odds against it are vanishingly small.

One could of course make the equations come out right by inserting the required matter density as an "initial condition," but this amounted to invoking the guiding hand of God, which in science is rather like playing tennis without a net.* Alternately, one could "explain" the flatness of the universe by identifying it as a prerequisite of human existence. This argument, called the anthropic principle, went as follows: Were the cosmic matter density only slightly higher, the universe would have stopped expanding and have collapsed before enough time had elapsed for stars and planets and life to form; were it only slightly lower, the universe would have expanded too rapidly for stars and planets to have congealed from the rapidly thinning primordial gas. Therefore, the argument goes, the fact that we are here constrains certain cosmological parameters, among them the value of omega. The anthropic principle "explains" the miracle of the flat universe if we

---

*"Initial" conditions in cosmology are seldom absolutely initial, since nobody yet knows how to calculate the state of matter and space-time prior to the Planck time, which culminated at about 10^{-43} second ABT. One instead designates as "initial" some point subsequent to the Planck epoch. For most purposes this is regarded as quite initial enough.

imagine the creation of many universes, only a fraction of which chance to have the values requisite for life to appear in them. But the explanation cannot be tested unless the creation of other universes can be established, something that may well be impossible by definition. In that sense, the anthropic principle is a dead-end street. The English physicist Stephen Hawking, whose work is said to have contributed to the formulation of the principle, nonetheless called it "a counsel of despair."[7]

But where there is enigma there is also the promise of discovery: A paradox may signal an inadequacy in the way we are looking at a question, thereby suggesting a new and more fruitful way of approaching it. This, I think, is what Bohr meant when he exclaimed, "How wonderful that we have met with paradox. Now we have some hope of making progress."[8] And it was in this spirit that the flatness conundrum was resolved, by the invention of a new cosmological hypothesis, the inflationary universe.

The inflation hypothesis was first proposed by a young American physicist named Alan Guth. He learned of the flatness problem one November afternoon in 1978 at Cornell, in a talk by Robert Dicke, a resourceful Princeton relativist whose thoughts on the cosmic background radiation recalled those of Gamow. Trained as a physicist, Guth at the time knew little of cosmology, and, with the fierce conservatism of the young, dismissed ideas about the early evolution of the universe as "too speculative." Dicke's point about the oddity of omega equaling 1 struck Guth as "amazing," he recalled, but at the time he had no idea what to do about it.

The physics community, however, was at the time commencing its mating dance with cosmology, and Guth soon found himself working on the question of how magnetic monopoles might have been produced in the early universe. Guth found monopoles intriguing: First conceived in Dirac's austere imagination in 1931, they were purported to be massive particles with a unipolar magnetic charge. The grand unified theories indicated that they would have been created out of knots in space-time, by the same symmetry-breaking event that split the electroweak and strong nuclear forces asunder. Anachronistically, each magnetic monopole would harbor trapped W and Z bosons, as well as a tiny region at its core where the unified, electronuclear force still functioned.

The problem that engaged the attention of Guth, and of his Cornell colleague Henry Tye, was that the grand unified theories predicted the production of far too many magnetic monopoles—roughly one hundred times more monopoles than there are atoms. Given that most of the matter in the universe is invisible—the "dark matter" question—cosmologists generally welcome the suggestion that massive subatomic particles might make up the deficit, but this was an embarrassment of riches. Searches for monopoles had turned up null results: One event had been recorded, on Valentine's Day, 1982, on a device built by Blas Cabrera in a basement laboratory at Stanford, but Cabrera's result had never been repeated, at Stanford or anywhere else. This plus several other lines of inquiry suggested that the cosmic monopole population was either negligible or zero. The disagreement between theory, which predicted many monopoles, and observation, which permitted few, could be resolved, Guth and Tye found, if the fabric of space-time had been smoother than expected at the time of the grand unified phase transition. Smoother space-time meant fewer space-time knots, resulting in fewer monopoles. It also meant an omega equal or close to 1.

On the evening of December 6, 1979, Guth wrote the words EVOLUTION OF THE UNIVERSE atop a blank page that he then went on to fill with calculations. His hypothesis was that the universe initially had expanded much faster than at the linear rate it evinces today—that, as Guth would later put it, there had been an "inflationary epoch," during which the universe expanded exponentially. This meant that space was flatter and smoother by the time of the grand unified phase transition, and that far fewer monopoles therefore were produced. Here, too, was the solution to the flatness problem Dicke had outlined: Since the universe would have been much larger at the end of an inflationary period than was envisioned in the old, linear-expansion model, space would be much flatter—just as, say, an acre of the surface of the earth is flatter than is an acre of a spherical asteroid only ten miles in diameter.

SPECTACULAR REALIZATION, the young Guth wrote in his notebook the following day, drawing a box around the words. The hypothesis was not unprecedented; its revised picture of phase transitions had been arrived at independently by Katsuhiko Sato in Japan and Martin Einhorn in the United States, and the "pumping" of the expansion rate up to an exponential rate by a symmetry-breaking mechanism had been proposed by Demosthenes Kazanas of NASA. Nor did it work very well in its original form; it had to

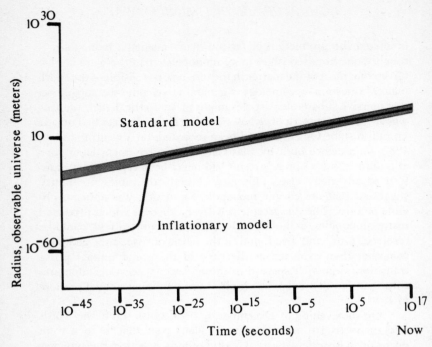

The inflationary model hypothesizes that the universe underwent a brief period of very rapid expansion, after which it settled into the linear expansion rate that has characterized it ever since.

be refined, by A. D. Linde in Moscow and by Andreas Albrecht and Paul Steinhardt at the University of Pennsylvania. But Guth came up with the idea on his own, and in its finished form it enlightened and illuminated the study of the very early universe.

According to the inflationary scenario, the radius of the universe increased by some 10^{50} times, from smaller than a proton to larger than a softball, during the first 10^{-30} second of time. During this brief but critical period the universe was but a vacuum. Its potential mass and energy could not yet manifest itself as particles, because space was expanding too fast for the particles to congeal out of the vacuum. Technically, one described this condition by saying that the vacuum was hung up in a symmetrical state during a phase transition. A Taoistic simile may be drawn from water. Liquid water is more symmetrical than ice, and the change in water when it cools from a liquid to a solid state marks a phase transition that breaks the symmetry. If liquid water is cooled very rapidly to below its freezing point it will not congeal into ice at once, but instead will linger in a liquid state for a while. Similarly, in the inflationary universe account, the cosmic vacuum remains empty even after falling below the temperature at which particle production ordinarily would take place. Indeed, it is this hang-up that drives the expansion: The latent energy is tied up in what is called a zero-value Higgs field, and the field acts as an engine that inflates the dimensions of cosmic space, driving the expansion so that the empty universe balloons in perfect, Platonic sphericity.*

Eventually (meaning after about 10^{-30} second) the quantum instability of the situation catches up with it, and the expansion abruptly slows to a linear rate. When that happens the energy latent in the vacuum precipitates out as particles and antiparticles. (Thus was new life lent to the much ridiculed steady-state picture of atoms congealing out of a vacuum.) The particles mutually annihilate, and the resulting flood of energy inaugurates the big bang. The grand unified theories, the composition of which requires attention to Higgs fields, even demonstrated how symmetry-breaking at the end of the inflationary epoch could have delivered up a small imbalance of matter over antimatter, leaving a residue, after the fireworks were over, from which to build the material universe.

Inflation resolved not only the flatness problem, revealing why omega is equal to or nearly unity, but also another major cosmological mystery, the *horizon* problem. The observable universe, taken as a whole, is remarkably homogeneous. In every star, in every direction, we find identical atoms functioning in accord with the same physical laws, and the cosmic background radiation, too, is everywhere the same. This, strange to say, had never been explained by the standard big bang model. The trouble was that the linearly expanding universe of the old model expanded too rapidly for all the quanta of the very early universe to have ever been in causal contact with one another: 90 percent of the universe in the old model lay beyond the causal horizon of any one observer, meaning that there was insufficient time for information, even if traveling at the speed of light, to permeate the universe.

*The sphere can be in many dimensions; that is another question, addressed by the supersymmetry theories, which did not as yet prescribe a timetable describing when the young universe allegedly collapsed its ten or so dimensions into three of space and one of time.

This omission mocked the universality of natural law. How could atoms and photons on one side of the universe behave exactly like atoms and photons on the other side, if they had never communicated with one another? To visualize the problem, imagine a marching band gathered on a greensward, ready to start playing as soon as the drummer standing at the center delivers a downbeat. At the moment that time begins, the band members march rapidly away from the drummer in all directions, at nearly the speed of sound. The result will be chaos. Only a few musicians will hear the downbeat; most will go hurrying away, unable to hear it, and so will not know when to start playing or what to play. In cosmological terms, the speed of sound is replaced by the speed of light, the fastest velocity at which information can be exchanged. The standard model required that the particles of the early universe depart before they could get their marching orders: Without hearing the drumbeat, then, how did the first quarks "know" how to be quarks, and all the photons learn the rules that govern photons? Had such been the true tale of genesis, nearly every cluster of galaxies would be made of different stuff and would obey different laws. Instead, the observable universe is a lawful unity. How so?

At first blush, inflation would seem only to make matters worse, since it postulates an even speedier cosmic expansion rate. But actually it resolves the dilemma, by permitting the material of the very early universe to remain together, in causal contact, for a relatively long period before inflation began. The band members now have time to listen for the downbeat before leaving; then they board the inflationary express, which goes so fast that they soon catch up with the linear expansion rate. Now they all have their marching orders when they go, and all, consequently, can play the same tune. Inflation thus explained why the cosmic background radiation is isotropic, and why the quarks and electrons of the earth are identical to those of the Coma cluster of galaxies.*

---

*Inflation theory indicates that the universe is many billions of times greater in volume than had been estimated in the old big bang model. The *observable* universe, however, is thought to constitute but a fraction of the universe as a whole: Its limits are determined less by space than by time, in that we can see only those events the light from which has had time to reach Earth. If, for instance, the first stars began to shine eighteen billion years ago, then no observer will see stars any farther than eighteen billion light-years away, regardless of how large the universe as a whole may be.

All of which was cheering to Ed Tryon and his little cadre of vacuum genesis enthusiasts. The inflationary hypothesis made vacuum genesis appear more plausible, by admitting the possibility that the universe could have started as a relatively modest, cold particle, with the heat of the big bang coming later, in the blast of fire released by latent vacuum energy when the inflationary epoch ended. And inflation painted the vacuum in new and more vibrant colors. Once one entertained the idea that all the matter and energy in the universe erupted from a vacuum at a brief but finite interval after the beginning of time, it no longer seemed quite so preposterous to imagine that the whole affair might have *begun* as a vacuum.

Guth, for his part, became an aficionado of the vacuum, regarding it less as emptiness than as a cornucopia. He calculated that only a small amount of vacuum flux might, if sufficiently concentrated, have been enough to set off inflation. If, then, our universe began as a quantum flux—a sort of bubble—in a primordial vacuum, other universes might reasonably be imagined to have formed from other bubbles. Moreover, Guth conjectured, creation need not necessarily be relegated solely to the past, but might happen again: If a vacuum instability in our universe were to blister in such a way as to form another universe, we would never know it. From our perspective, the only trace of the new creation event would be a pinpoint of infinite spatial curvature. As it happens, there appear to be such places here and there, in the infinitely curved regions of space surrounding black holes. Conceivably, every time a giant star goes supernova and its remnant collapses to form a black hole it might give birth to a new universe, on the other side of space and time.

If so, Guth speculated, the artificial creation of a black hole through application of an advanced technology could create another universe. Nor would such a custom-made black hole have to be terribly massive. "You might even be able to start a new universe using energy equivalent to just a few pounds of matter," Guth suggested, in a 1987 interview. "Provided you could find some way to compress it to a density of about 10^{75} grams per cubic centimeter, and provided you could trigger the thing, inflation would do the rest." And if *we* could do it, so, perhaps, could someone else have done it long ago. "For all we know," said Guth, who had a gift for the laconic statement of radical ideas, "our own universe may have started in someone's basement."[9]

If only to get our feet back on the ground, let it be noted that there were problems with both the inflationary universe and vacuum genesis. Inflation smoothed out the early universe, all right, but did so with such a vengeance that theorists had trouble coaxing enough lumpiness out of the equations to allow for the formation of galaxies and of the superclusters (and, evidently, meta-super-clusters) in which they are gathered. Vacuum genesis suffered from a lingering suspicion that, if anything, it was just not crazy enough. The quantum vacuum is a characteristic of the universe we live in—virtual particles *today* boil in the space between real particles —but who was to say that the same was true of the "vacuum" that allegedly preceded the beginning of the expansion of the universe? *That* vacuum, after all, ought to have been very different from the one we encounter in the present-day universe: Its relativistic curvature was infinite and its matter content presumably zero, and neither is true of cosmic space today.

Some theorists proposed, instead, a set of even stranger but at least equally promising hypotheses. Together, these ideas went by the name of "quantum genesis." Their approach involved taking the random nature of quantum flux to heart and enshrining it as the ruling law of the extremely early universe. Here a pioneer was Stephen Hawking, holder of Newton's old chair as Lucasian Professor of Mathematics at Cambridge University. Described by colleagues as "the nearest thing we have to a living Einstein," Hawking carried on a productive career in physics despite suffering from ALS, a fatal disease that attacks the central nervous system. He worked from a wheelchair, writing and communicating by means of a computer controlled by a toggle that he manipulated with one finger. He expressed impatience less with his affliction than with people who worshiped him as a hero, pitied him as a sick man, or otherwise treated him as if he were any different from any other genius. In his postdoctoral days Hawking and his colleague Roger Penrose demonstrated that general relativity implies that the universe began in a "singularity," a state of infinitely curved space in which the laws of relativity break down; this proved, as Hawking put it, that "relativity predicts its own downfall."[10]

But quantum theory might function where relativity did not, and in later years Hawking began to explore the prospect of understanding the origin of the universe in terms of quantum probabilities. His tools included "imaginary time"—a kind of time measured in terms of imaginary numbers—and Richard Feynman's "sum over histories" method of doing quantum mechanics.

Imaginary numbers make no sense when handled by customary mathematical rules. An example is the square root of -1, which will produce an "error" message if demanded of an electronic calculator. They work quite well, however, according to their own rules; imaginary numbers have been employed to excellent effect, for instance, in hydrodynamics. Feynman's "sum over histories" strategy consists of calculating all the possible past trajectories of a particle, and arriving, via quantum probabilities, at the most likely path by which the particle reached its observed state. Hawking, working with the American cosmologist James Hartle at the University of California at Santa Barbara, applied this method to the universe as a whole. Speaking via an interpreter, in a vaulted hardwood hall at Padua where Galileo used to lecture, Hawking announced that he had been able to derive the quantum wave function of the universe as a whole. "The universe today is accurately described by classical general relativity," he said.

> However, classical relativity predicts that there will be a singularity in the past, and near that singularity, the curvature [of space] will be very high, classical relativity will break down, and quantum effects will have to be taken into account. In order to understand the initial conditions of the universe, we have to turn to quantum mechanics, and the quantum state of the universe will determine the initial conditions for the classical universe. So today I want to make a proposal for the quantum state of the universe.[11]

What emerged was a tale of cosmic evolution possessed of a strangely alien beauty. All world lines diverge from the singularity of genesis, Hawking noted, like longitude lines proceeding from the north pole on a globe of the earth. As we travel along our world line we see the other lines moving away from us, as would an explorer sailing south along a given longitude; this is the expansion of the universe. Billions of years hence the expansion will halt and the universe will collapse, eventually to meld into another fireball at the end of time. There is, however, no meaning to the question

of when time began, or when it will end: "If the suggestion that
space-time is finite but unbounded is correct," said Hawking on
another occasion, "the big bang is rather like the North Pole of the
earth. To ask what happens before the big bang is a bit like asking
what happens on the surface of the earth one mile north of the
North Pole. It's a meaningless question."[12]

Imaginary time in Hawking's view was the once and future
time, and time as we know it but the broken-symmetry shadow of
that original time. When a hand calculator cries "error" upon being
asked the value of the square root of -1, it is telling us, in its way,
that it belongs to *this* universe, and knows not how to inquire into
the universe as it was prior to the moment of genesis. And that
is the state of all science, until we have the tools in hand to explore
the very different regime that pertained when time began.

Another quantum approach to genesis, championed by John
Wheeler, emphasized the quantization of space itself. Just as matter
and energy are made of quanta, went this line of reasoning, so space
itself ought to be quantized at its foundations. Wheeler liked to
compare quantum space to the sea: Viewed from orbit, the surface
of the ocean looks smooth, but if we set out in a rowboat on the
surface, "we see foam and froth and breaking waves. And that foam
and froth is how we picture the structure of space down at the very
smallest scales."[13]

In the present-day universe, the foamy structure of space manifests itself in the constant blooming forth of virtual particles. In
the extremely early universe—meaning prior to the Planck time—
space would have been a very rough sea indeed, and its storm-
tossed quantum flux might have dominated all particle interactions.
How, here, do we find our bearings?

Wheeler—an elder statesman who learned his science from
Einstein and Bohr and in turn educated a whole generation of
physicists—thought the answer lay in space-time geometry: "What
else is there out of which to build a particle except geometry itself?"[14] he asked. Wheeler compared the quantum flux of the early
universe to a complicated sailor's knot of a kind that looks impos-
sibly tangled, yet will fall apart if one can find the end of the rope
and give it a tug in the right way. The knot in his simile is the
hyperdimensional geometry of the original universe, the untangled
rope the universe we inhabit today. Penrose had said, "I do not
believe that a real understanding of the nature of elementary par-

COMING OF AGE IN THE MILKY WAY   365

ticles can ever be achieved without a simultaneous deeper understanding of the nature of spacetime itself."[15] For Wheeler, this was
true of the universe as a whole:

> "Space is a continuum." So bygone decades supposed from the
> start when they asked, "Why does space have three dimen-
> sions?" We, today, ask instead, "How does the world manage
> to give the impression it has three dimensions?" How can there
> be any such thing as a spacetime continuum except in books?
> How else can we look at "space" and "dimensionality" except
> as approximate words for an underpinning, a substrate, a "pre-
> geometry," that has no such property as dimension?[16]

To answer such questions, Wheeler argued, science would
somehow have to bootstrap itself into a new realm, a world of "law
without law," in which, as taught by the quantum indeterminacy
principle, the answer depends upon the question asked. Wheeler
recalled being the subject in a game of twenty questions. He left
the room for a period during which the answer was to be decided
upon by the other players, then returned and started asking questions. The answers were progressively slower in coming, until
Wheeler finally guessed, "Cloud," and was told, to general amusement, that he was right. When his friends stopped laughing they
explained that they had been playing a trick on Wheeler: There
had originally *been* no right answer; his friends had agreed to formulate their answers so that each would be consistent with the
answers given to his previous questions. "What is the symbolism
of the story?" asked Wheeler.

> The world, we once believed, exists "out there" independent
> of any act of observation. The electron in the atom we once
> considered to have at each moment a definite position and a
> definite momentum. I, entering, thought the room contained
> a definite word. In actuality the word was developed step by
> step through the questions I raised, as the information about
> the electron is brought into being by experiment that the ob-
> server chooses to make; that is, by the kind of registering
> equipment that he puts into place. Had I asked different ques-
> tions or the same questions in a different order I would have
> ended up with a different word as the experimenter would
> have ended up with a different story for the doings of the

electron. . . . In the game no word is a word until that word is promoted to reality by the choice of questions asked and answers given. In the real world of quantum physics, *no elementary phenomenon is a phenomenon until it is a recorded phenomenon.*[17]

We are left, then, with an image of genesis as a soundless and insubstantial castle, where our eyes cast innovative, Homeric beams and the only voices are our own. Having ushered ourselves in and having reverently and diligently done our scientific homework, we ask, as best we can frame the question, how creation came to be. The answer comes back, resounding through vaulted chambers where mind and cosmos meet. It is an echo.

19

MIND AND
MATTER

Life, like a dome of many-colored glass,
Stains the white radiance of Eternity.
                                —Shelley

A sad spectacle. If they be inhabited, what a
scope for misery and folly. If they not be
inhabited, what a waste of space.
                                —Thomas Carlyle

The scientific developments we have been discussing in this book have worked, however inadvertently, to implicate and involve our species in the wider universe. Astronomy, in shattering the crystalline spheres that had been said to seal off the earth from the aethereal realms above the moon, placed us *in* the universe. Quantum physics cracked the metaphorical pane of glass that had been assumed to separate the detached observer from the observed world; we are, we found, unavoidably entangled in that which we study. Astrophysics, in determining that matter is the same everywhere and that it everywhere obeys the same rules, laid bare a cosmic unity that extends from nuclear fusion in stars to the chemistry of life. Darwinian evolution, in indicating that all species of earthly life are related and that all arose from ordinary matter, made it clear that there is no wall dividing us from our fellow creatures on Earth, or from the planet that gave us all life—that we are such stuff as worlds are made of.

The conviction that we are in some sense at one with the universe had of course been promulgated many times before, in other spheres of thought. Yahweh fashioned Adam out of dust; Heraclitus the Greek wrote that "all things are one"; Lao-tzu in China depicted man and nature alike as ruled by a single principle ("I call it the Tao"); and a belief in the unity of humankind with the cosmos was widespread among preliterate peoples, as evidenced by the Suquamish Indian chief Seattle, who declared on his deathbed that "all things are connected, like the blood which unites one family. It is all like one family, I tell you." But there is something striking about the fact that the same general view has arisen from sciences that pride themselves on their clearheaded pursuit of objective, empirical fact. From the chromosome charts and fossil records that chart the interrelatedness of all living things on Earth to the similarity of the cosmic chemical abundance to that of terrestrial biota, we find indications that we really are a part of the universe at large.

This scientific verification of our involvement in the workings of the cosmos has of course many implications. One of them—the subject of this chapter—is that, if intelligent life has evolved on this planet, it may have also done so elsewhere.* Darwin's theory of evolution, though it does not explain away the ancient conundrum of why there is such a thing as life, does make it clear that life may arise from ordinary matter and evolve into an "intelligent" form, at least on an Earthlike planet orbiting a sunlike star. As there are plenty of sunlike stars (over ten billion of them in the Milky Way galaxy alone), and, presumably, more than a few Earthlike planets, we can speculate that we are not the only species ever to have studied the universe and wondered about our role in it.

Our comprehension of the relationship between mind and the universe may depend upon whether we can make contact with another intelligent species with which to compare ourselves. Seldom has science done very well at studying phenomena of which

---

*I will not dwell on the self-flattery we exhibit in describing ourselves as "intelligent," or on the enormous variations that might be embraced by the term when it is applied in a panstellar context. For the purposes of this discussion, "intelligent" creatures are defined merely as those with the means and inclination to engage in interstellar communication via electromagnetic ("radio") waves.

but a single example was available: Newton's and Einstein's laws would have been far more difficult—perhaps impossible—to formulate had there been only one planet to test them against, and it is often said that the central problem of cosmology itself is that we have but a single universe to examine. (The discovery of cosmic evolution eases this difficulty, by proffering for our consideration the very different state of the universe during the first moments of cosmic evolution.) The question of extraterrestrial life, then, goes beyond such issues as whether we are alone in the universe or may look forward to cosmic companionship or need fear alien invasion; it is also a way of examining ourselves and our relationship to the rest of nature.

Though much here is new, recent interest in extraterrestrial life can be viewed as resulting from the latest upturn in the fortunes of materialism, the philosophical doctrine that events can be explained solely in terms of material interactions, without recourse to insubstantial conceptions such as that of spirit. Darwinism engendered a new respect for the potential of ordinary matter: A lump of mud in a puddle of rainwater begins to look quite magical, if one appreciates that its like once managed to rear itself up into the whole panoply of earthly life, including that possessed by the individual who contemplates the mud. No longer could a thinking person, mindful that his or her ancestry stretches back through the mammals to the fishes to the amino acids and sugars of prebiotic matter, readily agree with Martin Luther that the earth is but "defiled and noxious," or accept the verdict of the Christian Science service that "there is no life, truth, substance, nor intelligence in matter."

Historically, materialists have been inclined to imagine that there is life on other worlds. Metrodorus the atomist wrote in the fourth century B.C. that "to consider the earth as the only populated world in infinite space is as absurd as to assert that in an entire field sown with millet only one grain will grow."[1] Five centuries later, Lucretius the Epicurean proposed that "there are infinite worlds both like and unlike this world of ours."[2] The Roman Catholic Church, convinced that humans are essentially immaterial spirits, felt threatened by the materialistic point of view: When Giordano Bruno, the Renaissance doyen of pop mysticism, asserted that matter "is in truth all nature and the mother of all living things,"[3] and declared that God "is glorified not in one, but in countless suns;

not in a single earth, but in a thousand, I say, in an infinity of worlds,"[4] he was tied to an iron stake and burned alive, on February 19, 1600, in the Piazza Campo dei Fiori in Rome.

Nevertheless, as science ascended so did materialism, and with it the belief in a plurality of populated worlds. In England in 1638, a Protestant clergyman named John Wilkins published a book proposing that the moon was habitable. Descartes, whose theory of cosmic vortices foreshadowed aspects of Newton's universal gravitation, wondered whether "elsewhere there exist innumerable other creatures of higher quality than ourselves."[5] But no writer did more to infuse the conception of a diverse, fertile universe with a sense of delight than the young French Cartesian Bernard de Fontenelle, whose *Conversations on the Plurality of Worlds* was published in 1686 and has enchanted readers ever since. The book takes the form of a dialogue between Fontenelle and a beautiful, unnamed countess with whom he walks in the gardens each evening at twilight, discussing the stars as they wink into view in the darkening sky. "Who can think long of the moon and stars, in the company of a pretty woman!" Fontenelle exclaims, but he soon gets down to business. "The earth swarms with inhabitants," he tells the countess. "Why then should nature, which is fruitful to an excess here, be so very barren in the rest of the planets?"[6] The moon, he thinks, may be inhabited, but he does not know by what sort of beings: "Put the case that we ourselves inhabited the moon, and were not men, but rational creatures; could we imagine, do you think, such fantastical people upon the earth, as mankind is?"[7]

The countess has her doubts: "You have made the world so large," she says, "that I know not where I am, or what will become of me. . . . I protest it is dreadful."

"Dreadful, Madam?" Fontenelle replies. "I think it very pleasant. When the heavens were a little blue arch, stuck with stars, I thought the universe was too strait and close, I was almost stifled for want of air. But now it is enlarged in height and breadth. . . . I begin to breathe with more freedom, and think the universe to be incomparably more magnificent than it was before."[8]

Not until the latter half of the twentieth century did it become possible to actually begin searching for life on other worlds. One way to do this was to send spacecraft to other planets in the solar system. This endeavor began with the American Pioneer and Soviet Venera missions to Venus in the 1960s, and continued with unmanned American missions to Mars and the Jovian planets in the following decades. The results of these and other preliminary reconnaissances were negative: Photographs taken by unmanned Soviet landers turned up no trace of life on Venus, which has a thick atmosphere but is hotter than Dante's hell, and two landers dispatched to the surface of Mars by the American Viking project recorded no obvious sign of Martian life, either. But these, of course, were insufficient grounds upon which to reach a conclusion about the prevalence of extraterrestrial life in general, since the solar system harbors fewer than one ten-billionth of the total number of planets estimated to exist in our galaxy alone.

One might search for life beyond the solar system by traveling to the stars, but to do so within any reasonable amount of time is a very tall order indeed. The stars are just too far away: A spacecraft capable of traveling a million miles per hour—and this would be a stunningly fast ship, one that could fly from Earth to Mars in less than two days—would take nearly *three thousand years* to reach Alpha Centauri, the nearest star. If the expeditionaries proceeded to the next promising star—Delta Pavonis, spectral class F8, would be a reasonable choice—and then hastened on to, say, Beta Hydri, and then kept going to Zeta Tucanae before stopping for a well-earned rest, they would have succeeded in visiting about one one-hundred billionth of the stars in the galaxy—a sample statistically less significant than attempting to understand all Shakespeare's writings by examining only two letters from one of his sonnets. What is more, the trip would have taken over thirty thousand years, which is a very long time; thirty thousand years ago, our Paleolithic ancestors were carving the world's first wooden drums.

There is, however, a better method of searching for intelligent life beyond the solar system. It is to employ radiotelescopes to listen for electromagnetic signals—radio or television transmissions—beamed into space by alien civilizations. Such a signal, transmitted using only a few cents' worth of electricity, travels at the speed of light and could be intercepted by radiotelescopes on Earth across distances of many thousands of light-years. This was the realization behind what came to be called SETI—the search for extraterrestrial intelligence.

SETI was first proposed in 1959 by two scientists, Giuseppe Cocconi and Philip Morrison. "The probability of success is difficult to estimate," they noted, "but if we never search, the chance of

success is zero."[9] The first SETI experiment, Project Ozma, was
conducted in the early 1970s by the American astronomer Frank
Drake. Drake observed a total of 659 stars over a three-year period,
listening at but a single frequency with radio dish antennae of 300
and 140 feet in diameter. He detected no artificial extraterrestrial
signals, but was not disappointed, given that the total number of
stars in the Milky Way is so large; even if there were, say, a thousand
civilizations beaming signals our way at exactly the listened-for
frequency, the odds against Project Ozma's having detected one of
them would have been nearly a million to one. When one factored
in the many other uncertainties—guessing at the right frequency,
allowing for Doppler shifts introduced by the motion of the sun
and earth, and so forth—the chances became even slimmer. If SETI
were to succeed, it would have to be a continuous, long-term endeavor.

     In the meantime, however, the very existence of even a few
modest SETI projects in the United States and the Soviet Union
spurred fresh reaction against the extraterrestrial-life hypothesis.
When NASA proposed spending two million dollars a year to divert
a small amount of radiotelescope time to SETI, Senator William
Proxmire of Wisconsin scotched the idea, bestowing upon it his
"Golden Fleece" award for wasteful public spending and declaring
that extraterrestrial civilizations, "even if they once existed, are now
dead and gone."[10] Thus ridiculed, SETI had to proceed in fits and
starts, and by the mid-1980s only a few thousand hours of listening
time had been accumulated on major radiotelescopes.

     The opponents to SETI marshaled two central arguments.
One was probablistic in nature, and like many statistical arguments
was dogged by confusion. The other, called Fermi's question, raised
important issues that, if they did not dismiss the SETI case, did
hold interesting implications for its search strategy.

     The probabilistic argument consisted of adding up all the con-
ditions thought to have been necessary for intelligent life to have
evolved on Earth, then calculating that it was highly unlikely that
the same thing had happened elsewhere. Its proponents began with
the size of the earth's orbit—were the earth slightly closer to the
sun all its water would boil, and were it slightly farther away all
the water would freeze—and tallied all the twists and turns in
evolutionary history thought to have led to the emergence of *Homo*

sapiens. Were all these variables ascribed to chance (as both sides presumed they should be) the result was a vanishing small likelihood of intelligent life appearing *anywhere.* There are more than one million species of life on Earth today, the evolution of each of which is estimated to have involved perhaps a thousand unsuccessful mutations that led nowhere; the probability, therefore, that another planet has evolved a similar biology can be estimated at a thousand million to one. Allow for all the cultural and biological variables involved in the advent of human beings and their civilization, and the odds go up to perhaps 10^{15} or 10^{18} to one—a number that exceeds the likely total of all the planets in the galaxy. Therefore, went the argument, we are almost certainly alone.

This line of thought resembled the old argument from design, as succinctly summarized by Bertrand Russell: "You all know the argument from design: Everything in the world is made just so that we can manage to live in the world, and if the world was ever so little different, we could not manage to live in it. This is the argument from design."[11] The poverty of this argument is that one cannot reliably calculate the odds of a particular thing having happened unless one either understands the process—that is, can properly identify and quantify all the variables involved—or has an adequate experimental data base from which to draw phenomenological information about it. If, for instance, we want to predict how close an intercontinental ballistic missile will land to its intended target, we can calculate all the variables—the flight characteristics of the missile, the influences of environment on its navigational system, etc.—or we can test real missiles, as often as possible, in order to generate a data base about how they perform. In practice one does both, since both approaches may err. But when the question involves intelligent life arising on other planets, we can with confidence do neither, since we have only a rudimentary understanding of the variables involved, and none whatsoever regarding the statistics. To reason probabilities without them is to fall victim to the post hoc fallacy, by the lights of which almost every event may be calculated to be unique. If, for instance, we were to ask how likely it is for you to be reading this page at this moment, we might add up all the twists and turns of your life and mine, beginning with our births and running down through a billion variables to the circumstances in which I wrote these words and you read them, and conclude that the thing is so nearly impossible that it would almost certainly never have happened, anywhere in the universe. Yet here we are.

In fairness, though, sentience might, indeed, be but an improbable accident, in which case we humans with our poor little radiotelescopes represent the highest form of intelligence in the universe. A SETI project could never prove that this were so, but if it went on listening for many decades and found nothing, the implication surely would be that we have few if any cosmic compatriots. But until that happens, as Michelangelo used to say, permit me to doubt.

The other argument is credited to Fermi, who is said to have asked, at Los Alamos one day in the late 1940s, "Where are they?" His reasoning was that if technically advanced extraterrestrial civilizations are prevalent, some of them ought to be able to migrate to other star systems, colonizing new planets as they go; the colonists subsequently could launch additional interstellar missions, until virtually every star system in the galaxy was occupied. Yet they are not here. Therefore, they do not exist.

On the face of it, Fermi's question is easy enough to answer: Advanced extraterrestrials are not here because, for one reason or another, they are not able or do not want to be here. Perhaps interstellar travel is too expensive and time-consuming an enterprise for them to bother making the trip to other stars, or at least to *our* star, a yellow dwarf in the galactic suburbs orbited by a small blue planet the atmosphere of which is contaminated by the notoriously poisonous gas oxygen. Or perhaps they know we are here, but scruple not to interfere with our development (the "zoo hypothesis"). One can think of many such explanations; the point is, as they say, that absence of evidence is not evidence of absence. Lobster exist on earth, but I can set a place at my dinner table and wait a very long time before a lobster will come in the front door and climb up onto my plate; it's just not worth the lobster's trouble.

Fermi's question, however, returns with far greater force when applied to the question of whether the solar system has been visited, not by a gigantic ark full of intrepid aliens, but by an automated, instrumented, self-replicating probe. Suppose that a nonhuman society were to send ten such machines to ten stars, and that each, upon arrival, were to mine an asteroid or planet in the system for materials and fuel from which to construct and dispatch ten more probes, while itself staying behind to keep an eye on things. In this fashion it would be possible for an advanced civilization to establish a remote-sensing presence in many star systems. If, for instance, the average time required for a probe to reach a new star, replicate, and send forth ten new probes were ten thousand years, there would be probes in orbit around half the stars in the galaxy within one hundred thousand years after the project began. The galaxy is more than ten billion years old, so presumably there has been plenty of time for someone to have tried such a thing. The cost is manageable, the return in data enormous. So where are the probes?

One obvious answer is that they may already be here. The first rule of interstellar space-flight is to make everything as light as possible, in order to save fuel. Consequently we would expect a probe dispatched to the solar system from another star to be small. How small? A study conducted at NASA's Jet Propulsion Laboratory in 1980 concluded that an interstellar spacecraft with sensing devices and an antenna capable of calling home could be encapsulated, using near-future technology, in a package weighing less than three hundred pounds. The example of the human brain suggests that it should be possible to miniaturize considerable sensory and data-processing power into an even smaller package, perhaps the size of a grapefruit and weighing only a few pounds. A probe *that* small could be in orbit around the sun right now—adrift in space or sitting on one of the millions of asteroids orbiting the sun or on a satellite of Mars, or Jupiter, or Saturn, quietly watching and transmitting its findings back home—and we would not know it. We have as yet no means of detecting such a probe, unless it had been programmed to make itself conspicuous, and there are good reasons for it not to call attention to itself, the foremost among which is that if we discovered its presence we might very well go get it and take it apart. The further exploration of the solar system might, therefore, eventually turn up evidence of intelligent extraterrestrial life. This of course was the thesis of Arthur C. Clarke's science fiction tale *2001*, in which an astronaut investigates an alien probe discovered in orbit around Jupiter, the sun's largest planet.

The possibility of automation also raises an intriguing prospect with regard to the search for intelligent interstellar radio signals: It suggests that the first signal acquired by a SETI receiver might very well have been dispatched, not by the inhabitants of another planet, but by an intelligent machine. To see how this could be so, we need only to consider the practical exigencies that an advanced civilization would encounter once it had been in the interstellar communications business for a while.

Suppose that yours is one of among a hundred and one worlds in the Milky Way galaxy that have established radio communication with one another. You now have a minimum of one hundred antennae in action, each maintaining contact with a different planet thousands of light-years away. This arrangement has two drawbacks. First, it is inefficient; for the sake of economy, you would prefer to be using as few antennae as possible. Second, far more serious, is the Q and A time; if you ask a question it takes thousands of years to get an answer.

The way to alleviate both problems is to network the system. You install a single, automated station in space to handle all the radio traffic, and you link it to your planet via a single antenna system. Getting out your galactic map, you then determine strategic locations for siting other such automated stations, and you transmit an appeal to the worlds located at those junctures to please build them. Soon—meaning in a matter of a few dozen millennia or so—everybody is sending and receiving data to and from all the other worlds through local junction terminals, which may be in their own star system or in one next door. This way they need not employ separate antennae for each planet with which they communicate, any more than Earthlings maintain a separate telephone for every person they call.

The network, once established, could be endowed with a number of salutary features. For one thing, it could be assigned the task of acquiring signals from new worlds and bringing them on-line. Indeed, its stations might broadcast acquisition signals for the purpose of attracting the attention of such relatively undeveloped planets. If so, the first signal intercepted by a SETI radiotelescope might come from an automated station sited far from any inhabited planet. In order to accomplish this and other tasks efficiently, the network should be made capable not only of repairing itself but also of expanding as the growing body of data requires. Here the technology of the self-replicating probes comes in handy; the network could dispatch probes to strategically favorable star systems in the galaxy, where each would build itself into a new junction station that would in turn hook up with the rest of the network.

Most important, the network should be equipped with a commodious and self-expanding memory, one that is replicated and regularly updated at every station. The great advantage of this is that it alleviates the Q and A problem. What one wants from interstellar communication is not really conversation, which takes too long, but *information:* One wants to know who else lives in the galaxy, what they look like and how they think and what they do, and about their history and that of the species that preceded them. To make this and all other information available to everyone interested, one should have the network *remember* everything that it conveys.

The network, then, would be not only a telephone or television system, but also a computer and a library, access to which would be as near as the nearest junction. If a species of intelligent birds on one side of the galaxy were interested in the biology of a species

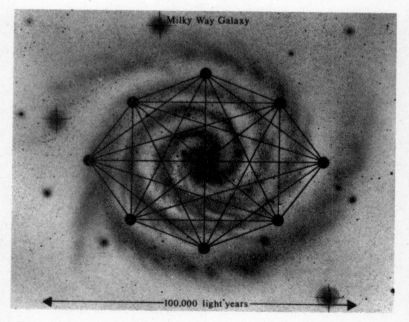

Direct radio communication between intelligent species would be relatively slow and inefficient. Here, eight inhabited worlds scattered around the galaxy are communicating directly. The average Q&A time is one hundred thousand years.

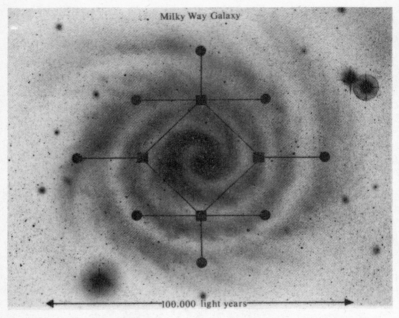

Milky Way Galaxy

←————————100,000 light years————————→

Networking of interstellar communication greatly improves the efficiency of the system. This rudimentary network, consisting of only four junctions, cuts the Q&A time (for communication with the terminals' memory banks) in half, to fifty thousand years. By adding more junction stations to the network, the Q&A time can be reduced to a century or less.

of intelligent reptiles on the opposite side of the galaxy, they would not have to send a direct message and wait two hundred thousand years for a reply. Instead, the information would already be stored in the memory banks of the network itself, and the requisite Q and A time would be little longer than the light travel time to and from the nearest network junction. Nor would the information be hostage to the fate of any particular world; once submitted to the network, it could survive indefinitely.

We arrive, then, at the prospect of an immortal system, constantly expanding and continually acquiring and storing information from all the worlds that choose to subscribe to it. In the long run, the network itself might reasonably be expected to evolve into the single most knowledgeable entity in the galaxy. It alone could survey the full sweep of galactic history and experience the development of knowledge on a panstellar scale. Growing in sophistication and complexity with the passage of aeons, forever articulating itself among the stars, the network would come to resemble nothing so much as the central nervous system of the Milky Way.

Which, perhaps, is the ultimate purpose of intelligence, if life and intelligence may be said to have a purpose. We often find that our deepest yearnings have less to do with ourselves than with the wider scheme of things. (Love, which makes the world go round, is a highly individual experience, but its ultimate function is to perpetuate and advance the species.) Perhaps this is true as well of our deep but seemingly inexplicable desire to learn whether we are alone in the galaxy. Life might be the galaxy's way of evolving a brain.

The process could extend beyond the galaxy, too, through contact with similar networks in other galaxies. Intergalactic Q and A times go to many millions of years—too long a wait for mortal beings, even if they are as longevous as Joshua trees, but perfectly manageable for an interstellar network. The network could afford to fashion giant antennae, use them to broadcast powerful acquisition signals to the Andromeda galaxy, to Centaurus A, even to the populous heart of the Virgo Supercluster, sixty million light-years distant, and then wait for a reply. Every world on every network would stand to benefit as galaxy after galaxy established contact, spinning electromagnetic threads across the expanding universe and exchanging the wealth of galactic libraries. The human species is only about two million years old, a time equal to that required for a message to travel one-way from the Andromeda galaxy to ours; we cannot very well expect to start up a meaningful dialogue with a society in Andromeda. But if information about Andromeda and the history of its worlds were already stored in our galaxy's network, we might be able to begin accessing it within a matter of decades after making contact.

All this may be a dream. Certainly it is no more than a conjecture, and a materialist's conjecture at that; what could be more materialistic than a galactic "intelligence" composed of communicating computers that rear themselves up out of the ash-gray rocks of dead asteroids? Yet it points to an idealistic vision of worlds by the thousands, some in their youth and some in their graves, linked by an intergalactic mechanism devoted to pure thought. And, it suggests a cosmic role for intelligence—that the combination of intelligence and technology could awaken the universe to its own life and thought and history. That would make us all the substance of a cosmic mind.

THE PERSISTENCE OF MYSTERY

Drawn by my eager wish, desirous of seeing the great confusion of the various strange forms created by ingenious nature, I wandered for some time among the shadowed cliffs, and came to the entrance of a great cavern. I remained before it for a while, stupefied, and ignorant of the existence of such a thing, with my back bent and my left hand resting on my knee, and shading my eyes with my right, with lids lowered and closed, and often bending this way and that to see whether I could discern anything within; but this was denied me by the great darkness inside. And after I stayed a while, suddenly there arose in me two things, fear and desire—fear because of the menacing dark cave, and desire to see whether there were any miraculous thing within.

—Leonardo da Vinci

A great truth is a truth whose opposite is also a great truth.

—Niels Bohr

In this book I have discussed how we inhabitants of this one world pieced together a credible picture of the (much) larger universe. I have described this process as a "coming of age," by which I mean that we have, through centuries of fitful effort, finally begun to comprehend a few of the fundamental facts about the universe an acquaintanceship with which presumably is prerequisite to the most modest claim of cosmological maturity. We now know, for example, *where* we are—that we live on a planet orbiting a star located out toward one edge of a spiral galaxy, which in turn lies near the outskirts of a supercluster of galaxies, whose position has been determined relative to several neighboring superclusters that altogether harbor some forty thousand galaxies arrayed across a million billion cubic light-years of space. We also know, more or less, *when* we have come upon the scene—at about five billion years since the sun and its planets formed, in an expanding universe that is probably between twice and four times that old. We have discerned the basic mechanisms responsible for the evolution of life on Earth, found evidence of chemical evolution on the cosmic scale as well, and learned enough physics to investigate nature on a wide range of scales, from the jitterbugging of the quarks to the waltz of the galaxies.

These are accomplishments in which humanity can with justice take pride. Since the ancient Greeks first set the Western world on the path of science, our mensuration of the past has deepened from a few thousand years to over ten billion years, while that of space has expanded from a low-roof sky not much higher than the real distance of the moon to the more than ten-billion-light-year radius of the observable universe. We have reason to hope that our age will be remembered (if there is anyone around to remember it) for its contributions to the supreme intellectual treasure of any society, its concept of the universe at large.

And yet the more we know about the universe, the more we come to see how little we know. When the cosmos was thought to be but a tidy garden, with the sky its ceiling and the earth its floor and its history coextensive with that of the human family tree, it was still possible to imagine that we might one day comprehend it in both plan and detail. That illusion can no longer be sustained. We might eventually obtain some sort of bedrock understanding of cosmic structure, but we will never understand the universe in detail; it is just too big and varied for that. If we possessed an atlas of our galaxy that devoted but a single page to each star system in the Milky Way (so that the sun and all its planets were crammed on one page), that atlas would run to more than ten million volumes of ten thousand pages each. It would take a library the size of Harvard's to house the atlas, and merely to flip through it, at the rate of a page per second, would require over ten thousand years. Add the details of planetary cartography, potential extraterrestrial biology, the subtleties of the scientific principles involved, and the historical dimensions of change, and it becomes clear that we are never going to learn more than a tiny fraction of the story of our galaxy alone—and there are a hundred billion more galaxies. As the physician Lewis Thomas writes, "The greatest of all the accomplishments of twentieth-century science has been the discovery of human ignorance."[1]

Our ignorance, of course, has always been with us, and always will be. What is new is our awareness of it, our awakening to its fathomless dimensions, and it is *this*, more than anything else, that marks the coming of age of our species. Space may have a horizon and time a stop, but the adventure of learning is endless. As the philosopher of science Karl Popper writes:

> The more we learn about the world, and the deeper our learning, the more conscious, specific, and articulate will be our knowledge of what we do not know, our knowledge of our ignorance. For this, indeed, is the main source of our ignorance—the fact that our knowledge can be only finite, while our ignorance must necessarily be infinite.[2]

It is widely though erroneously supposed that science has to do with explaining everything, and that unexplained phenomena therefore upset scientists by threatening the hegemony of their world view. The technician in the white lab coat in the low-budget movie slaps palm to forehead when confronted with something novel, gasping, "But . . . there's no *explanation* for this!" Actually, of course, any worthy scientist will rush to embrace the unexplained, for without it science would get nowhere. It is the grand, mystical systems of thought, couched in terminologies too vague to be wrong, that explain everything and seldom err and do not grow.

Science is inherently open-ended and exploratory, and makes mistakes every day. Indeed that will always be its fate, according to the bare-bones logic of Kurt Gödel's second incompleteness theorem. Gödel's theorem establishes that the full validity of any system, including a scientific one, cannot be demonstrated within that system itself. In other words, the comprehensibility of a theory cannot be established unless there is something outside the frame against which to test it—something beyond the boundary defined by a thermodynamics equation, or by the collapse of the quantum wave function, or by any other theory or law. And if there *is* such a wider reference frame, then the theory by definition does not explain everything. In short, there is not and never will be a complete and comprehensive scientific account of the universe that can be proved valid. The Creator must have been fond of uncertainty, for He (or She) has given it to us for keeps.

Which is, I would argue, a salutary finding and cause for good cheer. Hell would be a small universe that we could explore thoroughly and fully comprehend. Alexander the Great may have wept upon being told that there were infinite worlds ("And we have not conquered even one," he sobbed) but the situation looks more sanguine to those inclined to untie rather than to cut nature's gordian knot. No thinking man or woman ought really to want to know everything, for when knowledge and its analysis is complete, thinking stops.

René Magritte in 1926 painted a picture of a pipe and wrote beneath it on the canvas, in a careful schoolboy script, the words *"Ceci n'est pas une pipe"*—"This is not a pipe."[3] His painting might suitably be made the emblem of scientific cosmology. The word "universe" is not the universe; neither are the equations of supersymmetry theory or the Hubble law or the Friedmann-Walker-Robinson metric.* Nor, more generally, is science very good at

---

*Numbers get closer to the rationally intelligible reality than do words—as Bohr said, "When it comes to atoms, language can be used only as in poetry"—but this is true only because mathematics is less ambiguous and more logical in structure than is ordinary language. The efficacy of mathematics in scientific research does not in itself establish that God is a geometer (if such a statement can have any meaning) or that the universe is a mathematical puzzle.

explaining what anything, much less the entire universe, actually "is." Science describes and predicts events, but it pays for this power in the coin of the *ding an sich*—the thing in itself.

Why, then, does science work? The answer is that nobody knows. It is a complete mystery—perhaps *the* complete mystery— why the human mind should be able to understand anything at all about the wider universe. As Einstein used to say, "The most incomprehensible thing about the universe is that it is comprehensible."[4] Perhaps it is because our brains evolved through the workings of natural law that they somehow resonate with natural law. Nature exhibits a number of self-similarities—patterns of behavior that recur on different scales, making it possible to identify principles, such as the conservation laws, that apply universally—and these may provide the link between what goes on inside and outside the human skull. But the mystery, really, is not that we are at one with the universe, but that we are to some degree at odds with it, different from it, *and yet* can understand something about it. Why is this so?

In search of an answer, let us pause to slake our thirst one last time at symmetry's bubbling spring. Symmetry, we recall, implies not only the existence of an invariance under a transformation, the basis of all natural law, but also a "due proportion" between the invariance at hand and some larger, more comprehensive frame of reference. In this relationship may be found parallels with the process of scientific thought. The mind with its inherent limitations makes a frame within which our ideas can cavort; even the most expansive theory is "framed" in a specific mathematical or verbal or visual vocabulary. We then test our ideas against a piece of the outer world, which, however, itself has a frame around it. This process will work so long as we never reach an unframed, limitless arena. Gödel's theorem suggests that we never will—that a theory by its very nature requires for its verification the existence or contemplation of a larger reference frame. It is the boundary condition, then, that provides the essential distinction between mind and the universe: Thoughts and events are bounded, even if the totality is not.*

And where did the boundaries come from? Quite possibly from

---

*Similar ideas appeared quite early in Greek thought, as when Philolaus of Tarentum wrote, in about 460 B.C., "Nature in the cosmos was fitted together of Unlimit and Limit, the order of the all as well as all things in it."[5]

the breaking of cosmic symmetries at the moment of genesis. We look out across a cosmic landscape riven by the fractal lines of broken symmetries, and draw from their patterns metaphors that aspire to be as creative, if not always quite as flawed, as the universe they purport to describe. (All metaphors are imperfect, said the poet Robert Frost, that is the beauty of them.)

It may be, then, that the universe is comprehensible because it is defective—that because it forsook the perfection of nonbeing for the welter of being, it is possible for us to exist, and to perceive the jumbled, blemished reality, and to test it against the ghostly specter of the primordial symmetry thought to have preceded it. We are, therefore we think. (Or, as the fabulist Jorge Luis Borges put it, "In spite of oneself, one thinks.")[6]

Science is a process, not an edifice, and sheds old concepts as it grows. "Theories," said Ernst Mach, "are like withered leaves, which drop off after having enabled the organism of science to breathe for a time."[7] The process depends upon error—as Popper notes, a theory is valuable only if it is capable of being disproved —as if to testify to the ubiquity and efficacy of cosmic imperfection. "Error can often be fertile," remarked the historian A.J.P. Taylor, "but perfection is always sterile."[8] Taken as a whole, the scientific endeavor is as open-ended as the expansion of the universe—which, I think, is what Bohr had in mind when on his deathbed he complained of the philosophers that they too often "have not that instinct that it is important to learn something, and that we must be prepared to learn."[9] Every answer opens up new questions: Like Atalanta stooping to gather up the golden apples, we pause to marvel at each new discovery, only to realize that we have fallen behind in the race and must hurry on to the next turn in the path, where another golden apple awaits us.

Our explications of nature will always be inadequate, if only because it is the difference between the idea and the reality that makes the idea possible. Nature may be counted upon forever to retain the mysterious, magical quality that arises from the contrast between her innumerable splendors and the limitations of our metaphors. As Wheeler put it:

> There is nothing deader than an equation. Imagine that
> we take the carpet up in this room, and lay down on the floor
> a big sheet of paper and rule it off in one-foot squares. Then

I get down and write in one square my best set of equations for the universe, and you get down and write yours, and we get the people we respect the most to write down their equations, till we have all the squares filled. We've worked our way to the door of the room.

We wave our magic wand and give the command to those equations to put on wings and fly. Not one of them will fly. Yet there is some magic in this universe of ours, so that with the birds and the flowers and the trees and the sky it flies! What compelling feature about the equations that are behind the universe is there that makes them put on wings and fly?

. . . If I had to produce a slogan for the search I see ahead of us, it would read like this: That we shall first understand how simple the universe is when we realize how strange it is.[10]

Science is young. Whether it will survive long enough to become old depends upon our sanity and courage and vigor, and, as one always must add in this nuclear age, upon whether we blow ourselves up first. "Nothing that is vast enters into the life of mortals without a curse," as Sophocles said, and the knowledge of how the stars shine is very great, and its dark side is very dark indeed. Needless to say, science in itself will not deliver us from the dangers to which its knowledge has exposed us. "Scientific statements of facts and relations, indeed, cannot produce ethical directives," wrote Einstein, though he allowed that "ethical directives can be made rational and coherent by logical thinking and empirical knowledge."[11]

Viewed from so cold a perspective, we may esteem ourselves less but will know ourselves better, as creatures of darkness and light, in love with death as well as with life, as eager to destroy as to create. Our lives are suspended like our planet in a gimbals of duality, half sunlight and half shadow. If we plead with nature, it is in vain; she is wonderfully indifferent to our fate, and it is her custom to try everything and to be ruthless with incompetence. Ninety-nine percent of all the species that have lived on Earth have died away, and no stars will wink out in tribute if we in our folly soon join them.

Epictetus the former slave remarked that

every matter has two handles, one of which will bear taking hold of, the other not. If thy brother sin against thee, lay not

hold of the matter by this, that he sins against thee; for by this
handle the matter will not bear taking hold of. But rather lay
hold of it by this, that he is thy brother, thy born mate; and
thou wilt take hold of it by what will bear handling.[12]

Therefore, we say—speaking as living and (we think) thinking beings,
as carriers of the fire—therefore, choose life.

Three philosophers came together to taste vinegar, the Chinese symbol for the spirit of life. First Confucius drank of it. "It is sour," he said. Next, Buddha drank. He pronounced the vinegar bitter. Then Lao-tzu tasted it. He exclaimed, "It is fresh!"

—Traditional Chinese tale,
repeated by Niels Bohr

For all my pains, I only beg this favor, that whenever you see the sun, the heavens, or the stars, you will think of me.

—Bernard de Fontenelle

GLOSSARY

The breaking of a wave cannot explain the
whole sea.

—Vladimir Nabokov

ABT. Abbreviation employed in this book to mean "after the beginning of time," which is here defined as the beginning of the expansion of the universe.

Absolute luminosity. See *luminosity*.

Absolute magnitude. See *magnitude*.

Absolute space. Newtonian space, hypothesized to define a cosmic reference frame independent of its content of matter or energy. The existence of absolute space, enshrined in *aether* theory, was denied in *relativity*.

Aberration of starlight. Displacement in the apparent location of stars in the sky, introduced by the motion of the earth.

Absorption lines. Dark lines in a *spectrum*, produced when light or other electromagnetic radiation coming from a distant source passes through a gas cloud or similar object closer to the observer. Like *emission lines*, absorption lines betray the chemical composition and velocity of the material that produces them.

Acceleration. An increase in velocity over time.

Accelerator. A machine for speeding subatomic particles to high velocity, then colliding them with a stationary target or with another beam of particles moving in the opposite direction. (In the latter instance, the machine may be called a *collider*.) At velocities approaching that of light, the mass of the particles increases dramatically, adding greatly to the energy released on impact. The resulting explosion promotes the production of exotic particles, which are analyzed according to their behavior as they fly away through a particle *detector*.

Aether. (1) In *Aristotelian physics*, the fifth element, of which the stars and planets are made. (2) In *classical physics*, an invisible medium that was thought to suffuse all space.

Alchemy. Art of bringing parts of the universe to the perfect state toward which they were thought to aspire—e.g., gold for metals, immortality for human beings.

Andromeda galaxy. Major spiral galaxy, 2.2 million *light-years* from Earth. Gravitationally bound to the *Milky Way galaxy*, with which it shares membership in the *Local Group*, it is currently approaching us, rather than receding as is the case for most galaxies.

Angular momentum. The product of mass and angular velocity for an object in rotation; similar to linear momentum. In *quantum mechanics*, angular momentum is quantized, i.e., is measured in indivisible units equivalent to *Planck's constant* divided by 2π.

Anisotropy. The characteristic of being dependent upon direction. (Light coming with equal intensity from all directions is *isotropic;* a spotlight's beam is *an*isotropic.) The *cosmic background radiation* is generally isotropic—i.e., its intensity is the same in all parts of the sky—but small anisotropies have been detected which are thought to reflect the earth's proper motion relative to the framework of the universe as a whole.

Anthropic principle. The doctrine that the value of certain fundamental constants of nature can be explained by demonstrating that, were they otherwise, the universe could not support life and therefore would contain nobody capable of worrying about why they are as they are. Were the *strong nuclear force* slightly different in strength, for instance, the stars could not shine and life as we know it would be impossible.

Antimatter. Matter made of *particles* with identical *mass* and *spin* as those of ordinary matter, but with opposite charge. Antimatter has been produced experimentally, but little of it is found in nature. Why this should be so is one of the questions that must be answered by any adequate theory of the early universe.

Apparent magnitude. See *magnitude*.

Aristotelian physics. Physics as promulgated by Aristotle; includes the hypothesis that our world is comprised of four elements, and that the universe beyond the moon is made of a fifth element and so is fundamentally different from the mundane realm.

Asteroids. Low-mass, solid objects that orbit the sun and shine by reflected light. Most belong to the "asteroid belt," a zone located between the orbits of Mars and Jupiter. Though they number in the millions, their total mass is but a tiny fraction of the earth's. Also called *minor planets*.

Astrolabe. Sighting instrument employed since antiquity to determine the elevation above the horizon of celestial objects. Eventually replaced by the *sextant*.

Astrology. The belief that human affairs and people's personalities and characters are influenced by (or encoded in) the positions of the planets.

Astronomical unit. The mean distance from the earth to the sun, equal to 92.81 million miles or 499.012 light-seconds.

Astronomy. The science that studies the natural world beyond the earth.

Astrophysics. The science that studies the physics and chemistry of extraterrestrial objects. The alliance of physics and astronomy, which began with the advent of *spectroscopy*, made it possible to investigate *what* celestial objects are and not just *where* they are.

Asymmetry. A violation of *symmetry*.

Asymptotic freedom. Orwellian liberty enjoyed by *quarks*, which move freely when close together but are reined in by an increasingly powerful *strong nuclear force* whenever they begin to drift apart.

Atoms. The fundamental units of a chemical element. An atom consists of a *nucleus*, which may contain *protons* and *neutrons*, and *electrons*, which occupy shells that surround the nucleus and are centered on it.

Avoidance. The fact that *galaxies* appear to "avoid" the *Milky Way*, and are most numerous in other parts of the sky. When galaxies were known as spiral *nebulae* and their nature was not yet understood, avoidance was thought by some researchers to indicate a connection between them and the Milky Way. Now the effect is understood to be due to dark clouds of dust and gas in our galaxy, which obscure our view of the universe beyond in those quarters of the sky.

Background radiation. See *cosmic background radiation*.

Baryon number. The total number of *baryons* in the universe, minus the total number of antibaryons. An index, therefore, of the cosmic matter-antimatter *asymmetry*.

Baryons. Massive elementary particles with half-integral spin that experience the strong nuclear force. Protons and neutrons are baryons. See *hadrons*.

BeV. One billion (10^9) *electron volts*. See *GeV*.

Big bang theory. Model of cosmic history in which the universe begins in a state of high density and temperature, both of which decrease as the universe expands. Less a theory than a school of theories that attempt to trace how the universe evolved.

Billion. This book employs the American billion, equal to one thousand million or 10^9.

Binary star. A double star system, in which the two stars are bound together by their mutual *gravitational force*.

Biology. The scientific study of life and living matter.

Black-body curve. Plot of energy level against wavelength for heat or other radiation emitted by an object capable of absorbing all the energy that strikes it. The curve has a pronounced hump that moves toward shorter wavelengths as the temperature increases. The *cosmic background radiation*, thought to consist of photons emitted during the *big bang*, conforms to a black-body curve.

Black holes. Objects with a gravitational field so intense that their *escape velocity* exceeds that of light. No macroscopic object inside the black hole, therefore, can escape it. In terms of general relativity, space surrounding a black hole is said to reach infinite curvature, making it a *singularity*.

Bosons. Elementary particles with integer spin that do not obey the *Pauli exclusion principle*. They include the *photons* and the *W and Z particles*, carriers of the *electromagnetic* and the *electroweak* forces respectively.

Boundary condition. Restriction on the limits of applicability of an equation. Examples include the definition of a "closed system" in thermodynamics, and the theater within which one collapses the *wave function* in quantum mechanics. Every equation in physics may in principle be reduced to two fundamentals— the *initial conditions* and the boundary conditions.

Broken symmetry. In cosmology and particle physics, a state in which traces of an earlier *symmetry* may be discerned. A broken symmetry condition differs from chaos in that its parts can in theory be united in a symmetrical whole, like the pieces of a jigsaw puzzle.

Caltech. The California Institute of Technology, in Pasadena.

Carbon reaction. An important *nuclear fusion* process that occurs in stars. Carbon-12 both initiates it and, following interactions with nuclei of nitrogen, hydrogen, oxygen, and other elements, reappears at its conclusion.

Catastrophism. Nineteenth-century hypothesis that depicted the many changes evinced by the geological record as having resulted from cataclysms occurring during a relatively brief period of history. Compare *uniformitarianism*.

Causation, causality. The doctrine that every new situation must have resulted from a previous state. Causation underlay the original atomic hypothesis of the Greeks, and was popular in *classical physics*. It is eroded in *quantum mechanics* and has, in any case, never been proved essential to the scientific world view. See *chance, determinism*.

Centaurus A. Giant elliptical galaxy, located between the *Local Group* and the center of the *Virgo Supercluster*.

Cepheid variable. A pulsating *variable star* whose periodicity—i.e., the time it takes to vary in brightness—is directly related to its absolute *magnitude*. This correlation between brightness and period makes Cepheids useful in measuring intergalactic distances.

CERN. The Center for European Nuclear Research, located outside Geneva, Switzerland.

Chance. Characteristic of a regime in which predictions cannot be made exactly, but only in terms of probabilities. In *classical physics*, chance was thought to pertain only where ignorance limited our understanding of an underlying mechanism of strict *causation*. But in the Copenhagen interpretation of *quantum mechanics*, chance is portrayed as inherent to all observations of nature.

Charm. The fourth *flavor* of quarks. Predicted by theory, charmed quarks were discovered in 1974.

Chromatic aberration. Introduction of spurious colors by a lens. This defect flawed the performance of refracting *telescopes* for centuries, until attenuated by the introduction of corrective elements into a compound lens.

Chronometer. A highly accurate timepiece.

Circle. An *ellipse* possessing but one focus.

Classical physics. Physics prior to the introduction of the quantum principle. Classical physics incorporates Newtonian mechanics, views energy as a continuum, and is strictly causal.

Closed universe. Cosmological model in which the universe eventually stops expanding and begins to collapse, presumably to end in a fireball like that of the big bang. Compare *open universe*.

Cloud chamber. A glass-walled enclosure containing a vapor in which *particles* can be detected by photographing the tracks of water droplets they leave behind when they pass through the chamber.

Collider. See *accelerator*.

Color. Property of *quarks* that expresses their behavior under the *strong force*. Analogous to the concept of charge in electromagnetism, except that, whereas there are two electrical charges (plus and minus), the strong force involves three color charges—red, green, and blue. The term is whimsical, and has nothing to do with color in the conventional sense, any more than quark "*flavor*," which determines the weak force behavior of quarks, has anything to do with taste.

Comets. Minor members of the solar system, thought to be lumps of dirt and ice left over from the formation of the solar system. Millions of comets are believed to reside in the *Oort Cloud*, a spherical region with a radius of some thirty to one hundred thousand *astronomical units* centered on the sun. Comets falling in from the Oort Cloud are heated by the sun and grow glowing tails, which can make them conspicuous in the skies of Earth.

Confinement. The inability of *quarks* to escape the bonds that hold them in pairs and triplets at the energy levels found in the universe today. See *gluon lattice, asymptotic freedom*.

Conservation laws. Laws that identify a quantity, such as *energy*, that remains unchanged throughout a transformation. All conservation laws are thought to involve *symmetries*.

Copernicanism. Broadly, the hypothesis that the earth and the other planets orbit the sun.

Cosmic background radiation. Microwave radio emission coming from all directions and corresponding to a *black-body curve*; its properties coincide with those predicted by the *big bang theory* as having been generated by *photons* released from the big bang when the universe was less than one million years old. The big bang theory suggests the existence of *neutrino* and *gravitational* background radiations as well, though the means to detect such do not yet exist.

Cosmic matter density. The average number of *fermions* per unit volume of space throughout the universe. Since matter is depicted in general relativity as bending space, the value of the cosmic matter density, if known, could reveal the overall curvature of cosmic space. See *critical density, omega*.

Cosmic rays. Subatomic particles, primarily *protons*, that speed through space and strike the earth. The fact that they are massive, combined with their high velocities, means that they pack considerable energy—from 10^8 to more than 10^{22} *electron volts*.

Cosmogony. The study of the origin of the universe.

Cosmology. (1) The science concerned with discerning the structure and composition of the universe as a whole. Combines astronomy, astrophysics, particle physics, and a variety of mathematical approaches including geometry and topology. (2) A particular cosmological theory.

Cosmological constant. A term sometimes employed in cosmology to express a force of "cosmic repulsion," such as the energy released by the false vacuum thought to power exponential expansion of the universe in the *inflationary universe* models. Whether any such thing as cosmic repulsion exists or ever played a role in cosmic history remains an open question.

Coulomb barrier. Electromagnetic zone of resistance surrounding protons (or other electrically charged particles) that tends to repel other protons (or other particles of like charge).

Creationism. Belief that the universe was created by God in the relatively recent past, as implied by literal interpretations of biblical chronology, and that the species of terrestrial life did not arise through Darwinian evolution but, rather, all came into existence at once.

Critical density.   The cosmic density of matter required to "close" the universe and so eventually to halt cosmic expansion. Its value amounts to about ten hydrogen atoms per cubic meter of space. The observed density is so close to the critical value that the question of whether the universe is open or closed has not yet been resolved by observation. See *open universe, closed universe*.

Dark matter.   Matter whose existence is inferred on the basis of dynamical studies—e.g., the orbits of stars in galaxies—but which does not show up as bright objects such as stars and nebulae. Its composition is unknown: It might consist of subatomic particles, or of dim dwarf stars or black holes, or a combination of various sorts of objects.

Darwinism.   Theory that species arise through the *natural selection* of random mutations that better fit changing conditions in a generally *uniformitarian* Earth.

Dead reckoning.   Navigation by recording one's heading, velocity, and elapsed time, with little or no reference to the stars.

Deceleration parameter.   Quantity designating the rate at which the *expansion of the universe* is slowing down, owing to the braking effect of the galaxies' gravitational tug on one another. It is a function of the *cosmic matter density*.

Declination.   Location on the sky in a north-south direction. Lines of declination are the celestial equivalent of latitude on Earth. Compare *right ascension*.

Decoupling.   Separation of classes of particles from regular interaction with one another, as in the decoupling of photons from particles of matter that produced the *cosmic background radiation*.

Deduction.   Process of reasoning in which a conclusion is derived from a given premise or premises, without a need for additional information. Compare *induction*.

Degree.   (1) A measure of temperature: Unless otherwise specified, all temperatures in this book are in degrees Kelvin. (To convert to Celsius, subtract 273.) (2) An angle subtended in the sky: From the zenith to the horizon is 90 degrees; the distance between the pointer stars of the Big Dipper is 5 degrees.

Detector.   Device for recording the presence of *subatomic particles*. A typical modern detector consists of an array of electronic sensors connected to a computer, capable of recording the paths of the particles as they fly out from the collision site in a particle *accelerator*.

Determinism.   The doctrine that all events are the predictable effects of prior causes. See *causation*.

Deuterium.   An isotope of hydrogen, the nucleus of which comprises one neutron plus one proton.

Dimension.   A geometrical axis.

Dirac equation.   Mathematical description of the electron, derived by Paul Dirac, that incorporates both *quantum mechanics* and *special relativity*.

DNA.   Deoxyribonucleic acid, the macromolecule that carries the genetic information requisite to life on Earth.

Doppler shift.   Change in the apparent wavelength of radiation (e.g., light or sound) emitted by a moving body. A star moving away from the observer will appear to be radiating light at a lower frequency than if at rest; consequently, lines in the star's spectrum will be shifted toward the red (lower frequency) end of the spectrum. The existence of a direct relationship between the *redshift* of light from galaxies and their distances is the fundamental evidence for the *expansion of the universe*.

Double star. See *binary star*.

Dwarf stars. Main-sequence stars with masses equal to or less than that of the sun. More generally, any star on or below the *main sequence* in the *Hertzsprung-Russell diagram*.

Dynamics. Study, in physics, of the motion and equilibrium of systems under the influence of *force*.

Dynamo. An electric generator that employs a spinning magnetic field to produce electricity.

Eccentrics. In Ptolemaic cosmology, displacement of the center of a rotating celestial sphere from the center of the universe.

Eclipse. Obscuration of one astronomical object (such as the sun) by another such object (such as the moon).

Electrodynamics. Study of the behavior of *electromagnetic force* in motion.

Electromagnetic force (or interaction). Fundamental force of nature that acts on all electrically charged *particles*. Classical electromagnetics is based on Maxwell's and Faraday's equations, quantum electromagnetics on the theory of quantum electrodynamics (*QED*).

Electrons. Light elementary particles with a negative electrical charge. Electrons are found in shells surrounding the nuclei of *atoms;* their interactions with the electrons of neighboring atoms create the chemical bonds that link atoms together as *molecules*.

Electron shells. Zones in which the electrons in *atoms* reside. Their radius is determined by the quantum principle, their population by the *exclusion principle*.

Electronuclear force. Single fundamental force thought to have functioned in the very early universe and to have combined the attributes thereafter parceled out to the *electromagnetic* and the *strong* and *weak nuclear* forces. See *grand unified theory*.

Electron volt. Measure of energy, equal to 1.6×10^{-12} erg.

Electroweak theory. Theory demonstrating links between the *electromagnetic* and the *weak nuclear* forces. Indicates that in the high energies that characterized the very early universe, electromagnetism and the weak force functioned as a single, electroweak force. Also known as the Weinberg-Salam theory.

Ellipse. A plane curve in which the sum of the distances of each point along its periphery from two points—its "foci"—are equal.

Emission lines. Bright lines produced in a *spectrum* by a luminous source, such as a star or a bright nebula. Compare *absorption lines*.

Empiricism. An emphasis on sense data as a source of knowledge, in opposition to the rationalist belief that reasoning is superior to experience.

Energy. (1) The capacity to do work. (2) Manifestation of a particular variety of *force*.

Epicycles. In Ptolemaic cosmology, a circular orbit around a point that itself orbits another point.

Escape velocity. The speed at which an object can leave another object behind,

without being recalled by its gravitational force. The escape velocity of Earth
—which must, for instance, be attained by a spacecraft if it is to reach another
planet—is 25,000 miles per hour.

Euclidean geometry. See *geometry*.

Evolution. (1) In biology, the theory that complex and multifarious living things
developed from generally simpler and less various organisms. (2) In astronomy,
the theory that more complex and varied atoms develop from simpler ones, as
through the synthesis of heavy atomic nuclei in stars.

Exclusion principle. The rule that no two *fermions* can occupy the same quantum state.

Expansion of universe. Constant increase, with time, in the distance separating
distant galaxies from one another. Expansion does not take place within individual galaxies or clusters of galaxies, which are bound together gravitationally,
but evidences itself on the supercluster level.

Fermilab. The Fermi National Accelerator Laboratory, in Batavia, Illinois.

Fermions. Particles with half-integral spin. Fermions obey the *exclusion principle*,
which says that no two fermions can exist in an atom in the same quantum
state; in practice this restricts the number of electrons, which are fermions,
permitted in each electron shell.

Fermi's question. The question of why, if spacefaring extraterrestrial civilizations exist, their representatives haven't visited Earth.

Feynman diagram. Schematic representation of an *interaction* between *particles*.

Field. Domain or environment in which the real or potential action of a *force*
can be described mathematically at each point in space.

Fission, nuclear. Interaction in which *nucleons* previously united in an atomic
nuclei are disjoined, releasing energy. Fission powers "atomic" bombs. Compare
fusion.

Flatness problem. The riddle of why the universe is neither dramatically *open*
nor *closed*, but appears to be almost perfectly balanced between these states.

Flavor. Designation of quark types—up, down, strange, charmed, top, and
bottom. Flavor determines how the *weak nuclear force* influences *quarks*.

Force. Agency responsible for a change in a system. In Newtonian mechanics,
gravitational force bends the moon away from the straight trajectory it would
otherwise pursue.

Fossils. Geological remains of what was once a living thing.

Fraunhofer lines. Dark lines in a *spectrum*.

Fusion, nuclear. Interaction in which *nucleons* are forged together, creating new
atomic nuclei and releasing energy. Fusion powers "hydrogen" bombs.

Galactic disk. The plate-shaped component of a spiral *galaxy*, in which the
spiral arms are found.

Galactic halo. A spherical aggregation of stars, globular star clusters, and thin
gas clouds, centered on the nucleus of the galaxy and extending beyond the
known extremities of the galactic disk.

Galaxy. A large aggregation of stars, bound together gravitationally. There are
three major classifications of galaxies—spiral, elliptical, and irregular—and several subclassifications. The sun belongs to a spiral galaxy, the *Milky Way galaxy*.

Gamma rays.   Extremely short-wavelength electromagnetic energy.

Gauge theory.   Account of forces that views them as arising from broken *symmetries*.

Geocentric cosmology.   School of ancient theories that depicted the earth as standing, immobile, at the center of the universe.

Geology.   Scientific study of the dynamics and history of the earth, as evidenced in its rocks, chemicals, and fossils.

Geometry.   The mathematics of lines drawn through space. In euclidean geometry, space is postulated to be "flat," i.e., to be the three-dimensional analog of a plane. In noneuclidean geometry, space is "curved," i.e., is the three-dimensional analog of a sphere or a hyperbola.

GeV.   One billion (10^9) *electron volts*. Sometimes written as one BeV.

Giant stars.   High-luminosity stars that lie above the *main sequence* on the *Hertzsprung-Russell diagram*.

Globular clusters.   See *star clusters*.

Glueballs.   Theoretical particles made exclusively of *gluons*. Tentative evidence of the existence of glueballs had been found in accelerator experiments by the mid-1980s.

Gluon lattice.   Force field generated by the *strong nuclear force* that holds quarks together. See *gluons*.

Gluons.   Quanta that carry the *strong nuclear force*. Like photons, vector bosons, and gravitons—the carriers respectively of electromagnetism, the weak force, and gravitation—gluons are massless bosons. Consequently, for simplicity's sake, some physicists lump together all the force-carrying quanta under the term "gluons."

Grand unified theories.   Class of theories that purport to reveal identities linking the *strong* and *electroweak* forces. The differences between these forces in nature today is attributed to the breaking of symmetrical relationships among force-carrying particles as the very early universe expanded and cooled.

Gravitational force (or interaction).   Fundamental force of nature, generated by all particles that possess *mass*. Interpreted by means of Newtonian mechanics or by the general theory of *relativity*.

Gravitinos.   Hypothetical force-carrying particles predicted by *supersymmetry* theories. The gravitino's spin would be ½. Its mass is unknown.

Gravitons.   The quanta thought to convey *gravitational force;* analogous to the photons, gluons, and intermediate vector bosons of electromagnetism and the strong and weak nuclear forces. Predicted by quantum theory of gravity, gravitons have not yet been detected.

Gravity.   (1) In Aristotelian physics, an innate tendency of the elements earth and water to fall. (2) In Newtonian physics, the universal, mutual, attraction of all massive objects for one another; its force is directly proportional to the mass of each object, and decreases by the square of the distance separating the objects involved. (3) In Einstein's general *relativity*, gravity is viewed as a consequence of the curvature of space induced by the presence of a massive object. In quantum mechanics the gravitational field is said to be conveyed by quanta called *gravitons*.

Great year. Ancient concept of a celestial and historical cycle, its duration roughly a thousand or ten thousand years, at the end of which there is universal destruction and a new great year begins.

GUT. Acronym for *grand unified theory*.

Hadrons. Elementary particles that are influenced by the *strong nuclear force*. There are two sorts of hadrons—mesons, which have integral spin, and baryons, which have spin ½ or ³⁄₂.

Half-life. The time it takes for half of a given quantity of radioactive material to decay.

Halo, galactic. See *galactic halo*.

Heliocentric cosmology. School of models in which the sun was portrayed as standing at the center of the universe.

Hermetic. Of or relating to Hermes Trismegistus, a mythical philosopher beloved of the Neoplatonists and usually identified with ancient Egypt.

Hertz. A unit of frequency equal to one cycle (or wave) per second.

Hertzsprung-Russell diagram. Plot that reveals a relationship between the colors and absolute *magnitudes* of stars.

Higgs field. Mechanism operating in symmetry-breaking events; in *electroweak theory*, the Higgs field is said to have imparted mass to the *W and Z particles*.

High-energy physics. See *particle physics*.

Horizon problem. A quandary in standard *big bang theory*, which indicates that few of the particles of the early universe would have had time to be in causal contact with one another at the outset of cosmic expansion. It appears to have been resolved in the *inflationary universe theory*.

Hubble constant. The rate at which the universe expands, equal to approximately fifty kilometers of velocity per *megaparsec* of distance.

Hubble diagram. Plot of galaxy *redshifts* against their distances. This was the first evidence of the *expansion of the universe*.

Hubble law. That distant galaxies are found to be receding from one another at velocities directly correlated to their distances apart.

Hyperbolic space. See *geometry*.

Hyperdimensional. Involving more than the customary four dimensions (three of space plus one of time) of relativistic space-time.

Hypothesis. A scientific proposition that purports to explain a given set of phenomena; less comprehensive and less well established than a *theory*.

Indeterminacy principle. Quantum precept indicating that the position and trajectory of a particle cannot both be known with perfect exactitude. Indeterminacy thus indicates the existence of a basic quantum of knowledge of the particle world. And, since information about one quantity can be extracted at the expense of another, it demonstrates that the answers we obtain about natural events result to some extent from the questions we choose to ask about them.

Induction. System of reasoning in which the conclusion, though implied by the premises and consistent with them, does not necessarily follow from them.

Inertia. Quality of *mass*, such that any massive particle tends to remain at rest relative to a given reference frame, and to remain in constant motion once in motion, unless acted upon by a *force*.

Inflationary universe. Theory that the expansion of the very early universe proceeded much more rapidly than it does today—at an exponential rather than a linear rate.

Infrared light. Electromagnetic radiation of a slightly longer wavelength than that of visible light.

Infrared slavery. Inability of *quarks* to escape the bonds of the strong force that confines them to the company of other quarks.

Initial condition. (1) In physics, the state of a system at the time at which a given interaction begins—e.g., the approach of two electrons that are about to undergo an *electromagnetic* interaction. (2) In cosmology, a quantity inserted as a given in cosmogonic equations describing the early universe.

Intelligence. Defined in *SETI* as the ability and willingness to transmit electromagnetic signals across interstellar space.

Interaction. Event involving an exchange between two or more *particles*. Since the fundamental forces are portrayed by quantum theory as involving the exchange of force-carrying particles (the *bosons*), the forces are more correctly described as interactions.

Interferometer. A device for observing the interference of waves of light or similar emanations caused by a shift in the phase or wavelength of some of the waves.

Intermediate vector bosons. See *W, Z particles*.

Inverse-square law. In Newtonian mechanics, the rule that the measured intensity of light diminishes by the square of the distance of its source—so that, e.g., if stars A and B are of equal absolute *magnitude* but star B is twice as distant, it will appear to be one quarter as bright as star A.

Invisible astronomy. The study of celestial objects by observing their radiation at wavelengths other than those of visible light.

Island universe hypothesis. Assertion that the sun belongs to a *galaxy* and that the *spiral nebulae* are other galaxies of stars, which in turn are separated from one another by vast voids of space. Compare *nebular hypothesis*.

Isotopes. Atoms having the same number of *protons* in their nuclei but different numbers of *neutrons*, with the result that their mass differs though they may have the same number of *electrons*.

Isotropy. Quality of being the same in all directions. Compare *anisotropy*.

Jet Propulsion Laboratory. NASA installation near Pasadena, operated by *Caltech* and specializing in unmanned space exploration.

Jovian. Giant planets that have a gaseous surface; the sun's known Jovian planets are Jupiter, Saturn, Uranus, and Neptune.

Kaluza-Klein theory. Five-dimensional relativity theory that played a role in the development of *unified theory*.

Latitude. On Earth, distance north or south of the equator along a line connecting the poles.

Lattice. See *gluon*.

Law. A theory of such wide and invariable application that its violation is thought to be impossible.

Leptons. Elementary *particles* that have no measurable size and are not influenced by the *strong nuclear force*. Electrons, muons, and neutrinos are leptons.

Light.   Electromagnetic radiation with wavelengths of or close to those detectable by the eye.

Light-year.   The distance light travels in one year, equal to 5.8×10^{12} (about six trillion) miles.

Local Group.   The association of galaxies to which the *Milky Way galaxy* belongs.

Longitude.   On Earth, distance east or west of Greenwich, England, measured along lines drawn parallel to the equator.

Lookback time.   Phenomenon that, owing to the finite velocity of light, the more distant an object being observed, the older is the information received from it. A galaxy one billion light-years away, for instance, is seen as it looked one billion years ago.

Lorentz contraction.   Diminution in the observed length of an object along the axis of its motion, as perceived by an external observer who does not share its velocity.

Luminosity.   The intrinsic brightness of a star. Usually defined in terms of absolute *magnitude*.

M.   Designation of objects in the Messier catalog of nebulae, star clusters, and galaxies, published in the eighteenth century.

Mach's principle.   Precept that the inertia of objects results not from their relationship to Newtonian *absolute space*, but to the rest of the mass and energy distributed throughout the universe. Though unproved and perhaps unprovable, Mach's principle inspired Einstein, who sought with partial success to incorporate it into the general theory of *relativity*.

Magellanic Clouds.   Two galaxies that lie close to the *Milky Way galaxy*. They are visible in the southern skies of Earth.

Magnetic monopole.   A massive particle with but one magnetic pole, the production of which is indicated in some theories of the early universe.

Magnitude.   The brightness of a star or planet, expressed on a scale in which lower numbers mean greater brightness. *Apparent magnitude* indicates the brightness of objects as we see them from Earth, regardless of their distance. *Absolute magnitude* is defined as the apparent magnitude a star would have if viewed from a distance of ten *parsecs*, or 32.6 light-years. Each step in magnitude equals a difference of 2.5 times in brightness: The brightest stars in the sky are apparent magnitude 1; the dimmest, 6. The magnitudes of extremely bright objects are expressed in negative values—e.g., the apparent magnitude of the sun is about -26.

Main sequence.   The curving path in the *Hertzsprung-Russell diagram* along which most stars lie.

Many body problem.   The difficulty of calculating the interactions—e.g., the Newtonian gravitational interactions—of three or more objects.

Mass.   Measure of the amount of matter in an object. Inertial mass indicates the object's resistance to changes in its state of motion. Gravitational mass indicates its response to the *gravitational force*. In the general theory of *relativity*, gravitational and inertial mass are revealed to be aspects of the same quantity.

Materialism.   Belief that material objects and their interactions constitute the complete reality of all phenomena, including such seemingly insubstantial phenomena as thoughts and dreams. Compare *spiritualism*.

Matter waves. Characteristic by virtue of which matter, like energy, displays the qualities of waves as well as of particles. See *wave-particle duality*.

Mechanics. The study, in physics, of the influence of *forces*.

Megaparsec. One million (10^6) *parsecs*.

Mesons. See *hadrons*.

Metals. In astrophysics, all elements heavier than helium.

MeV. One million (10^6) *electron volts*.

Micrometry. The measurement of the apparent sizes and separations of astronomical objects by use of knife blades or crosshairs in the eyepiece of a telescope. If the distance of an object is known, its size can be determined through micrometry.

Microwave background. See *cosmic background radiation*.

Microwaves. *Radio* radiation with wavelengths of about 10^{-4} to 1 meter, equal to 10^9 to 10^{13} *hertz*.

Mile. The mile employed in this book is the statute mile, equal to 5,280 feet.

Milky Way. A softly glowing band of light that bisects the skies of Earth, produced by light from stars and nebulae in the *galactic disk*.

Milky Way galaxy. The spiral galaxy in which the sun resides.

Million. A thousand thousand (10^6).

Minor planets. See *asteroids*.

Missing matter. Alternate term for *dark matter*.

MIT. Massachusetts Institute of Technology, in Cambridge, Mass.

Molecules. The smallest units of a chemical compound. A molecule is composed of two or more atoms, linked by interactions of their *electrons*.

Monopole. See *magnetic monopole*.

Muon. Short-lived elementary particle with negative electrical charge. Muons are *leptons*. They resemble *electrons*, but are 207 times more massive.

Natural philosophy. A term widely employed in the seventeenth century to mean what today is encompassed in the word *science*.

Natural selection. Tendency of individuals better suited to their environment to survive and perpetuate their species, leading to changes in the genetic makeup of the species and, eventually, to the origin of new species. See *evolution*.

Nautical mile. Equals 1.15 statute *miles*.

Nebulae. Indistinct, nonterrestrial objects visible in the night sky. "Bright" nebulae glow with light emitted by the gas of which they are composed ("emission" nebulae) or by reflected starlight ("reflection" nebulae) or both. "Dark" nebulae consist of clouds of gas and dust that are not so illuminated. "Planetary" nebulae are shells of gas ejected by stars. *Spiral nebulae* are *galaxies*.

Nebular hypothesis. Hypothesis, maintained in the nineteenth and early twentieth century, that the *spiral nebulae* are not galaxies but are instead whirlpools of gas from which new systems of stars and planets are condensing. Compare *island universe* theory.

Neutrinos. Electrically neutral, massless particles that respond to the *weak nuclear force* but not the *strong nuclear* and *electromagnetic* forces.

GLOSSARY

Neutrons. Electrically neutral, massive particles found in the nuclei of *atoms*. Each neutron is composed of one up *quark* and two down quarks; its mass is 939.6 MeV, slightly more than that of the proton. Stable within the nucleus, the neutron if isolated decays, with a *half-life* of fifteen minutes.

Neutron stars. Stars with gravitational fields so intense that most of their matter has been compressed into neutrons. They are formed when massive stars run out of nuclear fuel and collapse. Many rotate rapidly and generate radio pulses; when detected by *radiotelescopes*, they are known as *pulsars*.

NGC. Designates entries in the New General/Catalog of nonstellar objects.

Noneuclidean geometry. See *geometry*.

Nova. A star that brightens suddenly and to an unprecedented degree, creating the impression that a new star has appeared where none was before. Hence the name, from *nova* for "new." See *supernova*.

Nuclei. (1) The central part of atoms, composed of *protons* and *neutrons* (which are made of *quarks*) and containing nearly all of each atom's *mass*. (2) The central region of a galaxy.

Nucleons. Protons and neutrons, the constituents of atomic *nuclei*.

Nucleosynthesis, nucleogenesis. The fusion of *nucleons* to create the nuclei of new atoms. Nucleosynthesis takes place in stars, and, at an accelerated rate, in *supernovae*.

Observational cosmology. The application of observational data to the study of the universe as a whole.

Open clusters. See *star clusters*.

Omega. Index of the matter density of the universe, defined as the ratio between its actual density and the "critical density" required to "close" the universe and eventually halt its expansion. If omega equals more than 1, the universe is "open" and will expand forever. See *cosmic matter density*.

Oort Cloud. Home of most solar system *comets*.

Open universe. Cosmological model in which the universe continues to expand forever; its space-time geometry is hyperbolic, or "open."

Optics. The science of light.

Oscillating universe. Cosmological model in which the universe is "closed" and its expansion is destined to stop, to be succeeded by collapse and "then" (if ordinary temporal terms may be said to apply) by a rebound into a new expansion phase.

Panstellar. Of or pertaining to more than one star.

Paradox. A self-contradictory proposition. Paradoxes are most useful when they seem most likely to be true, for it is then that they best serve to expose flaws in the data or reasoning that led to their appearance.

Parallax. The apparent displacement in the position of a star or planet occasioned by its being viewed from two different locations—e.g., by observing it from two widely separated stations on Earth, or at intervals of six months, when the earth is at either extreme of its orbit around the sun. The resulting angle can be used, by *triangulation*, to determine the distance of the star or planet.

Parsec. Astronomical unit of distance, equal to 3.26 light-years.

Particle accelerator. See *accelerator*.

GLOSSARY 405

Particle physics. The branch of science that deals with the smallest known structures of matter and energy. As their experimental investigation usually involves the application of considerable energy, particle physics overlaps with *high-energy physics*.

Particles. Fundamental units of matter and energy. All may be classed as *fermions*, which have half-integral spin and obey the *exclusion principle*, and *bosons*, which have integral spin and do not obey the exclusion principle. The term *particle* is metaphoric, in that all subatomic particles also evince aspects of wavelike behavior.

Pauli exclusion principle. See *exclusion principle*.

Period-luminosity function. The relationship between the absolute *magnitude* and period of variability of *Cepheid variable* stars.

Phase transition. An abrupt change in the equilibrium state of a system, as evoked by the cooling of the early universe as it expanded.

Photon decoupling. The release of *photons* from constant collisions with massive particles as the universe expanded and its matter density diminished. See *decoupling*.

Photons. The *quanta* of the *electromagnetic force*. The name comes from the fact that light is a form of electromagnetism. Photons have zero rest mass and can therefore travel infinitely far.

Physics. The scientific study of the interactions of matter and energy.

Planck epoch, Planck time. The first instant following the beginning of the expansion of the universe, when the *cosmic matter density* was still so high that *gravitational force* acted as strongly as the other fundamental forces on the subatomic scale.

Planck's constant. The fundamental quantity of action in quantum mechanics.

Planet. An astronomical object more massive than an *asteroid* but less so than a *star*. Planets shine by reflected light; stars generate light of their own.

Plasma. A state in which matter consists of electrons and other subatomic particles without any structure of an order higher than that of atomic *nuclei*.

Platonic solids. The five regular polyhedrons—the tetrahedron, octahedron, hexahedron, icosahedron, and dodecahedron—esteemed by Plato as embodying aesthetic and rational ideals.

Plurality of worlds. Hypothesis that the universe contains inhabited planets other than Earth.

Pole star. The star—Polaris—that lies near the direction in the sky toward which the north pole of the earth points.

Positron. The *antimatter* twin of the electron.

Post hoc fallacy. The erroneous assumption that, because B follows A, B therefore was caused by A. More strictly, the fallacy of calculating, in retrospect, the odds of B's having occurred by adding up a long sequence of such putative causes.

Precession. The slow (once per twenty-six thousand years) gyration of the earth's axis.

Proper motion. Individual drifting of stars through space.

Protogalaxy. A galaxy in the process of formation. None are observed nearby, indicating that all or most galaxies formed long ago.

Proton. A massive *particle* with positive electrical charge found in the nuclei of atoms. Composed of two up *quarks* and one down quark. The proton's mass is 938.3 MeV, slightly less than that of the *neutron*.

Proton decay. Spontaneous disintegration of the proton, predicted by *grand unified theory* but never observed experimentally.

Proton-proton reaction. An important *nuclear fusion* reaction that occurs in stars. It begins with the fusion of two hydrogen *nuclei*, each of which consists of a single *proton*.

Pulsars. See *neutron stars*.

QCD. See *quantum chromodynamics*.

QED. See *quantum electrodynamics*.

Quadrant. An instrument, based on a quarter of a circle, employed to measure the altitude above the horizon of astronomical bodies. Eventually replaced by the *sextant*.

Quanta. Fundamental units of energy.

Quantum chromodynamics. The quantum theory of the *strong nuclear force*, which it envisions as being conveyed by quanta called *gluons*. The name derives from the assignment of a quantum number called *color* to designate how *quarks* function in response to the strong force.

Quantum electrodynamics. The quantum theory of the electromagnetic force, which it envisions as being carried by quanta called *photons*.

Quantum genesis. Hypothesis that the origin of the universe may be understood in terms of a quantum chance.

Quantum leap. The disappearance of a subatomic *particle*—e.g., an electron—at one location and its simultaneous reappearance at another. The counterintuitive weirdness of the concept results in part from the limitations of the particle metaphor in describing a phenomenon that is also in many respects a wave.

Quantum mechanics. See *quantum physics*.

Quantum physics. Physics based upon the quantum principle, that energy is emitted not as a continuum but in discrete units.

Quantum space. Vacuum with the potential to produce *virtual particles*.

Quantum tunneling. A *quantum leap* through a barrier.

Quarks. Fundamental particles from which all *hadrons* are made. According to the theory of quantum chromodynamics, *protons*, *neutrons*, and their higher-energy cousins are composed of trios of quarks, while the *mesons* are each made of one quark and one antiquark. Held together by the *strong nuclear force*, quarks are not found in isolation in nature today; see *asymptotic freedom*.

Quasars. Pointlike sources of light whose redshifts indicate that they lie at distances of billions of light-years. Thought to be the *nuclei* of young *galaxies*.

Radio. Long-wavelength electromagnetic radiation.

Radioactivity. Emission of *particles* by unstable elements as they decay.

Radio astronomy. Study of the universe at the *radio* wavelengths of electromagnetic energy.

Radiocarbon dating. Determination of the age of a substance containing radioactive carbon by means of its radioactive *half-life*.

Radiometric dating. Determination of the age of objects—e.g., earth and moon rocks—by means of the *half-life* of the unstable elements they contain.

Radiotelescopes. Sensitive radio antennae employed to detect the radio energy emitted by nebulae, galaxies, pulsars, etc.

Recombination. The capture of an electron by a proton. Numerous recombinations are thought to have occurred when the universe was a little less than one million years old, resulting in the formation of electron shells around helium and hydrogen nuclei to create complete *atoms*.

Red giants. Large stars with an atmosphere that is relatively cool, and therefore looks redder in color than does that of a *main sequence* star.

Redshift. Displacement of the spectral lines in light coming from the stars of distant galaxies, thought to be produced by the velocity of the galaxies outward in the expanding universe. See *Hubble law*.

Redshift-distance relation. The correlation between redshift in the spectra of galaxies and their distances. See *Hubble law*.

Relativistic. Approaching the velocity of light. Particles moving at these speeds demonstrate effects predicted by the special theory of relativity—increased mass, slowing of time, etc.—that must be taken into account by combining relativity with quantum theory if accurate predictions are to be made.

Relativity, general theory of. Einstein's theory of *gravitational force*.

Relativity, special theory of. Einstein's theory of the electrodynamics of moving systems.

Renaissance. Generally, the period of cultural awakening in the West beginning at about 1350 and ending with the death of Giordano Bruno in 1600 or of Shakespeare in 1616.

Renormalization. The removal of nonsensical infinities from quantum mechanics equations by a mathematical procedure in which other infinities are introduced in order to cancel them.

Retrograde. Apparent motion of a planet in a direction opposite to its normal progress across the sky, produced by the orbital motion of the earth.

Right ascension. Location in the sky along an east-west direction; the celestial equivalent of longitude. Compare *declination*.

Royal Society. English organization founded in the seventeenth century and dedicated to the advancement of science.

Satellite. An object in orbit around another, more massive object.

Scholastics. Adherents to the philosophy and cosmology of Aristotle. Their dominance in the universities, which had been founded largely to study Aristotle, constituted an obstacle to acceptance of the Copernican system advocated by Kepler and Galileo.

Science. Systematic study of nature, based upon the presumption that the universe is based upon rationally intelligible principles and that its behavior can therefore be predicted by subjecting observational data to logical analysis.

SETI. The Search for Extraterrestrial Intelligence, by using *radiotelescopes* to listen for signals transmitted by intelligent alien beings.

Sextant. Instrument employed to measure the elevation of astronomical objects above the horizon. Based upon an arc equal to a sixth of a circle, sextants are more compact and easier to use than are the *quadrants* that preceded them.

Shadow matter. Theoretical classes of particles, their existence intimated by *supersymmetry* theory, that participate in few if any of the four known fundamental *forces*. Planets, stars, and galaxies made of shadow matter could conceivably exist in the same space and time we occupy without our sensing their presence.

Singularity. A point of infinite curvature of space where the equations of general relativity break down. A *black hole* represents a singularity; so, perhaps, did the universe at the first moment of time.

Solar system. The sun, its planets, and the asteroids and comets that, like the planets, orbit the sun.

Space. Traditionally, the three-dimensional theater within which events transpire, explicable by means of euclidean geometry. In *relativity*, space is depicted in terms of noneuclidean geometries as well. In *quantum physics*, space may be constructed out of any of a variety of abstractions, such as a "charge space" employed in dealing with electrically charged particles or the "color space" in which *quarks* can for convenience be plotted. See *geometry*.

Space-time. Arena in which events are depicted in the theory of *relativity*. The orbit of a planet, for instance, can be described as a "world line" in a four-dimensional space-time continuum.

Spectrograph. A device, usually based on a finely etched grate that performs the function of a prism, for breaking up light into its constituent parts and making a photographic or electronic record of the resulting *spectrum*. When lacking a means for recording the spectrum, the device is called a spectroscope.

Spectroscopic binary. A double star in which the individual stars cannot be resolved, but can be detected through their effects on the *spectrum* of the system—e.g., the relative motions of the stars may be detected from *Doppler shifts* in the spectral lines of starlight.

Spectroscopy. Scientific investigation of an object by studying its *spectrum*.

Spectrum. A record of the distribution of matter or energy (e.g., light) by wavelength. Spectra can be studied to learn the chemical composition and motion of stars and galaxies.

Spherical space. See *geometry*.

Spin. The intrinsic angular momentum of an elementary *particle*, as by the particle's spinning on its axis. Spin is quantized in units of Planck's constant of action, h, so that, e.g., "spin 1," means spin = $1h$. Particles with integral spin (0, 1) are called *bosons;* those with half spin are *fermions*.

Spiral nebulae. See *nebulae*.

Spiritualism. Belief that material interactions alone cannot account for all phenomena, and that some—e.g., thought—are due to the fundamentally insensible actions of intangibles.

Standard model. The theories of the four *forces*, which, taken together, can predict the outcome of every known fundamental *interaction*.

Star. A celestial object that generates energy by means of nuclear fusion at its core. To do this it must have more than about 0.08 the sun's mass. If, for instance, the planet Jupiter were some fifty to one hundred times more massive than it is, fusion reactions would transpire in its core and it would be a star. See *planet*.

Star clusters. Gravitationally bound aggregations of stars, smaller and less massive than galaxies. "Globular" clusters are the largest category; they are old, and may harbor hundreds of thousands to millions of stars, and are found both within and well away from the *galactic disk*. "Open" clusters are smaller, have a wide range of ages, and reside within the disk.

Statute mile. See *mile*.

Steady state. Theory that the expanding universe was never in a state of appreciably higher density—i.e., that there was no "big bang"—and that matter is constantly being created out of empty space in order to maintain the *cosmic matter density*.

Stellar evolution. The building of complex atomic nuclei from simpler nuclei in stars, with the result that succeeding generations of stars and planets contain a greater variety of chemical elements than did their predecessors. See *evolution*.

Stochastic cooling. The gathering (i.e., focusing) of clouds of subatomic *particles* in an accelerator by monitoring their scattering vectors and altering the magnetic environment in an accelerator *storage ring* to keep them close together. First employed in storing particles of *antimatter*, which are expensive to manufacture and ought not to be wasted.

Storage ring. A ring in which particles are kept in a circular motion, suspended in a magnetic field, until they can be injected into the larger ring of an *accelerator*.

String theory. Theory that subatomic *particles* actually have extension along one axis, and that their properties are determined by the arrangement and vibration of the strings.

Strong nuclear force (or interaction). Fundamental force of nature that binds *quarks* together, and holds *nucleons* (which are comprised of quarks) together as the nuclei of *atoms*. Portrayed in *quantum chromodynamics* as conveyed by quanta called *gluons*.

Subatomic. Of a scale smaller than that of an atom.

Subatomic particles. See *particles*.

Sum over histories. Probabilistic interpretation of a system's past, in which quantum *indeterminacy* is taken into account and the history is reconstructed in terms of each possible path and its relative likelihood.

Sun. The star orbited by the earth.

Supercluster. A cluster of clusters of galaxies. Superclusters are typically about one hundred million (10^8) light-years in diameter and contain tens of thousands of galaxies.

Superconducting super collider. A proposed *accelerator* of great size and high energy.

Supergiants. The largest and brightest class of stars.

Supernovae. The explosions of giant stars.

Superstring theory. Alternate name for *string theory*.

Supersymmetry. Class of theories that seek to identify symmetrical relationships linking fermions and bosons—i.e., particles of half-integral spin, like electrons, protons, and neutrinos, with those of integral spin, like photons and gluons. If attainable, a fully realized supersymmetry theory would provide a unified account of all four fundamental forces, and might well shed light on the very early evolution of the universe as well.

Superunified theory. Hypothetical theory that presumably would show how all four fundamental forces of nature functioned as a single force in the extremely early universe. The best current candidates for such a potential achievement are thought to be *supersymmetry* and *string theory.*

Symmetry.   State of a system such that it has a significant quantity that remains invariant after a transformation. More generally, an apt or pleasing proportion based upon such a state.

Symmetry breaking.   The loss of symmetry in a transformation. See *broken symmetry.*

Symmetry group.   A mathematical group with a common property that unites its members and evinces a symmetry.

Telescope.   A device for gathering and amplifying light or other energy. Refracting telescopes gather light by means of a lens, reflecting telescopes by means of a mirror. *Radiotelescopes* gather radio energy, typically by using a metallic dish antenna. Telescopes have also been built that can gather *X rays, gamma rays,* and other forms of energy.

TeV.   Equal to one teraelectron volt, or 1,000 *GeV.*

Tevatron.   A particle *accelerator* capable of attaining an energy of 1 TeV.

Theory.   A rationally coherent account of a wider range of phenomena than is customarily accounted for by a *hypothesis.*

Thermodynamics.   The study of the behavior of heat (and, by implication, other forms of energy) in changing systems.

Thought experiment.   An experiment that cannot be or is not carried out in practice, but can—given sufficient imagination and rigor—be reasoned through by thought and intuition alone.

Time.   A dimension distinguishing past, present, and future. In *relativity,* time is portrayed as a geometrical dimension, analogous to the dimensions of space.

Transit.   The passage of a smaller, nearer astronomical object across the face of a larger object in the background, as in a transit of Venus across the sun.

Triangulation.   Measurement of the distance of a planet or nearby star by sighting its apparent position against background stars from two or more separate locations. See *parallax.*

Trillion.   A thousand billion (10^{12}).

Ultraviolet light.   Electromagnetic radiation of a wavelength slightly shorter than that of visible light.

Unified theory.   In *particle physics,* any theory exposing relationships between seemingly disparate classes of *particles.* More generally, a theory that gathers a wide range of fundamentally different phenomena under a single precept, as in Maxwell's discovery that light and magnetism are aspects of a single, *electromagnetic force.*

Uniformitarianism.   The hypothesis that the extensive changes in the earth, as evinced in the geological record, have resulted, not from massive catastrophes, but from the slow operation of wind, weather, volcanism, and the like over many millions of years. Compare *catastrophism.*

Vacuum genesis.   Hypothesis that the universe began as nothingness, from which matter and energy arose by a process analogous to the appearance of *virtual particles* from a vacuum.

Variable star.   A star that changes in brightness periodically.

Virgo Cluster. A nearby cluster of galaxies.

Virgo Supercluster. An aggregation of galaxies—roughly ten thousand of them—to which the *Virgo Cluster* and our own galaxy belong.

Virtual particles. Short-lived particles that arise from a vacuum. Their existence is permitted by the *indeterminacy principle*.

Voyager. Pair of unmanned American spacecraft launched in 1977 on missions to Jupiter, Saturn, and beyond.

Wave function. A quantum mechanical expression that describes all the relevant properties of a particle.

Wave-particle duality. Quantum realization that *particles* of matter and energy also exhibit many of the characteristics of waves.

Waves. Propagation of energy by means of coherent vibration.

Weak nuclear force (or interaction). Fundamental force of nature that governs the process of *radioactivity*. It is currently accounted for by the *electroweak theory*.

Weinberg-Salam theory. See *electroweak theory*.

World line. In *relativity*, the path traced out in four-dimensional *space-time* by a given object or *particle*.

W particles. Massive *bosons* thought to have been abundant in the early universe, when the unified *electroweak* force was manifest.

X rays. Short-wavelength electromagnetic energy. The X ray portion of the electromagnetic spectrum lies between the realms of *gamma rays* and that of *ultraviolet light*.

Yang-Mills theory. See *gauge theory*.

Zoo hypothesis. Hypothesis that life on Earth has been detected by intelligent extraterrestrials who scruple not to visit us because they do not wish to interfere with our development.

Z particles. Massive *bosons* thought to have been abundant in the early universe, when the unified *electroweak* force was manifest.

A BRIEF HISTORY OF THE

UNIVERSE

Time	Noteworthy Events*
0	Origin of time, space, and energy—of the universe as we know it.
10^{-43} second ABT[†]	End of Planck epoch; gravitational radiation comes out of thermal equilibrium with the rest of universe.
10^{-34} second	Universe, in vacuum state, begins "inflating"—i.e., expanding at an exponential rate, some 10^{50} times present expansion rate.
10^{-30} second	Inflationary epoch ends; particles precipitate out of the vacuum.
10^{-11} second	Symmetry-breaking phase transition shatters the electroweak force into the electromagnetic and weak nuclear forces.
10^{-6}–10^{-5} second	Quarks and antiquarks cease mutual annihilation. The survivors link up in trios as protons and neutrons, the components of all future atomic nuclei.
10^{-4} second	Universe 1/10,000 second old. Constant capture of electrons and positrons turns neutrons into protons and vice versa. As slightly more energy is required to make neutrons than protons, the process leaves the universe with five times as many protons as neutrons.

*Most dates—and, for that matter, events—are approximate.
†ABT = After the beginning of time.

10^{-2} second	Particles of matter and energy interact in thermal equilibrium.
1 second	Neutrinos, previously embroiled with other particles, decouple and go their own way.
3 minutes 42 seconds	Protons and neutrons have linked up, forming nuclei of helium. Universe now composed of about 20 percent helium nuclei, 80 percent hydrogen.
1 hour	Universe has cooled to the point that most nuclear processes have stopped.
1 year	Ambient temperature of universe about that of the center of a star.
10^6 years	Origin of cosmic background radiation. Photons decouple, leaving electrons free to combine with nuclei, forming stable atoms. Hereafter, matter can begin to congeal into galaxies and stars.
10^9 years ABT ($\approx$ 17 billion years BP)	Protogalaxies, globular clusters forming. Epoch of quasars begins.
4.5 billion years BP*	Sun and planets congeal from a cloud of gas and dust in a spiral arm of the Milky Way galaxy.
3.8 billion years BP	Earth has cooled sufficiently for solid crust to form; age of oldest dated terrestrial rocks.
3.5–3.2 billion years	Microscopic living cells evolve on Earth.
1.8–1.3 billion years	Plants appear. Oxygen poisons Earth's atmosphere, and aerobic ("oxygen-loving") organisms proliferate.
900–700 million years	Advent of sex accelerates the pace of biological evolution.
700 million years	Animals—mostly flatworms and jellyfish—appear.
600 million years	First crustaceans.
500 million years	First vertebrates.
425 million years	Life migrates to dry land.
395 million years	First insects.
325 million years	First land vertebrates.
200 million years	First mammals.
180 million years	North America separates from Africa; genesis of the Atlantic.

*BP = Before Present

100 million years	Half a galactic year ago; Earth looks out on the other side of the universe.
70 million years	Preprimates evolve.
55 million years	Early horses appear.
35 million years	Early cats, dogs.
24 million years	Appearance of grass.
21 million years	Apes, monkeys depart along separate evolutionary pathways.
20 million years	Atmosphere approaches modern composition.
15 million years	Antarctica freezes over.
11 million years	Grazing animals proliferate.
5 million years	Apeman diverges from chimpanzee family.
3.7 million years	Apemen walk upright.
3.5 million years	Onset of latest series of ice ages.
1.8–1.7 million years	*Homo erectus*, "first true man," in China.
600,000 years	*Homo sapiens* emerges.
360,000 years	Controlled use of fire common among genus *Homo*.
150,000 years	Woolly mammoth roam.
100,000 years	Stars take on the forms of the recognizable modern constellations.
40,000 years	Invention of complex language; modern humans flourish.
35,000 years	Neanderthal man disappears. First musical instruments are crafted.
20,000–15,000 years	Agriculture invented.
19,000 years	Peopling of the Americas begins.
18,000 years	Animals are herded by humans.
14,000 years	Invention of fishhooks.
13,000 years	Development of ceramic pottery.
10,000 years	Cultivation of wheat, rice begins.
6,700 years	Early Babylonian calendar in use.
6,200 years	Refined solar calendar employed.
6,500 years	Copper is smelted.
5,600 years	First taxes.

5,500 years BP (= 3,500 BC)	Development of writing.

3,600–3,400 BC	Cotton cultivated in Peru, Mexico.
2,500 years	Stonehenge built.
2,200 years	Systematic astronomy in Egypt, Babylonia, India, China.
1,500 years	Sundial invented, in Egypt.
1,000 years	Homer declaims the *Odyssey*.
800 years	Olmec culture in Mexico.
700 years	Hesiod, *Works and Days*.
650 years	Mayan culture in Guatemala.
600 years	Lao-tzu, Confucius, Buddha, Zoroaster; Old Testament in Hebrew.
540 years	Pythagoras teaches that "all is number" and that nature is harmonious.
450 years	Leucippus and Democritus propose that matter is made of indivisible entities, the atoms. Paradoxes of Zeno raise doubts about the concept of the infinitesimal.
400 years	Plato teaches that the material world is but a shadow of a geometrically perfect reality. Aristotle, Eudoxus, theorize that universe is composed of crystalline spheres centered on Earth.
300 years	Euclid's geometry marries mathematical perfection to the world of experience.
260 years	Aristarchus of Samos hypothesizes that the earth orbits the sun in a gigantic universe.
100 years	Claudius Ptolemy constructs a complex geocentric cosmological model that "saves the appearances"—i.e., makes reasonably accurate predictions at the expense of claims to represent physical reality.
	Chinese seafarers reach the east coast of India.
60 BC	Lucretius writes *De Rerum Natura (On the Nature of Things)*, espousing Epicurean cosmology.

AD 325	Eusebius, chairman of the Council of Nicaea convened by the emperor Constantine, estimates that the world was created 3,184 years prior to the birth of Abraham.

400	Middle or Dark Ages begin; science dormant in the West.
455	Vandals sack Rome.
963	Al Sufi, in his *Book of the Fixed Stars*, mentions nebulae.
1001	Leif Ericsson reaches New England.
1276–1292	Marco Polo in Hangchow.
1400	Renaissance of learning commences in Europe.
1492	Columbus (re)discovers America.
1521	Cortez takes Mexico.
1522	Survivors of Magellan's final expedition complete circumnavigation of the globe.
1523	Pizarro takes Peru.
1543	Copernicus's *On the Revolutions* published.
1572	Tycho sees a nova (or "new star") in the sky, evidence against Aristotle's theory that the realm of the stars is unchanging and therefore unlike that of the earth.
1576	Thomas Digges in England publishes a defense of the Copernican cosmology in which he portrays the stars as distributed throughout infinite space.
1604	Galileo proposes that bodies fall with a uniformly accelerated motion, thus enunciating the first of the laws of classical dynamics.
	Kepler and Galileo observe a supernova.
1609	Galileo first observes the night sky through a telescope.
	Kepler demonstrates that the orbits of the planets are elliptical.
1611	Edition of the King James Bible published containing an estimate by James Ussher, bishop of Armagh, that "the beginning of time . . . fell on the beginning of the night which preceded the 23rd day of October, in the year . . . 4004 B.C."
1616	Roman Catholic Church bans all books that maintain that the earth moves.
1639	Transit of Venus observed by two English amateur astronomers.
1662	Royal Society chartered in London.
1665–1666	Isaac Newton, age twenty-three, home from college, re-

alizes that gravitational force obeying an inverse-square law would account alike for falling bodies on earth and the motion of the moon in its orbit.

1666   Newton observes spectrum produced by sunlight when shown through a prism.

1672   Opposition of Mars widely observed, by Richer at Cayenne and Cassini in Paris among others, leading to estimates of the distance from the earth to the sun of some eighty-one to eighty-seven million miles—90 percent of the correct value.

1675   Olaus Römer determines, from studying the satellites of Jupiter, that light has a finite velocity.

1684   Edmond Halley visits Isaac Newton at Trinity College, resurrects line of research that leads Newton to write the *Principia.*

1686   Bernard de Fontenelle's *Entretiens sur la Pluralité des Mondes* popularizes the idea that the universe contains many inhabited worlds.

1687   Newton's *Principia* published.

1716   Halley urges that future transit of Venus may be observed and timed in order to triangulate interplanetary distances.

1718   Halley finds that the bright stars Sirius, Aldebaran, Betelgeuse, and Arcturus have changed their position in the sky since Ptolemy's *Almagest* was compiled—first evidence of the "proper motion" of stars.

1719   John Strachey in England publishes notes on strata in the coal-rich district of Somerset, an early step in the establishment of geological science.

1728   James Bradley finds aberration in starlight produced by the motion of the earth.

1750–1784   French amateur astronomer Charles Messier catalogs scores of indistinct celestial objects that might be mistaken for comets; many will prove to be star clusters and interstellar gas clouds, others external galaxies.

1755   Kant proposes that spiral nebulae are galaxies of stars.

1761, 1769 Transits of Venus observed by widely scattered scientific expeditions, permitting new determinations of the distance from the earth to the sun—the "astronomical unit."

| 1765 | John Harrison is acknowledged by the English Board of Longitude to have developed the marine chronometer, making possible accurate timekeeping and the determination of longitude at sea. |

1766   Henry Cavendish identifies hydrogen, the most abundant element in the universe.

1781   William Herschel discovers the planet Uranus.

1783   Herschel derives the general direction of the solar system's motion through space, by studying the proper motion of thirteen bright stars.

1793   William Smith, a canal surveyor and consulting engineer excavating the Somersetshire Coal Canal, finds evidence for a consistent sequence of geological strata throughout England.

1795   James Hutton's *Theory of the Earth* advances a uniformitarian hypothesis of geological change having taken place in the course of a lengthy past.

1800   William Herschel detects infrared light.

1801   Johann Ritter detects ultraviolet light.

             Georges Cuvier identifies twenty-three species of extinct animals in the fossil record, confounding the doctrine that all species were created simultaneously and are imperishable.

1802   William Wollaston discovers spectral lines in the spectrum of the sun.

1814   Joseph Fraunhofer, using the first grating spectroscope, rediscovers solar spectral lines and charts them, laying the basis for astrophysical spectroscopy.

1820   Hans Christian Örsted discovers that electric current produces a magnetic field, ushering in the study of electromagnetic force.

1823   John Herschel proposes that Fraunhofer lines may indicate the presence of metals in the sun.

1830   Charles Lyell publishes the first volume of his *Principles of Geology*, presenting evidence for the uniformitarian theory that the geological record can be explained in terms of the slow action, over aeons of time, of processes that continue in the world today.

1831   Charles Darwin, a copy of Lyell's book in hand, departs aboard the *Beagle* on a five-year voyage around the world.

1837	Darwin adduces the essential elements of his theory of evolution by natural selection, but does not publish the theory for another twenty-two years.
1838	First precise measurement, by means of parallax, of the distance to a star.
1842	Christian Johann Doppler points out that the wavelength of sound or other emissions from a moving source will appear to a stationary observer to be higher in frequency if the object is approaching, lower if it is receding—the "Doppler shift."
1847	Hermann von Helmholtz proposes the law of conservation of energy.
1849	Jean-Léon Foucault detects spectral emission lines.
1850	First astronomical photograph—a daguerreotype of the moon—is made, by W. C. Bond at Harvard.
1855–1863	Robert Bunsen and Gustav Kirchhoff work out the basics of spectral analysis, by which the spectra of laboratory materials can be compared with those of the sun and stars.
1859	Darwin's *Origin of Species* published.
1862	Foucault refines estimates of the velocity of light.
1864	William Huggins obtains the first spectrum of a nebula, finds that it is composed of gas.
	James Clerk Maxwell publishes a unified theory of electricity and magnetism, portraying both as aspects of electromagnetic force.
1865	Gregor Mendel announces results of his research in genetics, revealing key to persistence of *unchanging* traits in living things, a critical missing element in Darwinism.
1874, 1882	Transits of Venus observed with new, more precise instruments, improving estimates of the astronomical unit.
1877	David Gill measures parallax of Mars during its opposition, deduces distance to the sun of ninety-three million miles.
1879	Albert Michelson, employing Foucault's principle, determines velocity of light.
1883	Henry Rowland's diffraction grating greatly improves the resolution of spectrographs.
1884	Johann Balmer determines harmonic sequence of hydro-

gen lines, initiating line of inquiry that will lead to investigation of the electron shells of atoms.

1887   Albert Michelson and Edward Morley perform the final and most precise in a series of experiments showing that space cannot be filled with the aether that had been thought to be responsible for transmitting light. Their work clears the ground for the ascent of the Lorentz contractions.

1892   Hendrik Lorentz and George FitzGerald independently propose that contraction of length of measuring rods with velocity explains the Michelson-Morley experimental results, a concept essential to the special theory of relativity.

1895   E. E. Barnard photographs the Milky Way, notes that dark patches are too numerous to be empty space but must represent dark clouds of interstellar matter.

1897   J. J. Thomson discovers the electron.

1898   Marie and Pierre Curie isolate the radioactive elements radium and polonium.

1900   Max Planck proposes the quantum theory of radiation, the basis of quantum physics.

1904   Ernest Rutherford suggests that the amount of helium produced by the radioactive decay of minerals in rocks could be employed to measure the age of the earth.

1905   Albert Einstein publishes special theory of relativity, indicating that measurements of space and time are distorted at high velocity and implying that mass and energy are equivalent; in another paper he shows that light is composed of quanta.

Jacobus Kapteyn, studying the proper motions of twenty-four hundred stars, finds evidence of what he calls "star streaming"—that stars in our neighborhood move in a preferred direction—an early clue to the rotation of our galaxy.

1911   Ernest Rutherford determines that most of the mass of atoms is contained in their tiny nuclei.

1912   Henrietta Swan Leavitt discovers a correlation between the absolute magnitude and the period of variability of Cepheid variable stars, opening the door to their use as intergalactic distance indicators.

1913   Niels Bohr develops theory of atomic structure, in which electrons are said to orbit the nucleus in a manner somewhat akin to that of planets orbiting the sun.

Henry Norris Russell presents a plot of the luminosities and colors of stars, extending work done in 1911 by Ejnar Hertzsprung. The resulting Hertzsprung-Russell diagram will be fundamental to the understanding of the evolution of stars.

1914   Walter Adams and Arnold Kohlschutter determine the absolute luminosity of stars from their spectra alone, making it possible to estimate the distances of millions of distant stars.

1915   Annie Jump Cannon classifies stars into categories according to their spectral type, a major step in discerning order underlying the diversity of the stars.

Arnold Sommerfeld refines Bohr model of the atom.

1916   Albert Einstein publishes the general theory of relativity, portraying gravitation as an effect of curved space and delivering cosmology from the ancient dilemma of a finite versus an infinite universe.

1916–1917   Arthur Stanley Eddington demonstrates theoretically that stars are gaseous spheres; his work lays the foundation for his later assertion that gravitational contraction cannot be the mechanism that powers the stars.

1917   Heber Curtis and George Ritchey announce that they have found novae (stars that have suddenly increased tremendously in brightness) in the Andromeda spiral. Opinions differ on whether this means Andromeda is a galaxy of stars, or a gaseous nebula from which new stars are condensing.

Vesto Slipher measures large Doppler shifts in the spectra of spirals, later found to be due to the motion of the spiral galaxies in the expanding universe.

1918   Harlow Shapley determines, by studying the distances of globular clusters, that the sun lies toward one edge of a galaxy of stars.

The 100-inch telescope at Mount Wilson, then the world's largest, begins operation.

1919   English expedition to observe a solar eclipse confirms Einstein's prediction that space, in a gravitational field, is strongly curved.

1920   The controversy over whether spiral nebulae are gaseous clouds or "island universes"—i.e., galaxies—comes to a head in a debate between Heber Curtis and Harlow Shapley.

1922   Ernst Öpik deduces, from rotation velocities and the mass to luminosity ratio of the Andromeda spiral, that it is a galaxy in its own right.

Aleksandr Friedmann shows that general relativity is consistent with an expanding-universe cosmology.

1923   Cecilia Payne demonstrates, from solar spectra, that the relative abundance of elements in the sun approximates that in the crust of the earth.

1924   Louis de Broglie develops wave theory of matter.

1925   Max Born, Pascual Jordan, and Werner Heisenberg develop quantum mechanics.

Wolfgang Pauli announces the exclusion principle, essential to understanding spectral lines of stars and nebulae.

Bertil Lindblad demonstrates that the motion of stars called "star streaming" by Kapteyn in 1905 can be explained as being due to the rotation of the Milky Way galaxy.

Edwin Hubble announces that he has identified Cepheid variable stars in the Andromeda galaxy, confirming that it is a galaxy of stars rather than a gaseous nebula and making it possible to measure its distance.

1926   Erwin Schrödinger proposes wave-mechanical theory of the atom.

Lindblad produces theory of rotation of the Milky Way galaxy.

1927   Jan Oort detects evidence of the rotation of the Milky Way galaxy, by examining the radial velocities of stars.

Georges Lemaître publishes an expanding-universe cosmology.

Werner Heisenberg discovers the quantum indeterminacy principle.

1927–1929   Relativistic quantum electrodynamics theory developed.

1928   George Gamow applies the uncertainty principle to the problem of how protons combine to build nuclei in stellar interiors, a signal step in establishing that nuclear fusion provides the energy that powers stars.

Ira Bowen determines that the spectra of nebulae are produced by doubly ionized oxygen and not by an unknown element called "nebulium," as had been thought. This strengthens the hopes of astrophysicists that the rest of the universe is made of the same elements and obeys the same natural laws as here on Earth.

Dirac publishes the "Dirac equation," a relativistic quantum theory of electromagnetism.

1929   Edwin Hubble announces a relationship between the redshift in the spectra of galaxies and their distances, indicating that the universe is expanding.

1930   Robert Trumpler's studies of open star clusters enable him to measure the extent to which interstellar clouds dim and redden starlight, greatly improving estimates of the distances of stars.

1931   Dirac predicts the existence of the positron, the antimatter equivalent of the electron.

Wolfgang Pauli, studying beta decay, predicts the existence of the neutrino.

Kurt Gödel's second incompleteness theorem indicates that the consistency of any system, including scientific systems, cannot be proved internally—i.e., that mathematics, and science are inherently open-ended.

1932   James Chadwick discovers the neutron.

Carl Anderson, without knowing of Dirac's 1931 paper postulating its existence, discovers the positron.

Karl Jansky finds that the Milky Way emits radio waves, opening door on the science of radio astronomy.

1935   Hideki Yukawa predicts existence of the meson.

1939   Niels Bohr and John Archibald Wheeler develop the theory of nuclear fission.

Hans Bethe and Carl Friedrich von Weizsäcker independently arrive at theory of the carbon and proton-proton reactions in stars.

1940   Grote Reber constructs a backyard radio telescope, makes the first radio map of the Milky Way.

1943   Carl Seyfert identifies Seyfert galaxies, the first of a larger class of galaxies found to have bright nuclei that emit abnormal amounts of energy.

1944   Walter Baade resolves the central region of the Androm-
                       eda galaxy into stars, establishing a fundamental dis-
                       tinction between the older, redder stars characteristic of
                       the centers of spiral galaxies, and the younger, bluer stars
                       found in their spiral arms.

1945   Hendrik van de Hulst predicts that clouds of interstellar
                       hydrogen emit radio energy at the 21-centimeter wave-
                       length.

1946   James Hey, S. J. Parsons, and J. W. Phillips identify a
                       powerful radio source in Cygnus, initiating research that
                       leads to finding galaxies that emit enormous amounts of
                       energy at radio wavelengths.

1948   Dedication of the 200-inch telescope on Palomar Mountain.

                       Ralph Alpher and George Gamow theorize about the
                       physics of the early universe; Alpher and Robert Her-
                       man, correcting Gamow's arithmetic, then predict that
                       the big bang should have produced a cosmic background
                       radiation.

1948–1949   "Renormalization" of quantum electrodynamics removes
                       unwanted infinities from the equations.

1948–1950   Willard Frank Libby develops technique of radiocarbon
                       dating.

1949   John Bolton, Gordon Stanley, and O. B. Slee use radio
                       interferometry to identify three radio sources with visible
                       objects; two of them are galaxies, suggesting that what
                       had been thought to be radio "stars" are actually objects
                       lying much farther away in space.

1951   Harold Ewen and Edward Purcell, closely followed by
                       C. Alex Muller and Jan Oort, detect 21-centimeter radio
                       radiation emitted by interstellar clouds.

1952   Baade clears up serious discrepancies in the cosmic dis-
                       tance scale when he finds that the Cepheid variable stars
                       used in measuring intergalactic distances actually come
                       in two varieties, with different magnitude-periodicity
                       relationships.

1953   Murray Gell-Mann proposes a new quantum number
                       called strangeness, notes that it is conserved in strong
                       interactions.

1954   Walter Baade and Rudolph Minkowski identify the radio
                       source Cygnus A with a distant galaxy.

Chen Ning Yang and Robert Mills develop a gauge symmetrical field theory, a major step toward viewing the universe in terms of underlying symmetries that were broken in early cosmic evolution.

1956 Yang and Tsung Dao Lee theorize that parity is not conserved in weak interactions—i.e., that the weak force does not function symmetrically. Experiments conducted by Chien-Shiung Wu and collaborators the same year confirm their prediction.

1957 Julian Schwinger proposes that the electromagnetic and weak forces are but aspects of a single variety of interaction.

1958 Oort and colleagues use radio telescopes to map the spiral arms of the Milky Way galaxy.

1960 Allan Sandage and Thomas Matthews discover quasars.

1961 Gell-Mann and Yuval Ne'eman independently arrive at the "eightfold way" scheme of classifying subatomic particles that react to the strong nuclear force.

1963 Maarten Schmidt finds redshift in the spectral lines of a quasar, indicating that quasars are the most distant class of objects in the universe.

1964 Murray Gell-Mann and George Zweig independently propose that protons, neutrons, and other hadrons are composed of still smaller particles, which Gell-Mann dubs "quarks."

The omega-minus particle is detected at Brookhaven National Laboratory, confirming a prediction of the Gell-Mann–Ne'eman "eightfold way."

1967 Chia Lin and Frank Shu show that the spiral arms of galaxies may be created by density waves propagating across the galactic disk.

Jocelyn Bell and Antony Hewish discover pulsars, leading to verification of the existence of extremely dense "neutron stars."

1968 Experiments at the Stanford Linear Accelerator Center support the theory that hadrons are made of quarks.

1981 Alan Guth postulates that the early universe went through an "inflationary" period of exponential expansion.

1983 Electroweak unified theory verified in collider experiments at CERN. Attempts accelerate to arrive at a unified theory of all four forces.

1987   Proton-decay experiments in the United States and Japan
               detect neutrinos broadcast by a supernova in the Large
               Magellanic Cloud, ushering in the new science of ob-
               servational neutrino astronomy.

1988   Quasars are detected near the outposts of the observable
               universe; their redshifts indicate that their light has been
               traveling through space for some seventeen billion years.

Notes

Preface and Acknowledgments

1. Carlyle, *On History*, in Pais, 1986, p. 129.

Chapter One: The Dome of Heaven

1. Copernicus, *On the Revolutions*, Wallis translation, p. 510.
2. Hesiod, *Works and Days*, Wender translation, p. 78.
3. In Williamson, 1984, p. 297.
4. Ibid., p. 210.
5. In Morison, 1963, p. 362.
6. In Wycherley, 1978, p. 222.
7. Aristotle, *On Youth*, 2, 14, in Barnes, 1984.
8. In Duhem, 1969, p. 19.
9. Ibid., p. 17.
10. Plato, *Phaedrus* 230a, Hackford translation, in Hamilton and Cairns edition, 1969.

Chapter Two: Raising (and Lowering) the Roof

1. Archimedes, "The Sand Reckoner," in *The Works of Archimedes*, Heath translation, 1952, p. 520.
2. Ibid.
3. Ibid.
4. Plutarch, *Lives*, Vol. II, p. 282.
5. In Pappus, *Collectio*, Book VIII, Prop. 10, Sect. XI.
6. Plutarch, *Lives*, Vol. II, p. 283.
7. Ibid., p. 286.
8. Euclid, Heath translation, p. 3.
9. The Gospel According to Saint John, Chapter 18.

10. In Adams, 1938, pp. 52–53.
11. In Alic, 1986.
12. In Nasr, 1964, p. 182.
13. Boethius, Watts translation, pp. 46–47.
14. Ibid., p. 57.
15. Ibid., p. 73.

CHAPTER THREE: THE DISCOVERY OF THE EARTH

1. Virgil, *Aeneid*, from the Delabere-May, Fitzgerald, and Rhoades translations.
2. In Morison, 1963.
3. In Garnet, 1970, p. 49.
4. Su, Watson translation, 1965.
5. In Bell, 1974, p. 46.
6. In Parry, 1963, p. 247.
7. In Beazley, p. 213.
8. In Needham, 1954–1984, Vol. 2, p. 525.
9. Ibid., Vol. 4, Part 3, p. 514.
10. Encyclopaedia Britannica, 3rd ed., Vol. 4, p. 937.
11. In Newby, 1975, p. 67.
12. In Columbus, Ferdinand, 1979, p. 10.
13. Ibid.
14. In Pigafetta, 1969, p. 57.
15. Aristotle, *De Caelo*, 298a, J. L. Stocks translation, McKeon edition, 1968, p. 437.
16. See Heyerdahl, 1979.
17. In Morison, 1963, p. 62.
18. Ibid., p. 65.
19. In Heyerdahl, 1979, p. 147.
20. In Morison, 1963, p. 383.
21. In Mason, 1977, p. 243.
22. In MacCurdy, 1939, p. 295.

CHAPTER FOUR: THE SUN WORSHIPERS

1. Copernicus, *On the Revolutions*, Duncan translation, preface.
2. In Panofsky, 1969, p. 10.
3. Copernicus, *Commentariolus*, in Rosen, 1959.
4. Copernicus, *On the Revolutions*, John Dobson and Selig Brodetsky, translators, *Occasional Notes of the Royal Astronomical Society*, Vol. 2, No. 10, 1947, in Kuhn, 1979, p. 139.
5. Plutarch, *Moralia*, XII, p. 925; Cherniss and Helmbold translation, p. 75.
6. Nicole Oresme, "The Compatibility of the Earth's Diurnal Rotation With Astronomical Phenomena and Terrestrial Physics," in Grant, Edward, 1974, p. 505.

NOTES

tag7. Copernicus, *On the Revolutions*, Wallis translation, pp. 526–527.
8. In Kuhn, 1979, p. 130.
9. Copernicus, *On the Revolutions*, Wallis translation, pp. 526–527.
10. Ibid., p. 511.
11. In Bienkowska, 1973.
12. In Russell, Bertrand, 1945, p. 528.
13. Martin Luther, *Table Talk*, p. 69, in Fosdick, 1952, p. xviii.
14. Copernicus, *On the Revolutions*, Wallis translation, pp. 516, 549. The translation has been altered slightly.
15. Aristotle, *On the Heavens*, 270:14, J.L. Stocks translation, in McKeon, 1968.
16. Tycho, *Progymnasmata*, Chapter 3, in Clark and Stephenson, 1977, p. 174.
17. Tycho, *De Nova Stella*, in Clark and Stephenson, 1977, p. 172.
18. In Dryer, 1890, p. 27.
19. In Koestler, 1959, p. 273.
20. Ibid., p. 276.
21. In Baumgardt, 1951, p. 17.
22. Plato, *Republic*, VII: 530d, Paul Shorey translation, in Hamilton and Cairns edition, 1969.
23. Ibid., X:617c.
24. Aristotle, *On the Heavens*, 290b, J.L. Stocks translation, in Barnes, 1984, p. 479.
25. "On the Morning of Christ's Nativity," XIII, in Milton, 1952.
26. Shakespeare, *Merchant of Venice*, Act V, Scene 1.
27. Kepler, *The Harmonies of the World*, Wallis translation, pp. 1034, 1048.
28. Kepler, *Epitome of Copernican Astronomy*, p. 897.
29. In Koestler, p. 304.
30. Ibid, p. 278.
31. Ibid.
32. Ibid.
33. In Dryer, 1980, p. 386.
34. Kepler, *The New Astronomy*, in Koyré, 1973, p. 231.
35. Kepler, *The Harmonies of the World*, Wallis translation, p. 1009.
36. Ibid., p. 1009.
37. In Koestler, p. 381.
38. Ibid.
39. Ibid., p. 414.
40. Ibid., p. 421.

CHAPTER FIVE: THE WORLD IN RETROGRADE

1. Letter to Cosimo de' Medici, 1610, in Drake, 1957, p. 61.
2. In Fermi and Bernardini, 1961, p. 12.
3. Galileo, *The Assayer*, in Drake, 1957, p. 238.
4. Galileo, *The Starry Messenger*, in Drake, 1957, p. 29.
5. In Fermi and Bernardini, 1961, p. 51.

6. Brecht, 1966, Scene 3, p. 66. The Galileo quotations are of course Brecht's inventions.
7. Galileo, *The Starry Messenger*, Drake translation, p. 28.
8. Ibid., p. 57.
9. Ibid., p. 94.
10. Ibid., p. 49.
11. In Weaver, 1987, p. 683.
12. In Galileo, *Dialogue Concerning the Two Chief World Systems*, Drake translation, p. xix.
13. Galileo, *Dialogues Concerning Two New Sciences*, Crew and De Salvio translation, p. 63.
14. Galileo, *Dialogue Concerning the Two Chief World Systems*, pp. 186–187.
15. Galileo, *Letters on Sunspots*, in Drake, 1957, p. 113.
16. Galileo, *Dialogue Concerning the Two Chief World Systems*, p. 462.
17. I. Bernard Cohen, "An Interview With Einstein," in French, 1979, p. 41.
18. In Galileo, *The Sidereal Messenger*, Carlos translation, p. 111.
19. In Fermi & Bernardini, 1961, p. 72.
20. Galileo, *Letters on Sunspots*, in Drake, 1957, p. 62.
21. In Koestler, p. 440.
22. Bellarmine, letter to Paolo Foscarini, in Drake, 1957, pp. 163–164.
23. Ibid., p. 164.
24. Letter to the Grand Duchess Christina, in Drake, 1957, p. 166.
25. In Geymonat, 1965, pp. 85, 83.
26. Ibid., p. 73.
27. Koestler, p. 471.
28. Ibid., p. 472.
29. Galileo, *Dialogue Concerning the Two Chief World Systems*, p. 319.
30. In Geymonat, p. 146.
31. In Singer, 1917, p. 269.
32. *Science 81*, March 1981, p. 14.
33. In Kesten, 1945, p. 93.
34. Book VIII, ll. 167ff., in Milton 1952.

CHAPTER SIX: NEWTON'S REACH

1. J.M. Keynes, "Newton, The Man," *The Royal Society Newton Tercentenary Celebrations*, Cambridge University Press, 1947, p. 27.
2. Westfall, 1980, p. 354.
3. In Manuel, 1968, p. 26.
4. In Westfall, p. 65.
5. Ibid., p. 89.
6. Ibid., p. 22.
7. Ibid., p. 143.
8. Ibid., p. 188–189.
9. Ibid., p. 245.

10. In Spinoza, 1928, p. 80.
11. In Jones, 1981, p. 197.
12. Descartes, *Geometry*, p. 353.
13. William Stukeley, *Memoirs of Sir Isaac Newton's Life*, in Cohen, I. Bernard, 1971, p. 301.
14. In Manuel, pp. 27–28.
15. In Westfall, p. 141.
16. Ibid., p. 405.
17. Ibid., p. 406.
18. Ibid.
19. In Parton, 1882, Vol. 2, p. 213.
20. Newton, *Principia*, Cajori-Motte translation, p. 13.
21. Ibid.
22. In Westfall, p. 459.
23. Ibid.
24. Ibid., p. 581.
25. In Manuel, p. 216.
26. In Cohen, I. Bernard, 1958, p. 7.
27. Ibid., p. 284.
28. Newton, *Principia*, Cajori-Motte translation, p. 371.
29. Ibid., p. 547.
30. Einstein, "Autobiographical Notes," in Schilpp, 1969, pp. 32–33.
31. In Cohen, I. Bernard, 1958, p. 7.
32. In Dampier, 1949, p. 197.
33. In Cohen, I. Bernard, 1958, p. 284.

CHAPTER SEVEN: A PLUMB LINE TO THE SUN

1. Huygens, *Systema Saturnium*, 1659, in Van Helden, 1985, p. 123.
2. Richard Hakluyt, *Principal Navigations*, 2nd ed., Vol. 1, 1598, in Landes, 1983, p. 110.
3. In Howse, 1980, p. 12.
4. Edmond Halley, "A Unique Method by which the Parallax of the Sun, or its Distance from the Earth, may be Securely Determined by Means of Observing Venus Against the Sun," *Philosophical Transactions of the Royal Society*, No. 348, April–June, 1716, pp. 454–455, 460, Dave Fredrick, translator.
5. In Albert Van Helden, "The Importance of the Transit of Mercury of 1631," *Journal for the History of Astronomy*, Vol. 7, Part 1, No. 18, February 1976.
6. In Fernie, 1976, p. 10.
7. Ibid.
8. Ibid., p. 39.
9. Ibid. p. 49.
10. Ibid. pp. 52–53.
11. Ibid. p. 29.

CHAPTER EIGHT: DEEP SPACE

1. Kant, 1900, 1969, p. 30. The translation has been altered slightly.
2. Ibid., pp. 61–62.
3. Ibid., pp. 63–64.
4. In Lambert, 1976, p. 1.
5. Ibid., pp. 1–2.
6. Ibid., p. 53.
7. Ibid., pp. 106, 120–121. The translation has been altered somewhat.
8. In Lubbock, 1933, p. 10.
9. Ibid., p. 10.
10. Ibid., p 29.
11. Ibid., p. 31.
12. Ibid., pp. 61–62.
13. In Hoskin, 1963, p. 21.
14. In Lubbock, p. 66.
15. In King, 1979, p, 126.
16. In J.A. Bennett, "The Discovery of Uranus," *Sky and Telescope*, March 1981, p. 188.
17. In MacPherson, 1933, p. 101.
18. In Lubbock, p. 138.
19. Ibid., p. 16.
20. In King, p. 133.
21. In Lubbock, p. 228.
22. In Burnham, 1978, p. 1317.
23. In Lubbock, p. 355.

CHAPTER NINE: ISLAND UNIVERSES

1. In Smith, Robert, 1982, p. 4.
2. *Popular Astronomy*, Vol. 34, 1926, p. 379, quoting a Boston newspaper article, in Warner, 1968, p. 10.
3. In MacPherson, 1933, p. 161.
4. In Owen Gingerich, "Unlocking the Chemical Secrets of the Cosmos," *Sky and Telescope*, July 1981, p. 13.
5. In Abetti, 1952, p. 192.
6. William Huggins and Lady Huggins, *The Scientific Papers of Sir William Huggins* (London: Wesley & Son, 1909), p. 106, in Smith, 1982, pp. 2–3.
7. In Smith, 1982, p. 43.
8. In Lang and Gingerich, 1979, p. 523.
9. Ibid.
10. Ibid., p. 525.
11. Shapley, letter to Russell, March 31, 1920, in Smith, 1982, p. 66.
12. Edwin Hubble, "Cepheids in Spiral Nebulae," *Publications of the American Astronomical Society*, Vol. 5, 1925, pp. 261–264.

13. In Smith, 1982, p. 114.
14. Shapley to Hubble, February 27, 1924, in Smith, 1982, p. 119.
15. In Dick, 1984, p. 147.
16. Edwin Hubble, "NGC6822, A Remote Stellar System," *Astrophysical Journal*, Vol. 62, p. 432.

CHAPTER TEN: EINSTEIN'S SKY

1. Newton, *Principia*, Cajori-Motte translation, p. 6.
2. See Hoffmann, 1983, p. 60.
3. In Livingston 1973, p. 77.
4. Ibid., p. 132.
5. Ibid.
6. Poincaré, 1899, in Hoffmann, 1983, p. 86.
7. Dirac, 1971, pp. 13, 14.
8. In Hoffmann, 1972, p. 24.
9. Ibid. pp. 20, 25.
10. In Stachel, 1987, p. 334.
11. In Dukas and Hoffmann, 1979, p. 5.
12. Einstein, "Autobiographical Notes," in Schilpp, 1969, p 11.
13. Ibid., p. 5.
14. In French, 1979, p. 31.
15. In John Stachel, "Albert Einstein: The Man Behind the Myths," manuscript copy.
16. Einstein, "Autobiographical Notes," in Schilpp, 1969, p. 9.
17. Ibid., p. 35.
18. In Goldman, 1983, p. 138.
19. Ibid., p. 146.
20. Einstein, "Autobiographical Notes," in Schilpp, 1969, p. 33.
21. Ibid., p. 53.
22. Adams, 1931, p. 380.
23. In Miller, 1981, p. 145.
24. Sigmund Freud, "Memorandum on the Electrical Treatment of War Neurotics," in Strachey, 1955, p. 211.
25. Einstein, letter to Grossman, 1901, in Stachel, 1987, p. 290.
26. In Szilard, 1978, p. 12.
27. In Seelig, 1956, p. 71.
28. In Miller, 1981, p. 125.
29. Mach, 1960, p. 279.
30. In Pais, 1982, p. 201.
31. Einstein, "On the Electrodynamics of Moving Bodies," in Miller, 1981, p. 392.
32. In Hoffmann, 1972, p. 131.
33. In Pais, 1982, p. 179.
34. Ibid., p. 152.
35. In Rucker, 1984, pp. 66, 68.

36. In Davis and Hersh, 1981, p. 221.
37. In Hoffmann, 1983, p. 129.
38. Hoskin, 1982, p. 83.
39. In Needham, *Science and Civilization in China*, Vol. 2, p. 388.
40. Lucretius, *De Rerum Natura*, Book I, lines 1012ff., Cyril Bailey translation, p. 227. The translation has been altered slightly.
41. In Cohen and Seeger, 1970, p. 181.
42. In Rosenthal-Schneider, 1980, p. 74.
43. In French, 1979, p. 31.
44. In Holton and Elkana, 1982, p. 104.

CHAPTER ELEVEN: THE EXPANSION OF THE UNIVERSE

1. Albert Einstein, "Cosmological Considerations on the General Theory of Relativity," 1917, in Einstein, 1952, p. 188.
2. Einstein, 1923, p. 127.
3. In Smith, Robert, 1982, p. 173.
4. Hubble, 1985, p. 35.
5. Ibid.
6. Lemaître, quoted in *The New York Times Magazine*, February 19, 1933.
7. Andre Deprit, "Monsignor Georges Lemaître," in Berger, 1985, p. 370.
8. Ibid., p. 376.
9. In Ferris, 1983, p. 119.
10. In Berger, p. 373.
11. *Los Angeles Times*, January 12, 1933.
12. *The New York Times*, January 12, 1933; *Los Angeles Times*, January 12, 1933.
13. Lemaître, 1950, p. 140.
14. Ralph A. Alpher and Robert C. Herman, "Evolution of the Universe," *Nature*, Vol. 162, pp. 774ff., 1948, in Lang and Gingerich, 1979, p. 866.

CHAPTER TWELVE: SERMONS IN STONES

1. In Lyell, 1877, p. 29. The punctuation of the translation has been edited slightly.
2. S. Sambursky, "The Stoic Doctrine of Eternal Recurrence," in Capek, 1976, p. 170.
3. Aristotle, *Meteorology*, 352b, E.W. Webster translation, in Aristotle, 1984, p. 574.
4. J.D. North, "Chronology and the Age of the World," in Yourgrau and Breck, 1977, pp. 307ff.
5. In Toulmin and Goodfield, *The Discovery of Time*, 1982, p. 144.
6. In Ogburn, 1968, p. 32.
7. In Lovejoy, 1953, p. 184.
8. In Eiseley, 1970, p. 39.

9. In Lovejoy, p. 243.
10. In Loren Eiseley, "Charles Lyell," *Scientific American*, Vol. 201, pp. 1959, 98–106.
11. In Toulmin and Goodfield, *The Discovery of Time*, 1982, p. 157.
12. Burnet, Book II, p. 173.
13. Lyell, 1863, Vol. II, p. 101.
14. In Knedler, p. 10.
15. Ibid., p. 16.
16. Ibid., p. 41.
17. Ibid.
18. Ibid., p. 51.
19. Ibid., p. 55.
20. Ibid., p. 56.

CHAPTER THIRTEEN: THE AGE OF THE EARTH

1. In Keynes, 1979, p. 18.
2. In Loren Eiseley, "Charles Lyell," *Scientific American*, Vol. 201, August 1959, pp. 98, 106.
3. In W.W. Bartley III, "What Was Wrong With Darwin?" *The New York Review of Books*, September 15, 1977, p. 37.
4. In Keynes, p. 295.
5. Darwin, 1962, pp. 402–403.
6. Ibid., p. 480.
7. Ibid., p. 469.
8. In Keynes, p. 19.
9. In *Encyclopaedia Britannica*, 15th ed., Vol. 5, p. 492.
10. In Keynes, p. 19.
11. In Darwin, Francis, 1950, p. 57.
12. Darwin, Erasmus, 1818, Vol. 1, pp. 397, 400. Emphasis is Darwin's.
13. Darwin, in *Gardeners' Chronicle and Agricultural Gazette*, Vol. 45, November 8, 1862, p. 1052.
14. Darwin, in *Journal of Horticulture and Cottage Gardener*, Vol. 3, December 2, 1862, p. 696.
15. Darwin, "Notebooks on Transmutation of Species," series D, in Ruse, 1979, p. 173.
16. Darwin, 1872, p. 99.
17. Ibid., pp. 99–100.
18. Ibid., p. 371.
19. Ibid., p. 374.
20. In DeBeer, 1964, p. 253.
21. In W.W. Bartley III, "What Was Wrong with Darwin?" *The New York Review of Books*, September 15, 1977, p. 34.
22. In Patterson, 1978, p. 14.
23. Darwin, letter to Lyell, June 3, 1858, *Encyclopaedia Britannica*, 15th ed., Vol. 19, p. 530.

NOTES

438

24. In Marchant, Vol. 1, 1916, pp. 29–30.
25. Ibid., p. 110.
26. In DeBeer, p. 149.
27. In Darwin, Francis, 1888, Vol 2, pp. 179–204.
28. Darwin, 1872, p. 38.
29. In Price, 1956, p. 28.
30. In Darwin, Francis, 1950, p. 68.
31. Ibid., p. 67.
32. In Huxley, Julian, 1903, Vol. 1, pp. 265–266.
33. Ibid., p. 268. Accounts of the debate differ somewhat in the exact wording of the quotations, as Huxley *fils* describes in the work cited. This quotation was cited by the elder Huxley as the most nearly accurate of the several accounts proffered by witnesses to the debate.
34. In DeBeer, p. 167.
35. Darwin, 1872, p. 236.
36. Kant, 1969, p. 159.
37. In Eiseley, 1958, pp. 234, 240.
38. In Toulmin and Goodfield, *The Discovery of Time*, 1982, p. 222.
39. Thompson, William, Baron Kelvin, 1891, Vol. 1, p. 16.
40. Rutherford, 1904, p. 657.
41. In Eve, 1939, p. 107.
42. Ibid.
43. In Segrè, 1980, p. 42.
44. Arthur Stanley Eddington, "The Internal Constitution of the Stars," *Nature*, 1920, Vol. 106, pp. 14–20, in Lang and Gingerich, 1979, p. 281.
45. In Moss, 1968, p. 59.
46. P.M.S. Blackett, *New Statesman*, December 5, 1959.
47. In Dukas and Hoffmann, 1979, p. 81.

Chapter Fourteen: The Evolution of Atoms and Stars

1. Chamberlin, *Science*, Vol. 9, July 7, 1899, p. 12, in Albritton, 1980, p. 198.
2. Planck, Nobel Prize address, in Heathcote, 1954, p. 415.
3. Annie Jump Cannon, "Pioneering in the Classification of Stellar Spectra," from "The Henry Draper Memorial," *Journal of the Royal Astronomical Society of Canada*, Vol. 9, 1915, in Shapley, 1960, p. 158.
4. In Lang and Gingerich, 1979, pp. 14–20.
5. Ibid., p. 288.
6. Hans Bethe, Nobel Prize address, in Heathcote, 1954, p. 216.
7. In Bernstein, 1980, p. 53.
8. In Lang and Gingerich, p. 288.
9. Gamow, 1951, p. 73.
10. Hoyle, 1965, p. 102.
11. In Berger, 1985, p. 387.
12. Gamow, 1951, p. 49.

13. E. Margaret Burbidge, Geoffrey R. Burbidge, William A. Fowler, and Fred Hoyle, "Synthesis of the Elements in Stars," *Reviews of Modern Physics*, Vol. 29, 1957, pp. 547–650, in Lang and Gingerich, p. 383.
14. Ibid., p. 377.
15. Ibid., p. 386.
16. Shu, 1982, p. 157.

CHAPTER FIFTEEN: THE QUANTUM AND ITS DISCONTENTS

1. Planck, Nobel Prize address, 1918, in Weaver, 1987, Vol II., p. 292.
2. In Moore, 1985, p. 127.
3. In Born, Max, 1971, p. 82. Einstein's italics.
4. Ibid., p. 91.
5. Ibid., p. 158.
6. In Clark, 1971, p. 34.
7. In Segrè, 1970, p. 69.
8. Veltman, unpublished talk at Caltech, April 29, 1982.
9. Leon Lederman, interview with TF, Fermilab, February 24, 1985.

CHAPTER SIXTEEN: RUMORS OF PERFECTION

1. Poincaré, 1958, p. 19; Poincaré, 1908, p. 59. For a discussion of these remarks see Wechsler, 1978, p. 5.
2. Heisenberg, 1971, p. 68.
3. Dirac, "The Evolution of the Physicist's Picture of Nature," *Scientific American*, May 1963, p. 47, in Wechsler, 1978, p. 5.
4. Yang, 1983, p. 82.
5. In Judson, 1980, p. 198.
6. Wigner, 1967, p. 29.
7. Ibid., p. 5.
8. Behram Kursunoglu, in Mehra, 1973, p. 818.
9. Ibid.
10. Yang, 1961, p. 53.
11. Weinberg, "The Forces of Nature," *American Scientist*, Vol. 65, No. 2, March–April 1977, pp. 171–176.
12. Ibid.
13. Glashow, Nobel Prize address, *Review of Modern Physics*, Vol. 52, No. 3, July 1980, p. 543.
14. Schwinger, 1958, pp. xvii.
15. Glashow, Nobel Prize address, *Review of Modern Physics*, Vol. 52, No. 3, July 1980, p. 540.
16. In Crease and Mann, 1986, p. 224.
17. Ibid., p. 225.
18. Glashow, "Partial Symmetries of Weak Interactions," *Nuclear Physics*, Vol. 22, No. 4, February 1961, p. 579.

19. In Hassan and Lai, 1984, p. 17.
20. Weinberg, *Review of Modern Physics*, op. cit., p. 515.
21. Weinberg, interview with TF, Austin, Texas, February 28, 1985.
22. Weinberg, *Review of Modern Physics*, op. cit., p. 517.
23. Ibid., p. 518.
24. In Hilts, 1982, p. 86.
25. Wilson, interview with Linda Dackman, *Arts and Architecture*, Vol. 3, No. 1, 1984.
26. Abdus Salam, Nobel Prize address, *Review of Modern Physics*, op. cit., p. 530.
27. In Ne'eman and Kirsh, 1986, p. 250.
28. *CERN Courier*, January/February 1985, p. 5.
29. In Taubes, 1986, p. 39.
30. Rubbia, interview with TF, CERN, December 1984.
31. *Engineering and Science*, September 1985, p. 20.
32. In Crease and Mann, 1986, p. 417.
33. Green, Schwarz, and Witten, 1987, p. 55.
34. Gamow, 1966, p. 163.
35. Weinberg, interview with TF, Austin, Texas, April 5, 1982.
36. Sheldon Glashow and Paul Ginsparg, "Desperately Seeking Superstrings?" *Physics Today*, May 1986, p. 7.
37. Weinberg, interview with TF, Austin, Texas, April 5, 1982.

Chapter Seventeen: The Axis of History

1. Gell-Mann, interview with TF, Pasadena, February 2, 1985.
2. Weinberg, interview with TF, Austin, Texas, February 28, 1985.
3. Gell-Mann, interview with TF, Pasadena, February 2, 1985.
4. Turner, interview with TF, Fermilab, February 26, 1985.
5. John Archibald Wheeler, interview with TF, Austin, February 28, 1985.

Chapter Eighteen: The Origin of the Universe

1. In Rothenberg, 1969, p. 34. The translation has been somewhat edited and revised.
2. Sandage, interview with TF, Pasadena, February 2, 1985.
3. Richard Feynman, talk at the University of Southern California, December 6, 1983.
4. In Yourgrau and Breck, 1977, p. 95.
5. Edward Tryon, interview with TF, New York, May 1, 1984.
6. Ibid.
7. Hawking, interview with TF, Pasadena, April 4, 1983.
8. In Root-Bernstein, Robert Scott, *Discovering*, prepublication manuscript, 1985.
9. Guth, Alan, in *The New York Times*, April 14, 1987, p. 17.
10. Hawking, private communication, August 11, 1987.

11. Hawking, talk at the 10th International Conference on General Relativity, Padua, Italy, July 7, 1983.
12. Hawking, interview with TF, Pasadena, 1985.
13. Wheeler, interview with TF, Austin, Texas, February 28, 1985.
14. Misner, Thorn and Wheeler, 1973, p. 1202.
15. Roger Penrose, *Structure of Spacetime*, in DeWitt and Wheeler, 1968, p 121.
16. Wheeler, "Beyond the Black Hole," in Woolf, 1980, p. 351.
17. Ibid., p. 356. Wheeler's italics.

CHAPTER NINETEEN: MIND AND MATTER

1. In Von Puttkamer, Jesco, "Extraterrestrial Life: Where Is Everybody?" *Cosmic Search*, Summer 1979, p. 43. See also F.M. Cornford, "Innumerable Worlds in Presocratic Philosophy," *Classical Quarterly*, January 1934, p. 13.
2. In Bailey, Cyril, 1926, p. 25.
3. In Lange, 1925, p. 233.
4. In Debus, 1978, p. 87.
5. In Dick, 1984, p. 111.
6. Fontenelle, 1929, pp. 12, 82.
7. Ibid., p. 55.
8. Ibid., pp. 114–115.
9. Giuseppe Cocconi and Philip Morrison, "Searching for Interstellar Communications," *Nature*, September 19, 1959.
10. William Proxmire, press release dated February 16, 1978, p. 2. The senator did not reveal the source of his intelligence on this point.
11. In Barrow and Tipler, 1986, p. 26.

CHAPTER TWENTY: THE PERSISTENCE OF MYSTERY

1. Lewis Thomas, "Debating the Unknowable," *The Atlantic Monthly*, July 1981; address to the Mount Sinai School of Medicine, reprinted in *The New York Times*, July 2, 1978, p. 15.
2. Popper, *Conjectures and Refutations*, 1968, p. 28.
3. For a discussion of this painting, see Foucault, 1982.
4. In French, A.P., 1979, p. 53.
5. See, e.g., Santillana, 1970, p. 60.
6. Borges, interview with Amelia Barili, *The New York Times Book Review*, July 13, 1986, p. 28.
7. In Cohen and Seeger, 1970, p. 220.
8. In Mehta, 1965, p. 155.
9. Niels Bohr, interviewed by Thomas Kuhn, in Mackinnon, 1982, p. 375.
10. Wheeler, interview with TF, Austin Texas; remainder of quotation in Buckley and Peat, 1979.
11. Einstein, *Out of My Later Years*, 1979, p. 114.
12. Epictetus, *Encheiridion*, verse 43, Oldfather translation.

BIBLIOGRAPHY

A man will turn over half a library to make
one book.

—Samuel Johnson

The following lists many of the titles consulted in researching this book. In the interest of brevity only books are included, and not articles or technical papers.

Abbott, Edwin A. *Flatland: A Romance in Many Dimensions*. New York: Dover, 1952. Classic fictionalized exposition of dimensionality.

Abell, George. *Exploration of the Universe*. New York: Saunders, 1982. Leading undergraduate textbook.

————. and P.J.E. Peebles, eds. *Objects of High Redshift*. Boston: Reidel, 1980. Proceedings of a 1979 symposium on remote galaxies and quasars.

Abetti, Giorgio, *The History of Astronomy*, trans. Betty Burr Abetti. New York: Schuman, 1952.

Adamczewski, Jan. *Nicolaus Copernicus and His Epoch*. Philadelphia: Copernicus Society of America, 1978.

Adams, Frank Dawson. *The Birth and Development of the Geological Sciences*. Baltimore: Williams & Wilkins, 1938. Historical survey of geology from ancient times to about 1825.

Adams, Henry. *The Education of Henry Adams*. New York: The Modern Library, 1931.

Afnan, Soheil M. *Avicenna: His Life and Work*. London: Allen & Unwin, 1958.

Ager, D.V. *Principles of Paleontology: An Introduction to the Study of How and Where Animals and Plants Lived in the Past*. New York: McGraw-Hill, 1963.

Ahmad ibn Muhammad ibn 'Abd al-Ghaffar, al-Kazwini al-Ghifari. *Epitome of the Ancient History of Persia*, trans. W. Ouseley. London: Cadell, 1799. Contains statement of eternal geological cycles cited by the geologist Lyell.

Albritton, Claude C., Jr. *The Abyss of Time: Changing Conceptions of the Earth's Antiquity After the Sixteenth Century*. San Francisco: Freeman, 1980. Nontechnical introduction to the question of geological time and evolution.

Alexander, H.G., ed. *The Leibniz-Clarke Correspondence*. New York: Barnes & Noble, 1970. Early debate on the nature of Newtonian space.

Alic, Margaret. *Hypatia's Heritage: A History of Women in Science from Antiquity to the Late Nineteenth Century*. London: The Women's Press, 1986.

Allen, C.W. *Astrophysical Quantities*. London: University of London Press, 1963.

Allen, J.M., ed. *The Nature of Biological Diversity*. New York: McGraw-Hill, 1963.

Allen, Richard Hinckley. *Star Names: Their Lore and Meaning*. New York: Dover, 1963.

Aller, Lawrence H., and Dean B. McLaughlin, eds. *Stellar Structure*. Chicago: University of Chicago Press, 1965. Collection of technical papers of historical interest.

Alvarez, Luis W. *Alvarez: Adventures of a Physicist*. New York: Basic Books, 1987. Autobiography of the versatile researcher in accelerator physics and nuclear fusion.

Amaldi, Ginestra. *The Nature of Matter*, trans. Peter Astbury. Chicago: University of Chicago Press, 1961.

Andrade, Edward Neville da Costa. *Rutherford and the Nature of the Atom*. Garden City, N.Y.: Doubleday, 1964.

Anthony, H.D. *Science and Its Background*. New York: St. Martin's, 1961.

Apollonius of Rhodes. *The Voyage of Argo*, trans. E.V. Rieu. London: Penguin, 1959. The legend of Jason and the Golden Fleece.

Appleman, Philip, ed. *Darwin: A Norton Critical Edition*. New York: Norton, 1979. Includes excerpts from Lyell, Hooker, Huxley, etc.

Aquinas, Thomas. *Summa Theologica*, trans. Daniel J. Sullivan and the Fathers of the English Dominican Province. 2 vols. Chicago: University of Chicago Press, 1952.

Archimedes. *The Works of Archimedes Including the Method*, trans. Thomas L. Heath. Chicago: University of Chicago Press, 1952.

Aristarchus of Samos. *Distances of the Sun and Moon*, trans. Ivor Thomas. Cambridge, Mass.: Harvard University Press, 1980.

Aristotle. *The Basic Works of Aristotle*, ed. Richard McKeon. New York: Random House, 1968.

———. *The Complete Works of Aristotle*, ed. Jonathan Barnes. Princeton, N.J.: Princeton University Press, 1984.

Armitage, Angus. *Copernicus, the Founder of Modern Astronomy*, New York: Barnes, 1962.

———. *Sun, Stand Thou Still: The Life and Work of Copernicus*. London: Sigma, 1947.

———. *Edmond Halley*. London: Nelson, 1966.

———. *The World of Copernicus*. New York: Mentor, 1951.

Asimov, Isaac. *Understanding Physics*, Vol. III, *The Electron, Proton, and Neutron*. New York: Mentor, 1966. Introduction by a master of science popularization.

———. *The Neutrino*. New York: Avon, 1966.

Atkins, Kenneth R. *Physics*. New York: Wiley, 1970. Undergraduate textbook.

Atkins, P.W. *The Second Law*. San Francisco: Freeman, 1984. Illustrated explication of the second law of thermodynamics.

Atkinson, Geoffrey. *The Extraordinary Voyage in French Literature Before 1700.* New York: Columbia University Press, 1920.

Atkinson, R.J.C. *Stonehenge.* London: Pelican, 1960.

Augustine of Hippo. *City of God,* trans. Henry Bettenson. London: Pelican, 1972.

———. *Confessions.* trans. R.S. Pine-Coffin. London: Penguin, 1970.

Avicenna. *Metaphysics,* trans. Parviz Morewedge. New York: Columbia University Press, 1973.

Ayer, A.J. *Wittgenstein.* New York: Random House, 1985.

Baade, Walter. *Evolution of Stars and Galaxies.* Cambridge: Mass.: MIT Press, 1975.

Bailey, Cyril. *Epicurus: The Extant Remains.* London: Oxford University Press, 1926.

———. *The Greek Atomists and Epicurus.* London: Oxford University Press, 1928.

Bailey, James. *The God-Kings and the Titans.* New York: St. Martin's, 1973. Makes a case for cultural diffusion via pre-Columbian contacts among Asia, Europe, and the New World.

Banville, John. *Doctor Copernicus.* London: Grenada, 1984. A novel, light on science but strong on personality.

———. *Kepler: A Novel.* Boston: Godine, 1984.

Barnes, C.A., D.D. Clayton, and D.N. Schramm, eds. *Essays in Nuclear Astrophysics.* London: Cambridge University Press, 1982.

Barnett, Lincoln. *The Universe and Dr. Einstein.* New York: Sloane, 1948. Venerable popularization of relativity theory.

Barrow, John D., and Frank Tipler. *The Anthropic Cosmological Principle.* London: Oxford University Press, 1986.

Barut, Asim O., Alwyn van der Merwe, and Jean-Pierre Vigier, eds. *Quantum, Space, and Time—The Quest Continues.* London: Cambridge University Press, 1984. Essays in honor of de Broglie, Dirac, and Wigner.

Baumgardt, Carola. *Johannes Kepler, Life and Letters.* New York: Philosophical Library, 1951.

Beaglehole, J.C. *The Exploration of the Pacific.* London: Black, 1966.

Beazley, Charles Raymond. *Prince Henry the Navigator.* New York: Putnam's, 1904.

Beer, Arthur, and K.A. Strand, eds. *Copernicus: Yesterday and Today.* New York: Pergamon, 1975.

Beiser, Arthur. *Concepts of Modern Physics.* New York: McGraw-Hill, 1981. College-level textbook.

Bell, A.E. *Christiaan Huygens and the Development of Science in the Seventeenth Century.* London: Arnold, 1947. Emphasizes Huygens's contributions to a scientific world in which Newtonian and Cartesian mechanics still contended.

Bell, Christopher. *Portugal and the Quest for the Indies.* New York: Harper & Row, 1974.

Bell, E.T. *Men of Mathematics.* New York: Simon & Schuster, 1937. Biographical sketches of mathematicians from Pythagoras to Cantor.

Benacerraf, Paul, and Hilary Putnam, eds. *Philosophy of Mathematics.* London: Cambridge University Press, 1985.

Berendzen, Richard, ed. *Life Beyond Earth and the Mind of Man.* Washington, D.C.: NASA, 1973. Based on a 1972 SETI symposium held at Boston University.

————. Richard Hart, and Daniel Seeley. *Man Discovers the Galaxies*. New York: Science History Publications, 1976.

Berger, A., ed. *The Big Bang and Georges Lemaître*. Dordrecht: Reidel, 1985. Papers presented at a symposium at Louvain-la-Neuve, Belgium, 1983, honoring the originator of the big bang theory.

Bergmann, Peter G. *The Riddle of Gravitation*. New York: Scribner's, 1968. Semi-technical explication of general relativity.

Berkeley, George. *The Principles of Human Knowledge*. Chicago: University of Chicago Press, 1952.

Berkhuijsen, Elly M., and Richard Wielebinski, eds. *Structure and Properties of Nearby Galaxies*. Boston: Reidel, 1978. Proceedings of a 1977 astronomical symposium.

Bernal, J.D. *The Origin of Life*. London: Weidenfeld, 1967.

————. *Science in History*. London: Watts, 1954.

Bernstein, Jeremy. *Einstein*. New York: Viking, 1973. Mines ground left largely untouched by other biographies.

————. *Hans Bethe: Prophet of Energy*. New York: Basic Books, 1980. Anecdotal profile of the pathfinding physicist.

————. and Gerald Feinberg, eds. *Cosmological Constants*. New York: Columbia University Press, 1986.

Berry, Michael. *Principles of Cosmology and Gravitation*. London: Cambridge University Press, 1981.

Berry, W.B.N. *Growth of a Prehistoric Time Scale*. San Francisco: Freeman, 1968. Brief introduction to the development of stratigraphy and the emergence of a geochronological system.

Bertocci, P.A. *The Empirical Argument for God in Late British Thought*. Cambridge, Mass.: Harvard University Press, 1938.

Bertotti, B., F. de Felice, and A. Pascolini. *General Relativity and Gravitation*. Boston: Reidel, 1984. Papers presented at a 1983 conference.

Beyerchen, Alan D. *Scientists Under Hitler*. New Haven, Conn.: Yale University Press, 1977. Includes accounts of Max Planck's exemplary career and tragic life.

Bienkowska, Barbara, ed. *The Scientific World of Copernicus*. Boston: Reidel, 1973.

Billingham, John, ed. *Life in the Universe*. Cambridge, Mass.: MIT Press, 1981.

Blaauw, Adriaan, and Maarten Schmidt, eds. *Galactic Structure*. Chicago: University of Chicago Press, 1965.

Blacker, Carmen, and Michael Loewe, eds. *Ancient Cosmologies*. London: Allen & Unwin, 1975.

Blackmore, John T. *Ernst Mach*. Berkeley: University of California Press, 1972.

Blake, George. *British Ships and Shipbuilders*. London: Collins, 1946. Illustrated.

Blum, Harold F. *Time's Arrow and Evolution*. Princeton, N.J.: Princeton University Press, 1968. Study of evolution and thermodynamics.

Boas, Franz. *The Mind of Primitive Man*. New York: Macmillan, 1938.

Bochner, Salomon. *The Role of Mathematics in the Rise of Science*. Princeton, N.J.: Princeton University Press, 1981.

Boethius, Ancius. *The Consolation of Philosophy*, trans. V.E. Watts. New York: Penguin Books, 1976.

Bohm, David. *Causality and Chance in Modern Physics*. Philadelphia: University of Pennsylvania Press, 1957.

———. *The Special Theory of Relativity*. Reading, Mass.: Benjamin/Cummings, 1965. Some mathematics.

———. *Wholeness and the Implicate Order*. London: Routledge & Kegan Paul, 1981.

Bohr, Niels. *Atomic Physics and Human Knowledge*. New York: Wiley, 1960.

———. *The Theory of Spectra and Atomic Constitution*. London: Cambridge University Press, 1924.

Bok, Bart J., and Priscilla Bok. *The Milky Way*. Cambridge, Mass.: Harvard University Press, 1981.

Boltzmann, Ludwig. *Theoretical Physics and Philosophical Problems*, ed. Brian McGuinness. Boston: Reidel, 1974.

Bondi, Hermann. *Assumption and Myth in Physical Theory*. London: Cambridge University Press, 1967.

———. *Relativity and Common Sense*. New York: Dover, 1964. Innovative popularization.

Bonola, Roberto. *Non-Euclidean Geometry*, New York: Dover 1955. On the origins of the geometries employed by Einstein's general theory of relativity.

Boodin, John. *Cosmic Evolution*, New York: Macmillan, 1925. Expansive extrapolation of the evolutionary hypothesis, influential in its day.

Boorse, Henry A., and Lloyd Motz, eds. *The World of the Atom*. New York: Basic Books, 1966. Anthology of nontechnical papers and essays.

Boorstin, Daniel. *The Discoverers*. New York: Random House, 1983. Anecdotal history of the ages of exploration.

Born, Max. *The Born-Einstein Letters*. New York: Walker, 1971. Documents a decades-long debate over the philosophy of quantum physics, conducted in high spirits and with great mutual respect.

———. *Einstein's Theory of Relativity*. New York: Dover, 1965.

———. *My Life*, New York: Scribner's, 1978.

———. *The Restless Universe*, trans. Winifred M. Deans. New York: Dover, 1951. Illustrated introduction to quantum physics and field theory.

Bowker, J. *The Sense of God*. London: Oxford University Press, 1973.

Boxer, Charles R., *The Portuguese Seaborne Empire: 1415–1825*. New York: Knopf, 1969.

Bracewell, Ronald N. *The Galactic Club*. San Francisco: Freeman, 1975. Discusses interstellar probes.

Bradshaw, M.J. *A New Geology*. London: English Universities, 1968.

Braithwaite, R.B. *Scientific Explanation*. New York: Harper, 1960.

Brandon, S.G.F. *Time and Mankind: An Historical and Philosophical Study of Mankind's Attitude to the Phenomena of Change*. London: Hutchinson, 1951.

Brecher, Kenneth, and Michael Fiertag, eds. *Astronomy of the Ancients*. Cambridge, Mass.: MIT Press, 1980.

Brecht, Bertolt. *Galileo*, English version trans. by Charles Laughton. New York: Grove Press, 1966.

Bridgman, Percy, *The Logic of Modern Physics*. New York: Macmillan, 1961.

Brittan, Gordon. *Kant's Theory of Science*. Princeton, N.J.: Princeton University Press, 1978.

Broad, C.D. *Leibniz: An Introduction*. London: Cambridge University Press, 1975.
———. *Scientific Thought*. Paterson, N.J.: Littlefield, 1959.
Brodetsky, S. *Sir Isaac Newton*. London: Metheun, 1927.
Bronowski, Jacob. *The Ascent of Man*. Boston: Little, Brown, 1973.
———. *The Common Sense of Science*. Cambridge, Mass.: Harvard University Press, 1978.
———. *The Identity of Man*. Garden City, N.Y.: Natural History Press, 1966.
———. *Magic, Science, and Civilization*. New York: Columbia University Press, 1978.
———. *The Origins of Knowledge and Imagination*. New Haven, Conn.: Yale University Press, 1978.
———. *Science and Human Values*. New York: Harper & Row, 1965.
———. *A Sense of the Future*. Cambridge, Mass.: MIT Press, 1977.
———. *The Visionary Eye*. Cambridge, Mass.: MIT Press, 1978.
Brooks, Daniel R., and E.O. Wiley. *Evolution as Entropy*. Chicago: University of Chicago Press, 1986. Study of the relationship between biological evolution and the second law of thermodynamics.
Brown, Laurie M., and Lillian Hoddeson, eds. *The Birth of Particle Physics*. London: Cambridge University Press, 1983. Based on a 1980 Fermilab symposium.
Brown, Lloyd. *The Story of Maps*. New York: Crown, 1979. Survey history of mapmaking.
Bruno, Giordano. *The Ash Wednesday Supper*, ed. and trans. Edward Gosselin and Lawrence Lerner. Hamden, Conn.: Archon Books, 1977.
———. *The Ash Wednesday Supper*, ed. and trans. Stanley Jaki. Paris: Mouton, 1975.
Buchwald, Jed. *From Maxwell to Microphysics*. Chicago: University of Chicago Press, 1985. Semitechnical examination of British and continental approaches to electromagnetism in the late nineteenth century.
Bucke, Richard. *Cosmic Consciousness*. New York: Dutton, 1969.
Buckley, Paul, and F. David Peat. *A Question of Physics*. Buffalo: University of Toronto Press, 1979.
Buffon, Georges-Louis Leclerc de. *Selections from Natural History, General and Particular*, New York, Arno Press, 1977.
Bunbury, E.H. *A History of Ancient Geography*. 2 vols. New York: Dover, 1959.
Bunge, Mario. *Causality and Modern Science*. New York: Dover, 1979.
Burchfield, J.D. *Lord Kelvin and the Age of the Earth*. New York: Science History Publications, 1975. Study of Kelvin's geochronological ideas and their influence.
Burckhardt, Jacob. *The Civilization of the Renaissance in Italy*. New York: Oxford University Press, 1944.
Burger, Dionys. *Sphereland: A Fantasy About Curved Spaces and an Expanding Universe*. New York: Barnes & Noble, 1983. Expands upon Edwin Abbott's *Flatland*.
Burkert, Walter. *Greek Religion*. Cambridge, Mass.: Harvard University Press, 1985.
Burnet, Thomas. *Sacred Theory of the Earth*. London: Centaur Press, 1965. Reprint of 1691 edition.

Burnham, Robert, Jr. *Burnham's Celestial Handbook*. New York: Dover, 1978. Compendium of astronomical information and mythological lore on the major stars and nebulae.

Burtt, E. A. *The Metaphysical Foundations of Modern Science*. New York: Doubleday, 1932.

Bury, John Bagnell. *The Idea of Progress*. New York: Dover, 1960.

Butterfield, Herbert. *The Origins of Modern Science, 1300–1800*. New York: Macmillan, 1951. Overview of concepts implicated in the rise of the scientific world view.

Buttmann, Gunther. *The Shadow of the Telescope*, trans. B.E.J. Pagel. New York: Scribner's, 1970. Biography of John Herschel.

Bynum, W.F., E.J. Brown, and Roy Porter, eds. *Dictionary of the History of Science*. Princeton, N.J.: Princeton University Press, 1981.

Calder, Nigel. *The Comet Is Coming!*. New York: Viking, 1980.

———. *Einstein's Universe*. New York: Penguin, 1980.

———. *Timescale: An Atlas of the Fourth Dimension*. New York: Viking, 1983.

———. *Violent Universe*. New York: Viking, 1969. Popular survey of high-energy astrophysics.

Cameron, A.G.W. *Interstellar Communication: The Search for Extraterrestrial Life*. New York: Benjamin, 1963.

Campanella, Tommaso. *The Defense of Galileo of Thomas Campanella*, ed. and trans. Grant McColley. Merrick, N.Y.: Richwood, 1976. Bold statement of the case for Copernicus and Galileo, written by a Dominican friar in 1616 while imprisoned in a Neapolitan dungeon.

Campbell, Bernard. *Human Evolution: An Introduction to Man's Adaptations*. Chicago: Aldine, 1966.

Campbell, Lewis, and William Garnett. *The Life of James Clerk Maxwell*. London: Macmillan, 1882. Dated but still instructive biography of Maxwell, originally published within three years of his death.

Capek, Milic, ed. *The Concepts of Space and Time*. Boston: Reidel, 1976. Anthology of historical writings.

———. *The Philosophical Impact of Contemporary Physics*. New York: Van Nostrand, 1961.

———. *Philosophy of Space and Time*. New York: Dover, 1958.

Carnap, Rudolf. *Foundations of Logic and Mathematics*. Chicago: University of Chicago Press, 1939.

Carus, Paul. *The Religion of Science*. Chicago: Open Court, 1913.

Caspar, Max. *Kepler*, trans. C. Doris Hellman. New York: Abelard-Schuman, 1959.

Cassirer, Ernst. *Einstein's Theory of Relativity*. New York: Dover, 1953.

———. *The Individual and the Cosmos in Renaissance Philosophy*, trans. Mario Domandi. New York: Harper & Row, 1963.

———. *Kant's Life and Thought*, trans. James Haden. New Haven, Conn.: Yale University Press, 1981.

Chandrasekhar, Subrahmanyan. *Eddington, the Most Distinguished Astrophysicist of His Time*, London: Cambridge University Press, 1983. Memoir of Eddington by his eminent pupil.

————. *An Introduction to the Study of Stellar Structure*. New York: Dover, 1967.

Chapple, J.A.V. *Science and Literature in the Nineteenth Century*. London: Macmillan, 1986.

Chardin, Pierre Teilhard de. *Man's Place in Nature*. New York: Harper, 1956, 1966.

————. *The Phenomenon of Man*. New York: Harper, 1959.

Cheng, David C., and Gerard K. O'Neill. *Elementary Particle Physics*. Reading, Mass.: Addison-Wesley, 1979. Graduate-level textbook.

Child, J.M. *The Geometrical Lectures of Isaac Barrow*. Chicago: Open Court, 1916.

Choquet-Bruhat, Y., and T.M. Karade. *On Relativity Theory*. Singapore: World Scientific, 1984. Proceedings of an Arthur Eddington centenary symposium.

Christianson, Gale E., *In the Presence of the Creator: Isaac Newton and His Times*. New York: Free Press, 1984.

Cicero. *De Fato*, trans. H. Rackham, Cambridge, Mass.: Harvard University Press, 1982. Timeless critique of philosophical issues in cosmology.

————. *The Nature of the Gods*, trans. H.C.P. McGregory. London: Penguin, 1984. Translation of *De Natura Deorum*.

Clagett, Marshall, ed. *Critical Problems in the History of Science*. Madison: University of Wisconsin Press, 1969.

————. *The Science of Mechanics in the Middle Ages*. Madison: University of Wisconsin Press, 1959. Pre-Galilean dynamics.

Clark, David H., and F. Richard Stephenson. *The Historical Supernovae*. New York: Pergamon, 1977. Survey of scientific study and historical records of supernovae visible from Earth with the naked eye.

Clark, Grahame. *World Prehistory: An Outline*. London: Cambridge University Press, 1961.

Clark, Ronald W. *Einstein: The Life and Times*. New York: World, 1971. Non-technical biography with minimal stress on Einstein's scientific research.

————. *The Survival of Charles Darwin: Biography of a Man and an Idea*. New York: Random House, 1984. Popular biography with an emphasis on the fortunes of Darwinism following its founder's death.

Clark, Thomas Henry, and C.W. Stearn. *Geological Evolution of North America*. New York: Ronald, 1968.

Clayton, Donald C. *Principles of Stellar Evolution and Nucleosynthesis*. Chicago: University of Chicago Press, 1983. Graduate-level textbook.

Clerke, Agnes M. *A Popular History of Astronomy During the Nineteenth Century*. London: Adam & Charles Black, 1902.

Cloud, Preston. *Cosmos, Earth and Man*. New Haven, Conn.: Yale University Press, 1978.

Cohen, I. Bernard. *Album of Science: From Leonardo to Lavoisier*. New York: Scribner's, 1980. Pictorial history of science.

————. *The Birth of a New Physics*. New York: Norton, 1985.

————. *The Newtonian Revolution*. London: Cambridge University Press, 1983.

————. *Revolution in Science*. Cambridge, Mass.: Harvard University Press, 1985.

————. *Roemer and the First Determination of the Velocity of Light*. New York: The Burndy Library, 1944.

BIBLIOGRAPHY 451

——, ed. *Introduction to Newton's "Principia."* Cambridge, Mass.: Harvard University Press, 1971.

——. ed. *Isaac Newton's Papers and Letters on Natural Philosophy.* Cambridge, Mass.: Harvard University Press, 1958.

Cohen, Morris, and Ernest Nagel. *An Introduction to Logic and Scientific Method.* New York: Harcourt, Brace, 1934. Textbook.

——, and I.E. Drabkin. *A Source Book in Greek Science.* Cambridge, Mass.: Harvard University Press, 1948.

Cohen, Robert S., and Raymond J. Seeger, eds. *Ernst Mach, Physicist and Philosopher.* Boston: Reidel, 1970. Collection of papers on Mach and his thought.

Collingwood, R.G. *The Idea of Nature.* New York: Oxford University Press, 1960.

Columbus, Christopher. *Select Letters of Christopher Columbus,* ed. R.H. Major. London: Hakluyt Society, 1870.

Columbus, Ferdinand. *The Life of the Admiral Christopher Columbus,* trans. Benjamin Keen. New Brunswick, N.J.: Rutgers University Press, 1979.

Conant, J.B. *On Understanding Science.* New Haven, Conn.: Yale University Press, 1947.

Conway, J.H. *On Numbers and Games.* New York: Academic Press, 1976. Discusses how nonbeing implies being.

Cook, James. *The Journals of Capt. James Cook,* ed. J.C. Beaglehole. 4 vols. London: 1955.

Cook, Theodore Andrea. *The Curves of Life.* New York: Dover, 1979. Includes illustrations of natural spirals generated by invariances.

Cooper, Henry S.F., Jr. *The Search for Life on Mars.* New York: Holt, Rinehart & Winston, 1980. Report on the Viking mission to Mars.

Cooper, Lane. *Aristotle, Galileo, and the Tower of Pisa.* Ithaca, N.Y.: Cornell University Press, 1935. Study of the opposition between Galileo and the Scholastics on the falling bodies question.

Copernicus, Nicolaus. *Complete Works,* trans. and commentator, Edward Rosen, ed. Vol. 3 Pawel Czartoryski. 3 vols. London: Macmillan, 1985.

——. *On the Revolutions,* ed. Jerzy Dobrzycki and trans. Edward Rosen. Warsaw: Polish Scientific Publishers, 1978; Baltimore: Johns Hopkins University Press, 1978. Large format, with a commentary by Rosen.

——. *On the Revolutions,* ed. and trans. A.M. Duncan. New York: Barnes & Noble, 1976.

——. *On the Revolutions,* trans. Charles Glenn Wallis, Chicago: University of Chicago Press, 1952.

——. *Three Copernican Treatises: The Commentariolus of Copernicus, the Letter Against Werner, the Narratio Prima of Rheticus,* ed. and trans. Edward Rosen. New York: Dover, 1959. Shorter works of Copernicus not previously translated into English.

Cornell, James. *The First Stargazers.* New York: Scribner's, 1981.

Cornford, F.M. *Plato's Cosmology.* London: Routledge & Kegan Paul, 1977. Translation of the *Timaeus,* with running commentary.

Cornwall, I.W. *The World of Ancient Man.* New York: New American Library, 1965.

Craig, W.L. *The Cosmological Argument from Plato to Liebniz*. New York: Macmillan, 1980.

Crease, Robert, and Charles Mann. *The Second Creation*. New York: Macmillan, 1986. Nontechnical history of twentieth-century unified field theory, based upon extensive interviews and a survey of the scientific literature.

Crombie, A.C. *Augustine to Galileo: The History of Science A.D. 400–1650*. London: Falcon, 1952.

Crosland, M.P., ed. *The Science of Matter*. Harmondsworth, Eng.: Penguin, 1971.

Crowe, Michael J. *The Extraterrestrial Life Debate 1750–1900*. London: Cambridge University Press, 1986.

Cyrano de Bergerac, Savinien de. *The Comical History of the States and Empires of the Worlds of the Moon and Sun*. London: Rhodes, 1687.

D'Abro, A. *The Evolution of Scientific Thought from Newton to Einstein*. New York: Dover, 1950.

———. *The Rise of the New Physics*. 2 vols. New York: Dover, 1931.

D'Alembert, Jean. *Preliminary Discourse to the Encyclopedia of Diderot*, trans. Richard N. Schwab. Indianapolis: Bobbs-Merrill, 1981.

Dampier, William Cecil. *A History of Science and Its Relations with Philosophy and Religion*. London: Cambridge University Press, 1949.

———. *Readings in the Literature of Science*. New York: Harper, 1959.

Danto, Arthur, and Sidney Morgenbesser, eds. *Philosophy of Science*. New York: World, 1960.

Dantzig, Tobias. *Number: The Language of Science*. New York: Macmillan, 1954.

Darlington, C.D. *The Evolution of Man and Society*. New York: Simon & Schuster, 1969.

Darwin, Charles. *The Collected Papers of Charles Darwin*, ed. Paul H. Barrett. Chicago: University of Chicago Press, 1977.

———. *The Correspondence of Charles Darwin*, ed. Frederick Burkhardt and Syndey Smith. New York: Dover, 1955.

———. *The Origin of Species by Means of Natural Selection*, 6th ed. New York: Modern Library, 1936. Reprint of 1872 edition.

———. *The Variation of Animals and Plants Under Domestication*. London: Murray, 1868. Expands upon the first chapters of the *Origin of Species*.

———. *The Voyage of the Beagle*, ed. Leonard Engel. New York: Doubleday, 1962. Reprint of the 1860 edition.

Darwin, Erasmus. *Zoonomia: Or the Laws of Organic Life*, 4th American ed. Philadelphia: 1818. Darwin's grandfather's evolutionary theory.

Darwin, Francis, ed. *Charles Darwin's Autobiography*. New York: Schuman, 1950. Like Einstein's, more an outline than an autobiography, written late in life.

———, ed. *The Life and Letters of Charles Darwin*. 2 vols. London: John Murray, 1888; New York, Basic Books, 1959.

Darwin, George. *Scientific Papers*. 5 vols. London: Cambridge University Press, 1907–1916.

Davies, P.C.W. *The Forces of Nature*. London: Cambridge University Press, 1980. Semitechnical.

———. *The Physics of Time Asymmetry*. Berkeley: University of California Press,

1977. Technical investigation of the unidirectional motion of time from the perspective of physics.

————. *Space and Time in the Modern Universe*. London: Cambridge University Press, 1977. Semitechnical account of relativity and cosmology.

Davies, Paul. *God and the New Physics*. New York: Simon & Schuster, 1983.

————. *Superforce*. New York: Simon & Schuster, 1984. Popular account of grand unified theory.

Davis, Nuel Pharr. *Lawrence and Oppenheimer*. New York: Simon & Schuster, 1968. Report on two of the most influential figures in the development of American atomic policy.

Davis, Philip J., and Reuben Hersh. *The Mathematical Experience*. Boston: Birkhäuser, 1981. Popular exposition of mathematical concepts and methods.

DeBeer, Gavin. *Charles Darwin, Evolution by Natural Selection*. Garden City, N.Y.: Doubleday, 1964. Succinct account of Darwin's life.

de Broglie, Louis, *New Perspectives in Physics*. New York, Basic Books, 1962.

————, Louis-Armand, Pierre-Henri Simon et al. *Einstein*. New York: Peebles Press, 1979. Centenary observations.

Debus, Allen G. *Man and Nature in the Renaissance*. London: Cambridge University Press, 1978.

De Madariaga, Salvador. *Christopher Columbus*. London: Hollis & Carter, 1949.

Descartes, René. *Geometry*, trans. David Eugene Smith and Marcia L. Latham. Chicago: University of Chicago Press, 1952.

————. *Philosophical Works*, trans. Elizabeth S. Haldane and G.R.T. Ross. London: Cambridge University Press, 1979.

————. *Philosophical Writings*, ed. and trans. Elizabeth Anscombe and Peter Thomas Geach. Indianapolis: Bobbs-Merrill, 1971, 1981.

————. *Principles of Philosophy*, trans. Valentine Rodger Miller and Reese P. Miller. Boston: Reidel, 1983.

DeSober, Elliott. *The Nature of Selection: Evolutionary Theory in Philosophical Focus*. Cambridge, Mass.: MIT Press, 1985.

D'Espagnat, Bernard. *Conceptual Foundations of Quantum Mechanics*. Reading, Mass.: Benjamin, 1976.

DeWitt, C.M., and John Archibald Wheeler, eds. *Battelle Seattle Summer Rencontres in Mathematics and Physics, 1967*. New York: Benjamin, 1968.

Diaz, Bernal. *The Conquest of New Spain*, trans. J.M. Cohen. London: Penguin, 1975.

Dick, Steven J. *Plurality of Worlds: The Origins of the Extraterrestrial Life Debate from Democritus to Kant*. London: Cambridge University Press, 1984.

Dicks, D.R. *Early Greek Astronomy to Aristotle*. Ithaca, N.Y.: Cornell University Press, 1970.

Dickson, F.P. *The Bowl of Night*. Cambridge, Mass.: MIT Press, 1968. Accessible review of long-standing cosmological questions.

Dijksterhuis, E.J. *The Mechanization of the World Picture*. London: Oxford University Press, 1969.

Dingle, Herbert. *Through Science to Philosophy*. London: Oxford University Press, 1937.

Diogenes Laertius. *Lives of the Philosophers*, ed. and trans. A. Robert Caponigri. Chicago: Regnery, 1969.

———. *Lives of the Philosophers*, trans. R.D. Hicks. 2 vols. Cambridge, Mass.: Harvard University Press, 1979.

Dirac, Paul Adrien Maurice. *The Development of Quantum Theory*. New York: Gordon and Breach, 1971. Succinct summary, based on remarks delivered on accepting the J. Robert Oppenheimer Memorial Prize.

———. *The Principles of Quantum Mechanics*. London: Oxford University Press, 1981. Textbook that educated a generation of physicists.

Dobbs, Betty Jo Teeter. *The Foundations of Newton's Alchemy*. London: Cambridge University Press, 1976.

Dobrzycki, Jerzy, ed. *The Reception of Copernicus's Heliocentric Theory*. Boston: Kluwer, 1973.

Dobzhansky, Theodosius, et al. *Evolution*. San Francisco: Freeman, 1977. Standard textbook.

Dodd, J.E. *The Ideas of Particle Physics*. London: Cambridge University Press, 1984. Introduction intended for scientists in other fields.

Dodds, E.R. *The Greeks and the Irrational*. Berkeley: University of California Press, 1951.

Doig, Peter. *A Concise History of Astronomy*. New York: Philosophical Library, 1951.

Drachman, J.M. *Studies in the Literature of Natural Science*. New York: Macmillan, 1930. Includes discussion of Lyell and Darwin.

Drake, Frank. *Intelligent Life in Space*. New York: Macmillan, 1962.

Drake, Stillman. *Galileo*. New York: Hill & Wang, 1980.

———. *Galileo at Work: His Scientific Biography*. Chicago: University of Chicago Press, 1978.

———, ed. and trans. *Discoveries and Opinions of Galileo*. Garden City, N.Y.: Doubleday, 1957. Includes Galileo's *The Starry Messenger, Letters on Sunspots*, and *The Assayer*.

Draper, John William. *History of the Conflict Between Religion and Science*. New York: Appleton, 1879. Inquires into the historical background of the decline of religious faith and the rise of science in the late nineteenth century.

Dryer, J.L.E. *A History of Astronomy from Thales to Kepler*. New York: Dover, 1953. Intellectual history of early astronomy.

———. *Tycho Brahe*. Edinburgh: 1890; New York: Dover, 1963. Standard biography.

Drude, Paul. *The Theory of Optics*. New York: Dover, 1959. Book that influenced the young Einstein.

Duff, M.J., and C.J. Isham. *The Quantum Structure of Space and Time*. London: Cambridge University Press, 1982. Proceedings of a workshop held at Imperial College, London, in 1981.

Duhem, Pierre. *To Save the Phenomena*, trans. Edmund Doland and Chaninah Maschler. Chicago: University of Chicago Press, 1969.

Dukas, Helen, and Banesh Hoffmann, eds. *Albert Einstein: The Human Side*. Princeton, N.J.: Princeton University Press, 1979. Unique collection of Einstein memorabilia. Includes the German originals as well as their translations.

Duns Scotus. *Philosophical Writings*, trans. Allan Wolter. Indianapolis: Bobbs-Merrill, 1974.

Durham, Frank, and Robert D. Purrington. *Frame of the Universe*. New York: Columbia University Press, 1983. Brief history of cosmology.

Dyson, Freeman. *Perspectives in Modern Physics*. New York: Interscience, 1966.

Dyson, J.E., and D.A. Williams. *The Physics of the Interstellar Medium*. New York: Wiley, 1980.

Eddington, Alfred Stanley. *The Nature of the Physical World*. New York: Macmillan, 1929.

———. *The Philosophy of Physical Science*. New York: Macmillan, 1939.

Eicher, Don L. *Geologic Time*. Englewood Cliffs, N.J.: Prentice-Hall, 1969. Chronicles the establishment of modern geochronology.

Eigen, Manfred, and Ruthild Winkler. *Laws of the Game: How the Principles of Nature Govern Chance*. New York: Knopf, 1981.

Einstein, Albert. *Essays in Humanism*. New York: Philosophical Library, 1983.

———. *Ideas and Opinions*, trans. Sonja Bargmann. New York: Dell, 1979.

———. *The Meaning of Relativity*. Princeton, N.J.: University Press, 1923.

———. *Out of My Later Years*. Secaucus, N.J.: Citadel, 1979.

———. *The Principle of Relativity*, trans. W. Perrett and G.B. Jeffrey. New York: Dover, 1952. Useful but marred by errors in translation.

———. *Relativity: The Special and General Theory*, trans. Robert W. Lawson. New York: Crown, 1961.

———. *Sidelights on Relativity*. London, Methuen, 1922. Includes discussion of aether theory.

———. *The World As I See It*, trans. A. Harris. New York: Philosophical Library, 1935.

———, and Leopold Infeld. *The Evolution of Physics*, New York: Simon & Schuster, 1938.

Eiseley, Loren. *Darwin's Century*. Garden City, N.Y.: Doubleday, 1958. Study of Darwin's intellectual and scientific milieu.

———. *The Firmament of Time*. New York: Atheneum, 1970. Based on lectures given at the University of Cincinnati in 1959.

Eisenstein, Elizabeth L. *The Printing Press as an Agent of Change*. London: Cambridge University Press, 1980.

Eliade, Mircea. *The Myth of the Eternal Return*, trans. Willard Trask. New York: Harper, 1959. Treatise on the doctrine of cyclical time.

Elkana, Yehuda. *The Discovery of the Conservation of Energy*. Cambridge, Mass.: Harvard University Press, 1974.

Elliott, James P., and P.G. Dawber. *Symmetry in Physics*. London: Macmillan, 1979. Examination of symmetry in classical and quantum physics, with an emphasis on group theory.

Emmerson, John McLaren. *Symmetry Principles in Particle Physics*. London: Oxford University Press, 1972.

Epictetus. *Discourses*, trans. George Long. Chicago: University of Chicago Press, 1952.

———. *Discourses*, trans. W.A. Oldfather. Cambridge, Mass.: Harvard University Press, 1979.

Epstein, Lewis Carroll. *Relativity Visualized*. San Francisco: Insight, 1985. Amply illustrated, right-forebrain explication of the special and general theories.

Eratosthenes. *Measurement of the Earth*, trans. Ivor Thomas. Cambridge, Mass.: Harvard University Press, 1980.

Euclid. *The Elements*, trans. Isaac Barrow. London: Redmayne, 1705.

―――. *The Elements*, ed. and trans. Thomas L. Heath. 3 vols. New York: Dover, 1956.

Eve, A.S. *Rutherford*. London: Cambridge University Press, 1939.

Fakhry, Ahmed. *The Pyramids*. Chicago: University of Chicago Press, 1974.

Farrar, Glennys, and Frank Henyey, eds. *Problems in Unification and Supergravity*. New York: American Institute of Physics, 1984.

Feigl, Herbert, and May Brodbeck, eds. *Readings in the Philosophy of Science*. New York: Appleton, 1953.

Feinberg, Gerald. *What Is the World Made Of?* New York: Anchor, 1978. Popular account of nuclear physics.

―――, and Robert Shapiro. *Life Beyond Earth: An Intelligent Earthling's Guide to Life in the Universe*. New York: Morrow, 1980.

Ferguson, James. *Astronomy Explained upon Sir Isaac Newton's Principles*, 2nd ed. London: self-published, 1757. Popularization that helped kindle William Herschel's passion for astronomy.

Ferguson, Wallace, et al. *The Renaissance: Six Essays*. New York: Harper, 1953.

Ferm, Vergilius. *History of Philosophical Systems*. Paterson, N.J.: Littlefield, Adams, 1965.

Fermi, Enrico. *Collected Papers*. Chicago: University of Chicago Press, 1965.

Fermi, Laura, and Gilberto Bernardini. *Galileo and the Scientific Revolution*. New York: Basic Books, 1961.

Fernie, Donald. *The Whisper and the Vision: The Voyages of the Astronomers*. Toronto: Clarke, Irwin & Co., 1976. Account of transit expeditions undertaken to measure the dimensions of the solar system.

Ferris, Timothy. *Galaxies*. San Francisco: Sierra Club Books, 1980.

―――. *The Red Limit: The Search for the Edge of the Universe*. New York: Morrow, 1977, 1983.

―――. *SpaceShots: The Beauty of Nature Beyond Earth*. New York: Pantheon, 1984.

Feuer, Lewis S. *Einstein and the Generations of Science*. New York: Basic Books, 1974. Impressionistic study of Einstein's social milieu.

Feynman, Richard P. *QED*. Princeton, N.J.: Princeton University Press, 1985. Nontechnical introduction to quantum electrodynamics, by an author of the theory.

―――. *The Character of Physical Law*. Cambridge, Mass.: MIT Press, 1965. Feynman lectures on philosophy of science.

―――, Robert B. Leighton, and Matthew Sands. *The Feynman Lectures of Physics*. Reading, Mass.: Addison-Wesley, 1963. Classic introduction for scientifically inclined undergraduates, by one of the architects of contemporary quantum physics.

Finney, Ben R., and Eric M. Jones, ed. *Interstellar Migration and the Human Experience*. Berkeley: University of California Press, 1986.

Fiske, John. *Outlines of Cosmic Philosophy, Based on the Doctrine of Evolution*. 4 vols.

Boston: Houghton Mifflin, 1874. Concerns the philosophy of Herbert Spencer.

Flew, A. *God and Philosophy*. New York: Harcourt, Brace, 1966.

Flurry, Robert L. *Quantum Chemistry: An Introduction*. Englewood Cliffs, N.J.: Prentice-Hall, 1983.

Folse, Henry J. *The Philosophy of Niels Bohr*. New York: North-Holland, 1985. Scholarly though nontechnical discussion of Bohr's philosophy of complementarity.

Folsome, Clair Edwin. *The Origin of Life*. San Francisco: Freeman, 1979. Popular account of the origin of the solar system and of life on Earth.

Fontenelle, Bernard de. *A Plurality of Worlds*, trans. John Glanville. London: Nonesuch Press, 1929.

Fosdick, Harry Emerson, ed. *Great Voices of the Reformation*. New York: Random House, 1952.

Foster, J., and J.D. Nightingale. *A Short Course in General Relativity*. London: Longman, 1979.

Foucault, Michel. *This Is Not a Pipe*, ed. and trans. James Harkness. Berkeley: University of California Press, 1982. Study of symbolism and logic in the work of René Magritte.

Fox, Sidney, and Klaus Dose. *Molecular Evolution and the Origin of Life*. San Francisco: Freeman, 1972.

Frank, Philipp. *Einstein: His Life and Times*. New York: Knopf, 1970. Biography by its subject's friend and colleague. American edition unaccountably abridged.

———. *Foundations of the Unity of Science*. Chicago: University of Chicago Press, 1946.

Frankel, Charles. *The Faith of Reason: The Idea of Progress in the French Enlightenment*. New York: Octagon, 1948.

Franz, Marie-Louise von. *Creation Myths*. Dallas: Spring, 1972.

Fraser, J.T. *The Voices of Time*. Amherst: University of Massachusetts Press, 1981. Collection of scientific and humanistic essays.

Frazer, James George. *The Golden Bough*. New York: Macmillan, 1940.

Freeman, Kathleen. *The Presocratic Philosophers*. Oxford, Eng.: Blackwell, 1949.

Freeman, R.B. *Charles Darwin: A Companion*. Kent, Eng.: Dawson, 1978. Index of Darwinia.

French, A.P. *Newtonian Mechanics*. New York: Norton, 1971.

———. *Special Relativity*. New York: Norton, 1968.

———, and Edwin F. Taylor. *An Introduction to Quantum Physics*. New York: Norton, 1978.

———, ed., *Einstein: A Centenary Volume*. Cambridge, Mass.: Harvard University Press, 1979.

———, and P.J. Kennedy, eds. *Niels Bohr: A Centenary Volume*. Cambridge, Mass.: Harvard University Press, 1986.

Freud, Sigmund. *Civilization and Its Discontents*. New York: Harcourt, Brace, 1928.

———. *The Future of an Illusion*, trans. James Strachey. New York: Norton, 1961.

Freund, Philip. *Myths of Creation*. London: Allen & Unwin, 1964.

Friedman, Michael. *Foundations of Space-Time Theories: Relativistic Physics and Philosophy of Science*. Princeton, N.J.: Princeton University Press, 1983.

Fritzsch, Harald. *The Creation of Matter*. New York: Basic Books, 1984. Origin of matter in the early universe.

Funkenstein, Amos. *Theology and the Scientific Imagination from the Middle Ages to the Seventeenth Century*. Princeton, N.J.: Princeton University Press, 1986. Suggests a commonality between theological and scientific thinking by the seventeenth century.

Gale, Richard M., ed. *The Philosophy of Time*. Sussex, Eng.: Harvester, 1968. Anthology of historical writings.

Galileo. *Dialogue Concerning the Two Chief World Systems*, trans. Stillman Drake. Berkeley: University of California Press, 1967.

———. *Dialogues Concerning Two New Sciences*, trans. Henry Crew and Alfonso de Salvio. Chicago: University of Chicago Press, 1952.

———. *The Sidereal Messenger*, trans. Edward Stafford Carlos. London: Dawsons of Pall Mall, 1959.

———. *The Starry Messenger*, 1610, in Drake, Stillman, ed. and trans., *Discoveries and Opinions of Galileo*, Garden City, N.Y.: Doubleday, 1957.

Gamow, George. *Biography of Physics*. New York: Harper, 1961. Lighthearted history of physics.

———. *The Creation of the Universe*. New York: Mentor, 1951. Popular account of the big bang theory.

———. *Thirty Years That Shook Physics*. Garden City, N.Y.: Anchor, 1966.

Gardner, Martin. *The Ambidextrous Universe*. New York: Mentor, 1969. Account of the discovery of parity violation in the weak interaction.

Garnet, Jacques. *Daily Life in China on the Eve of the Mongol Invasion 1250–1276*. Stanford, Calif.: Stanford University Press, 1970. Describes Marco Polo's sixteen years in Hangchow.

Garnett, Christopher B. *The Kantian Philosophy of Space*. Port Washington, N.Y.: Kennikat, 1965.

Gell-Mann, Murray, and Yuval Ne'eman. *The Eightfold Way*. New York: Benjamin, 1964. Application of symmetry precepts to the study of the strong interaction.

George, Wilma. *Biologist Philosopher: A Study of the Life and Writings of Alfred Russel Wallace*. New York: Abelard-Schuman, 1964.

Geroch, Robert. *General Relativity from A to B*. Chicago: University of Chicago Press, 1978. Nonmathematical introduction.

Geymonat, Ludovico. *Galileo Galilei: A Biography and Inquiry into His Philosophy of Science*, trans. Stillman Drake. New York: McGraw-Hill, 1965.

Ghani, Abdul. *Abdus Salam*. Karachi: Ma'aref, 1982.

Ghyka, Matila. *The Geometry of Art and Life*. New York: Dover, 1977.

Gibbon, Edward. *The Decline and Fall of the Roman Empire*. 3 vols. New York: Modern Library, n.d.

Gibbons, G.W., S.W. Hawking, and S.T.C. Siklos. *The Very Early Universe*. London: Cambridge University Press, 1983. Papers presented at a 1982 Cambridge conference.

Gillispie, Charles Coulston. *Genesis and Geology*. Cambridge, Mass.: Harvard University Press, 1951.

————. *The Edge of Objectivity*. Princeton, N.J.: Princeton University Press, 1960. Essays on the historical structure of classical science.

Gingerich, Owen, ed. *The Nature of Scientific Discovery: Symposium Commemorating the 500th Anniversary of the Birth of Nicolaus Copernicus*. Washington, D.C.: Smithsonian Institution Press; New York, Braziller, 1975. Robust text, useful illustrations.

Gjertsen, Derek. *The Newton Handbook*. London: Routledge & Kegan Paul, 1986. Compendium of biographical information about Newton and his work.

Glass, Bentley, and Owsei Temkin, eds. *Forerunners of Darwin 1745–1859*. Baltimore: Johns Hopkins University Press, 1959.

Gödel, Kurt. *Collected Works*, ed. Solomon Feferman. New York: Oxford University Press, 1986.

Goldberg, Stanley. *Understanding Relativity: Origin and Impact of a Scientific Revolution*. Boston: Birkhäuser, 1984.

Goldman, Martin. *The Demon in the Aether: The Story of James Clerk Maxwell*. Edinburgh: Paul Harris, 1983. Includes ample excerpts from primary sources.

Goldman, T., and Michael Martin Nieto, eds. *The Santa Fe Meeting*. Philadelphia: World Scientific, 1985. Proceedings of a 1984 conference on particles and fields.

Goldsmith, Donald, ed. *The Quest for Extraterrestrial Life*. Mill Valley, Calif.: University Science Books, 1980.

————, and Tobias Owen. *The Search for Life in the Universe*. Reading, Mass.: Benjamin/Cummings, 1980. Undergraduate astronomy textbook that stresses exobiology.

Goldstein, Thomas. *Dawn of Modern Science: From the Arabs to Leonardo*. Boston: Houghton Mifflin, 1980.

Golino, Carlo L. *Galileo Reappraised*. Berkeley: University of California Press, 1966. Proceedings of a 1965 UCLA conference.

Good, Ronald. *The Philosophy of Evolution*. Stanbridge, Eng.: Dovecote Press, 1981. Brief overview.

Gooding, David, and Frank A.J.L. James, eds. *Faraday Rediscovered*. London: Stockton Press, 1985. Based on talks given at a 1984 Faraday conference.

Gorman, Peter. *Pythagoras: A Life*. London: Routledge & Kegan Paul, 1979.

Gottfried, Kurt, and Victor F. Weisskopf. *Concepts of Particle Physics*. London: Oxford University Press, 1984.

Gould, Rupert T. *John Harrison and His Timekeepers*. Greenwich, Eng.: National Maritime Museum, 1987.

Graham, Loren R. *Between Science and Values*. New York: Columbia University Press, 1981.

Grant, Edward. *In Defense of the Earth's Centrality and Immobility: Scholastic Reaction to Copernicanism in the Seventeenth Century*. Philadelphia: American Philosophical Society, 1984. Philosophical arguments deployed against Copernicanism.

————. *Much Ado About Nothing: Theories of the Infinite Void*. London: Cambridge University Press, 1981.

————. *Physical Science in the Middle Ages*. London: Cambridge University Press, 1977.

————, ed. *A Source Book in Medieval Science*. Cambridge, Mass.: Harvard University Press, 1974.

Grant, Michael. *Dawn of the Middle Ages*. New York: McGraw-Hill, 1981.

————. *From Alexander to Cleopatra*. New York: Scribner's, 1982.

Green, Michael, John Schwartz, and Edward Witten. *Superstring Theory*. London: Cambridge University Press, 1987. Introduction for physicists, written by three pioneers in the field.

————, and D. Gross, eds. *Unified String Theories*. Philadelphia: World Scientific, 1986.

Greenburg, Sidney. *The Infinite in Giordano Bruno*. New York: Octagon, 1978.

Greene, John. *The Death of Adam: Evolution and Its Impact on Western Thought*. Ames, Iowa: University of Iowa Press, 1959. The reaction to Darwin's theory.

Gregory, Joshua Craven. *A Short History of Atomism, From Democritus to Bohr*. London: Black, 1931.

Gribbin, John. *Genesis: The Origins of Man and the Universe*. New York: Delacorte, 1981.

Griggs, William. *The Celebrated "Moon Story," Its Origin and Incidents with a Memoir of Its Author*. New York: Bunnell & Price, 1852.

Guillemard, Francis Henry. *The Life of Ferdinand Magellan and the First Circumnavigation of the Globe, 1480–1521*. New York: AMS Press, 1971.

Guthrie, W.K.C. *A History of Greek Philosophy*. 2 vols. London: Cambridge University Press, 1962.

Haber, Francis. *The Age of the World, Moses to Darwin*. Baltimore: John Hopkins University Press, 1959. Nontechnical history of geochronology through the late nineteenth century.

Haeckel, Ernst. *Art Forms in Nature*. New York: Dover, 1974.

————. *The Riddle of the Universe at the Close of the Nineteenth Century*. New York: Harper, 1900.

Haile, H.G. *Luther*. Princeton, N.J.: Princeton University Press, 1980.

Haldane, E.S. *Descartes: His Life and Times*. New York: American Scholar Publications, 1966.

Hall, A.R. *The Scientific Revolution 1500–1800*. Boston: Beacon, 1966.

Halley, Edmond. *The Three Voyages of Edmond Halley in the Paramore 1698–1701*, ed. Norman J.W. Thrower. London: Hakluyt Society, 1981.

Hanson, Norwood Russell. *The Concept of the Positron*. London: Cambridge University Press, 1963.

————. *Observation and Explanation: A Guide to Philosophy of Science*. New York: Harper & Row, 1971.

————. *Patterns of Discovery*. London: Cambridge University Press, 1981.

Hapgood, Charles H. *Maps of the Ancient Sea Kings: Evidence of Advanced Civilization in the Ice Age*. New York: Chilton, 1966.

Hardin, Garrett. *Nature and Man's Fate*. New York: Holt Rinehart & Winston, 1959.

Hargittai, Istvan, ed. *Symmetry: Unifying Human Understanding*. New York: Pergamon, 1986. Collection of papers on symmetry in art and science.

Harman, P.M. *Energy, Force, and Matter*. London: Cambridge University Press, 1985. Study of development of nineteenth-century physics.

Harrison, Edward. *Cosmology: The Science of the Universe*. London: Cambridge University Press, 1981. Reflective study, on the undergraduate level.

———. *Darkness at Night: A Riddle of the Universe*. Cambridge, Mass.: Harvard University Press, 1987. Study of Olbers's paradox.

Harrison, Jane. *Prolegomena to the Study of Greek Religion*. London: Merlin Press, 1980.

Hartmann, William K. *Astronomy: The Cosmic Journey*. Belmont, Calif.: Wadsworth, 1978. Undergraduate textbook for nonscience majors.

Hassan, Z. and C.H. Lai, eds. *Ideals and Realities: Selected Essays of Abdus Salam*. Singapore: World Scientific, 1984.

Hawkes, Jacquetta. *The First Great Civilizations*. New York: Random House, 1973.

———, and Leonard Woolley. *History of Mankind*. New York: Harper & Row, 1963.

Hawking, Stephen W., and G.F.R. Ellis. *The Large Scale Structure of Space-Time*. London: Cambridge University Press, 1973.

———, and W. Israel, eds. *General Relativity: An Einstein Centenary Survey*. New York: Cambridge University Press, 1979.

———, and M. Rocek, eds. *Superspace and Supergravity*. London: Cambridge University Press, 1981. Proceedings of a 1980 Cambridge conference.

Hawkins, Gerald. *Beyond Stonehenge*. New York: Harper & Row, 1973.

———, with John B. White. *Stonehenge Decoded*. New York: Dell, 1965. Investigates astronomical alignments in the ancient megalith.

Hayek, F.A. *The Counter-Revolution of Science*. Indianapolis: Liberty, 1952.

Hazard, Cyril, and Simon Mitton. *Active Galactic Nuclei*. London: Cambridge University Press, 1979.

Healey, R. *Time, Reduction and Reality*. London: Cambridge University Press, 1981.

Heath, L.R. *The Concept of Time*. Chicago: University of Chicago Press, 1936.

Heath, Thomas. *Aristarchus of Samos*. New York: Dover, 1981.

———. *Greek Astronomy*. London: Dent, 1932.

———, *A History of Greek Mathematics*. London: Oxford University Press, 1921.

Heathcote, Niels Hugh de Vaudrey. *Nobel Prize Winners in Physics, 1901–1950*. New York: Schuman, 1954.

Heidel, William A. *The Frame of the Ancient Greek Maps*. New York: Arno Press, 1976.

Heilbron, J.L. *The Dilemmas of an Upright Man: Max Planck as Spokesman for German Science*. Berkeley: University of California Press, 1986. Concise, nonscientific biography, with minimal emphasis on Planck's research.

———. *Elements of Early Modern Physics*. Berkeley: University of California Press, 1982.

Heilbroner, Robert. *The Future as History*. New York: Harper, 1961.

Heisenberg, Elizabeth. *Inner Exile: Recollections of a Life with Werner Heisenberg*. Boston: Birkhäuser, 1984.

Heisenberg, Werner. *Across the Frontiers*, trans. Peter Heath. New York: Harper, 1974.

———. *Physics and Beyond*. New York: Harper & Row, 1971.

———. *Physics and Philosophy*. New York: Harper, 1962.

———. *Tradition in Science*. New York: Seabury Press, 1983.

Held, A., ed. *General Relativity and Gravitation*. New York: Plenum Press, 1979.

Helmholtz, Hermann von. *Epistemological Writings*, ed. R.S. Cohen and Y. Elkana, and trans. Malcolm F. Lowe. Boston: Reidel, 1977.

Hempel, Carl G. *Philosophy of Natural Science*. Englewood Cliffs, N.J.: Prentice-Hall, 1966.

Henderson, Linda Dalrymple. *The Fourth Dimension and Non-Euclidean Geometry in Modern Art*. Princeton, N.J.: Princeton University Press, 1983.

Heraclitus. *The Cosmic Fragments*, ed. G.S. Kirk. London: Cambridge University Press, 1962. Heraclitus on "the world as a whole," with extensive commentary.

Herbert, Nick. *Quantum Reality: Beyond the New Physics*. New York: Anchor, 1985. Sketch of philosophical questions raised by quantum mechanics.

Herbig, G.H., ed. *Spectroscopic Astrophysics*. Berkeley: University of California Press, 1970. Essays in honor of Otto Struve.

Herivel, John. *The Background to Newton's Principia*. London: Oxford University Press, 1965.

Hermann, A. *The Genesis of Quantum Theory*. Cambridge, Mass.: MIT Press, 1971.

"Hermes Trismegistus" *Hermetica: The Ancient Greek and Latin Writings Which Contain Religious or Philosophic Teachings Ascribed to Hermes Trismegistus*, ed. and trans. Walter Scott. London: Oxford University Press, 1924.

Herodotus. *The Histories*, trans. Aubrey de Sélincourt. London: Penguin, 1964.

Herschel, William. *Complete Works*, ed. J.L.E. Dryer. London: Royal Society, 1912.

Hesiod. *Works and Days*, trans. Dorothea Wender. Harmondsworth, Eng.: Penguin, 1977.

Heyerdahl, Thor. *Early Man and the Ocean*, Garden City, N.Y.: Doubleday, 1979.

Hilbert, David. *Geometry and the Imagination*. trans. P. Nemenyi. New York: Chelsea, 1952.

Hilts, Philip J. *Scientific Temperaments*. New York: Simon & Schuster, 1982. Includes a profile of Robert Wilson, builder of the Fermilab accelerator.

Himsworth, Harold. *Scientific Knowledge and Philosophic Thought*. Baltimore: Johns Hopkins University Press, 1986. Argues for wider application of scientific methods to philosophical problems.

Hitti, Philip K. *History of the Arabs*. New York: Macmillan, 1951.

Hobbes, Thomas. *English Works of Thomas Hobbes*, ed. William Molesworth. 11 vols. London: Oxford University Press, 1961. Reprint of 1839–1845 editions.

Hobson, E.W. *The Domain of Natural Science*. New York: Dover, 1968.

Hodge, Paul W. *Galaxies*. Cambridge, Mass.: Harvard University Press, 1986.

Hodson, F.R., ed. *The Place of Astronomy in the Ancient World*. London: Oxford University Press, 1974.

Hoffmann, Banesh. *Albert Einstein, Creator and Rebel*. New York: Viking, 1972. Nontechnical study of Einstein's life and work, by his former collaborator.

———. *Relativity and Its Roots*. San Francisco: Freeman, 1983. Einstein and the aether.

———. *The Strange Story of the Quantum*. New York: Dover, 1959. Nontechnical introduction to quantum physics.

Holton, Gerald, and Yehuda Elkana, eds. *Albert Einstein, Historical and Cultural Perspectives*. Princeton, N.J.: Princeton University Press, 1982.

———. *The Scientific Imagination*. London: Cambridge University Press, 1979.

———. *Thematic Origins of Scientific Thought, Kepler to Einstein*. Cambridge, Mass.: Harvard University Press, 1974.

Hook, Sidney, ed. *Determinism and Freedom in the Age of Modern Science*. New York: Collier, 1968.

Hooke, Robert. *An Attempt to Prove the Motion of the Earth by Observations*. London: 1674. Reprinted in R.T. Gunther, *Early Science in Oxford*, 8:1–28, Oxford, privately printed, 1923–1945.

Horace. *Epistles*, trans. H. Rushton Fairclough. Cambridge, Mass.: Harvard University Press, 1978.

Hoskin, Michael A. *Stellar Astronomy*. Giles, Bucks, Eng.: Science History Publications, 1982. Papers on Wright, Lambert, Newton, Herschel, Shapley, and others.

———. *William Herschel and the Construction of the Heavens*. London: Oldbourne, 1963.

Howells, William, ed. *Ideas on Human Evolution: Selected Essays 1949–1961*. Cambridge, Mass.: Harvard University Press, 1962.

Howse, Derek. *The Clocks and Watches of Captain James Cook, 1769–1969*. Ramsgate, Eng.: Thanet, 1970.

———. *Greenwich Time and the Discovery of the Longitude*. London: Oxford University Press, 1980.

———. *The Sea Chart: An Historical Survey Based on the Collections in the National Maritime Museum*. Newton Abbot, Eng.: David & Charles, 1973.

Hoyle, Fred. *Astronomy and Cosmology: A Modern Course*. San Francisco: Freeman, 1975.

———. *Encounter with the Future*. New York: Trident, 1965.

———. *From Stonehenge to Modern Cosmology*. San Francisco: Freeman, 1972.

———. *Frontiers of Astronomy*. New York: Harper, 1955.

———. *The Nature of the Universe*. New York: Harper, 1960.

———. *On Stonehenge*. San Francisco: Freeman, 1977.

———. *Steady-state Cosmology Re-visited*. Cardiff: University College Cardiff Press, 1980.

———, and Jayant Narlikar. *Action at a Distance in Physics and Cosmology*. San Francisco: Freeman, 1974.

———, and ———. *The Physics-Astronomy Frontier*. San Francisco, Freeman, 1980. Undergraduate astrophysics textbook.

Hubble, Edwin. *The Realm of the Nebulae*. New Haven, Conn.: Yale University Press, 1985. Popular account of galaxies by the discoverer of the expansion of the universe.

Humboldt, Alexander von. *Kosmos: A Sketch of a Physical Description of the Universe*. London: Bailliere, 1848.

Hurley, Patrick. *How Old Is the Earth?* Garden City, N.Y.: Anchor, 1959. On geological dating.

Hutton, James. *Theory of the Earth*. Edinburgh: Creech, 1795; Weinheim, Eng.:

Englemann, 1959 (2 vols.). Hutton's explication of how Earth's history may be traced in the geological record.

Huxley, Aldous. *Literature and Science*. New York: Harper, 1963.

Huxley, Julian. *Evolution in Action*. New York: Harper, 1953.

————. *Life and Letters of Thomas Henry Huxley*. 3 vols. London: Macmillan, 1903.

————, and H.B.D. Kettlewell. *Charles Darwin and His World*. New York: Viking, 1965. Popular biography.

Huygens, Christian, *The Celestial Worlds Discovered*. London: Cass, 1968.

————. *Treatise on Light*, trans. Silvanus P. Thompson. Chicago: University of Chicago Press, 1945.

Iltis, H. *Life of Mendel*, trans. Eden and Cedar Paul. New York: Norton, 1932.

Infeld, Leopold. *Quest: The Evolution of a Scientist*. New York: Doubleday, 1941. Autobiography of an Einstein collaborator.

Inge, William Ralph. *God and the Astronomers*. New York: Longman's Green, 1933.

Islam, J.N. *The Ultimate Fate of the Universe*. London: Cambridge University Press, 1983.

Jacob, Françoise. *The Possible and the Actual*. New York: Pantheon Books, 1982. Essays on the history and philosophy of science.

Jaki, Stanley L. *The Milky Way: An Elusive Road for Science*. New York: Science History Publications, 1972.

————. *The Road of Science and the Ways to God*. Chicago: University of Chicago Press, 1978.

Jammer, Max. *Concepts of Space*. Cambridge, Mass.: Harvard University Press, 1954.

————. *The Philosophy of Quantum Mechanics*. New York: Wiley, 1974.

Jardine, N. *The Birth of History and Philosophy of Science: Kepler's "A Defence of Tycho Against Ursus" with Essays on Its Provenance and Significance*. London: Cambridge University Press, 1984.

Jeans, James. *Physics and Philosophy*. Ann Arbor: University of Michigan Press, 1966.

————. *Science and Music*. New York: Dover, 1968.

Jerison, H.J. *Evolution of the Brain and Intelligence*. New York: Academic Press, 1973.

Jevons, W.S. *The Principles of Science*. London: Macmillan, 1900.

John, Laurie, ed. *Cosmology Now*. New York: Taplinger, 1973.

Johnson, Francis. *Astronomical Thought in Renaissance London*. Baltimore: Johns Hopkins University Press, 1937.

Jones, A.H.M. *The Later Roman Empire*. Oxford, Eng.: Blackwell, 1964.

Jones, Richard Foster. *Ancients and Moderns*. New York: Dover, 1981. The rise of scientific societies in seventeenth-century England.

Judson, Horace Freeland. *The Search for Solutions*. New York: Holt, Rinehart & Winston, 1980. Illuminating study of how science is practiced.

Jungnickel, Christa, and Russell McCormmach. *Intellectual Mastery of Nature*. 2 vols. Chicago: University of Chicago Press, 1986. Semitechnical history of physics from Ohm to Einstein.

Kahn, C.H. *Anaximander and the Origins of Greek Cosmology*. New York: Columbia University Press, 1960.

Kant, Immanuel. *Critique of Pure Reason*, trans. J.M.D. Meiklejohn. Chicago: University of Chicago Press, 1952.

———. *Universal Natural History and Theory of the Heavens*, trans. W. Hastie. Glasgow, 1900; Ann Arbor: University of Michigan Press, 1969. Includes text of the Hamburg review of Thomas Wright's book that launched Kant on his theory of galaxies.

———. *Metaphysical Foundations of Natural Science*, ed. James Ellington. New York: Bobbs-Merrill, 1970.

———. *Prolegomena to Any Future Metaphysics*, trans. Paul Carus, revised by James W. Ellington. Indianapolis: Hackett, 1977.

Kaplan, S.A., ed. *Extraterrestrial Civilizations*. Jerusalem: 1971. Based on a 1969 conference in Moscow.

Kardashev, N.S., ed., *Extraterrestrial Civilization*, trans. Z. Lerman. Jerusalem: Israel Scientific Translations, 1967.

Kastner, Joseph. *A Species of Eternity*. New York: Knopf, 1977. Recounts the life and researches of early American naturalists.

Kaufmann, William J. III. *Relativity and Cosmology*. New York: Harper & Row, 1977. Popular introduction.

———. *Stars and Nebulas*. San Francisco: Freeman, 1978. Popularized astrophysics survey.

Keller, Alex. *The Infancy of Atomic Physics: Hercules in His Cradle*. London: Oxford University Press, 1983.

Kelves, Daniel J. *The Physicists*. New York: Random House, 1979. History of the development of modern physics in America.

Kepler, Johann. *Epitome of Copernican Astronomy*, Part I. New York: Kraus Reprint Co., 1969.

———. *The Harmonies of the World*, trans. Charles Glenn Wallis. Chicago: University of Chicago Press, 1975.

———. *Kepler's Conversation with Galileo's Sidereal Messenger*, trans. Edward Rosen. New York: Johnson Reprint, 1965.

———. *Mysterium Cosmographicum: The Secret of the Universe*. A. trans. A.M. Duncan. New York: Abaris, 1981.

———. *Somnium: The Dream*, trans. Edward Rosen. Madison: University of Wisconsin Press, 1967.

Kerkut, G.A. *The Implications of Evolution*. Oxford, Eng.: Pergamon, 1960.

Kern, Stephen. *The Culture of Time and Space 1800–1918*. Cambridge, Mass.: Harvard University Press, 1983. Special relativity's social context.

Kesten, Hermann. *Copernicus and His World*. New York: Roy, 1945.

Keynes, Richard Darwin, ed. *The Beagle Record: Selections from the Original Pictorial Records and Written Accounts of the Voyage of H.M.S. Beagle*. London: Cambridge University Press, 1979. Illustrated compendium of Darwinia.

King, Henry C. *The History of the Telescope*. New York: Dover, 1979.

King, N.Q. *The Emperor Theodosius and the Establishment of Christianity*. Philadelphia: The Westminister Press, 1960. Relates the role of Theodosius in the Christian uprisings of the fourth century.

Kippenhahn, Rudolf. *Light from the Depths of Time*. New York: Springer-Verlag, 1984. Illustrated introduction to cosmology.

―――. *100 Billion Suns: The Birth, Life, and Death of the Stars*, trans. Jean Steinberg. New York: Basic Books, 1983. Nontechnical introduction to the physics of the stars.

Kirk, G.S., J.E. Raven, and M. Schofield. *The Presocratic Philosophers*. London: Cambridge University Press, 1983. Includes texts and translations of works by Thales, Heraclitus, Pythagoras, etc.

Kitcher, Philip. *Abusing Science: The Case Against Creationism*. Cambridge, Mass.: MIT Press, 1982. Discussion of creationist misrepresentations of Darwinism.

Kline, Morris. *Mathematics: The Loss of Certainty*. London: Oxford University Press, 1980.

Knedler, John Warren, ed. *Masterworks of Science*. New York: McGraw-Hill, 1973.

Koenigsberger, Leo. *Hermann von Helmholtz*, trans. Frances A. Welby. 3 vols. London: Oxford University Press, 1906. Spare narrative, with extensive quotations from Helmholtz's letters and talks.

Koenigswald, G.H.R. *The Evolution of Man*. Ann Arbor: University of Michigan Press, 1976.

Koestler, Arthur. *The Sleepwalkers*. New York: Grosset & Dunlap, 1959. On the life and work of Copernicus, Kepler, and Galileo.

Kokkedee, J.J.J. *The Quark Model*. New York: Benjamin, 1969.

Kolb, Edward, et al. *Inner Space/Outer Space: The Interface Between Cosmology and Particle Physics*. Chicago: University of Chicago Press, 1986. Proceedings of a 1984 Fermilab conference.

Koyré, Alexandre, *The Astronomical Revolution*, trans. R.E.W. Maddison. Ithaca, N.Y.: Cornell University Press, 1973.

―――. *Galileo Studies*, trans. John Mepham. Atlantic Highlands, N.J.: Humanities Press, 1978. What Galileo knew and when he knew it.

―――. *Metaphysics and Measurement: Essays in Scientific Revolution*. Cambridge, Mass.: Harvard University Press, 1968.

―――. *Newtonian Studies*. Cambridge, Mass.: Harvard University Press, 1965.

―――, and I. Bernard Cohen, eds. *Isaac Newton's Philosophiae Naturalis Principia Mathematica*. Mass.: Cambridge, Harvard University Press, 1972.

Kubler, George. *The Shape of Time*. New Haven, Conn.: Yale University Press, 1962.

Kühn, Ludwig. *The Milky Way: The Structure and Development of Our Star System*. New York: Wiley, 1982.

Kuhn, Thomas. *The Copernican Revolution*. Cambridge, Mass.: Harvard University Press, 1979.

―――. *The Structure of Scientific Revolutions*. Chicago: University of Chicago Press, 1970.

Lambert, Johann Heinrich. *Cosmological Letters on the Arrangement of the World-Edifice*, trans. Stanley Jaki. New York: Science History Publications, 1976.

Lanczos, Cornelius. *Albert Einstein and the Cosmic World Order*. New York: Interscience, 1965.

―――. *The Einstein Decade (1905–1915)*. New York: Academic Press, 1974.

Landes, David. *Revolution in Time*. Cambridge, Mass.: Harvard University Press, 1983. History of clocks and clockmaking.

Lang, Kenneth R., and Owen Gingerich, eds. *A Source Book in Astronomy and Astrophysics, 1900–1975*. Cambridge, Mass.: Harvard University Press, 1979.

Lange, Frederick Albert. *The History of Materialism*. London: Kegan Paul, 1925.

Laplace, Pierre-Simon de. *A Philosophical Essay on Probabilities*, trans. Frederick Wilson Truscott and Frederick Lincoln Emory. New York: Wiley, 1902.

Layzer, David. *Constructing the Universe*. New York: Freeman, 1984. Introduction to scientific cosmology.

Leach, Maria. *The Beginning: Creation Myths Around the World*. New York: Funk & Wagnalls, 1956.

Lear, John. *Kepler's Dream*, trans. Patricia Frueh Kirkwood. Berkeley: University of California Press, 1965.

Legge, James. *The Chinese Classics*. 5 vols. Hong Kong: Hong Kong University Press, 1960.

Leibniz, Gottfried Wilhelm. *A Collection of Critical Essays*, ed. Harry Frankfurt. Notre Dame, Ind.: University of Notre Dame Press, 1976.

———. *Monadology*, trans. Robert Latta. London: Oxford University Press, 1971.

———. *Philosophical Writings*, ed. G.H.R. Parkinson, and trans. Mary Morris and G.H.R. Parkinson. London: Dent, 1973.

Lemaître, Georges. *The Primeval Atom*. New York: D. Van Nostrand, 1950. Semipopular account of what was to become the big bang theory.

Leville, Jacques P., Lawrence R. Sulak, and David G. Unger, eds. *The Second Workshop on Grand Unification*. Boston: Birkhäuser, 1981.

Lewis, Wyndham. *Time and Western Man*. London: Chatto & Windus, 1927. Demystification of the concept of time.

Lindberg, David C., ed. *Science in the Middle Ages*. Chicago: University of Chicago Press, 1978.

Lindsay, Alexander Dunlop. *Kant*. London: Ernest Benn, 1934.

Livermore, H.V. *A New History of Portugal*. London: Cambridge University Press, 1966, 1976.

Livingston, Dorothy Michelson. *The Master of Light: A Biography of Albert A. Michelson*. Chicago: University of Chicago Press, 1973. By Michelson's daughter.

Lloyd, G.E.R. *Early Greek Science: Thales to Aristotle*. New York: Norton, 1970.

———. *Greek Science After Aristotle*. New York: Norton, 1973.

Lockyer, J. Norman. *The Dawn of Astronomy*. Cambridge, Mass.: MIT Press, 1964. Victorian study of the orientation of Egyptian pyramids and temples to the sun and stars.

Longair, M.S. *High Energy Astrophysics*. London: Cambridge University Press, 1981. Semitechnical.

———, and J. Einasto. *The Large-Scale Structure of the Universe*, Boston: Reidel, 1978. Proceedings of a 1977 astronomical symposium.

———, and J.W. Warner, eds. *Scientific Research with the Space Telescope*. Washington, D.C.: NASA, 1979. Papers presented at a colloquium on the Hubble Space Telescope.

Lorentz, H.A. *The Einstein Theory of Relativity*. New York: Brentano's, 1920.

Losee, J. *An Historical Introduction to the Philosophy of Science*. London: Oxford

University Press, 1972. Short survey, with an emphasis on the history of the concept of experimentation.

Lovejoy, Arthur O. *The Great Chain of Being*. Cambridge, Mass.: Harvard University Press, 1953. Classic study of a durable metaphor.

Lubbock, Constance A. *The Herschel Chronicle*. New York: Macmillan, 1933. Memoirs by William Herschel's granddaughter.

Lucretius. *De Rerum Natura*, trans. Cyril Bailey. London: Oxford University Press, 1947.

———. *On Nature (De Rerum Natura)*, trans. Russel Geer. Indianapolis: Bobbs-Merrill, 1965.

Lyell, Charles, *The Antiquity of Man*. London: Murray, 1863.

———. *Principles of Geology*, 11th ed. London: Appleton, 1877. The case for uniformitarian geology.

Lyell, Katherine, ed. *Life, Letters, and Journals of Sir Charles Lyell*. 2 vols. Westmead, Eng.: Gregg, 1970.

Maclagan, David. *Creation Myths*. London: Thames & Hudson, 1979.

Mach, Ernst. *The Science of Mechanics*, 6th ed. LaSalle, Ill.: Open Court, 1960. Influential work in the philosophy of science that helped inspire the young Einstein.

———. *Space and Geometry*. LaSalle, Ill.: Open Court, 1906.

MacCurdy, Edward, ed. and trans. *The Notebooks of Leonardo da Vinci*. New York: Braziller, 1939.

MacKay, D.M. *The Clockwork Image*. London: Inter-Varsity Press, 1974.

MacKinnon, E.M. *Scientific Explanation and Atomic Physics*. Chicago: University of Chicago Press, 1982.

MacPherson, Hector. *Makers of Astronomy*. London: Oxford University Press, 1933. Colorful, hyperbolic history.

MacPike, Eugene Fairfield. *Correspondence and Papers of Edmond Halley*. London: Oxford University Press, 1932.

McCrea, M.J. Rees et al. *The Constants of Physics*. London: Royal Society, 1983.

McCuster, Brian. *The Quest for Quarks*. London: Cambridge University Press, 1983. Semitechnical explication of experimental tests of quark theory.

McMullin, Ernan, ed. *The Concept of Matter in Greek and Medieval Philosophy*. Notre Dame, Ind.: University on Notre Dame Press, 1978.

McNeill, William H. *The Pursuit of Power: Technology, Armed Force, and Society Since AD 1000*. Chicago: University of Chicago Press, 1982.

———. *The Rise of the West: A History of the Human Community*. Chicago: University of Chicago Press, 1963.

Mahaffey, J.P. *Greek Life and Thought from the Age of Alexander to the Roman Conquest*. London: Macmillan, 1887.

Mainx, Felix. *Foundations of Biology*. Chicago: University of Chicago Press, 1955.

Malthus, Thomas Robert. *An Essay on the Principle of Population*, ed. Philip Appleman. New York: Norton, 1976.

Mandelbrot, Benoit B. *The Fractal Geometry of Nature*. New York: Freeman, 1983. Introduction to fractal geometry, by its founder.

Manier, E. *The Young Darwin and His Cultural Circle*. Boston: Reidel, 1978.

Manuel, Frank E. *A Portrait of Isaac Newton*. Washington, D.C.: New Republic, 1968. Psychological study.

Marchant, James. *Alfred Russel Wallace: Letters and Reminiscences*. 2 vols. London: Cassel & Co., 1916.

Marques, A.H. de Oliveira. *History of Portugal*. New York: Columbia University Press, 1972.

Martins, J.P. Oliveira. *The Golden Age of Prince Henry the Navigator*, trans. James Johnston Abraham and William Edward Reynolds. London: Chapman & Hall, 1914.

Mason, H.T. *The Liebniz-Arnauld Correspondence*. Manchester, Eng.: Manchester University Press, 1967.

Mason, Stephen F. *A History of the Sciences*. New York: Macmillan, 1977.

Matthews, Donald. *Atlas of Medieval Europe*. New York: Equinox, 1983.

Matthews, William. *Invitation to Geology*. Garden City, N.Y.: Natural History Press, 1971.

May, W.E. *How the Chronometer Went to Sea*. London: Antiquarian Horological Society, 1976.

Mayr, Ernst. *Principles of Systematic Zoology*. New York: McGraw-Hill, 1969. On living versus extinct species.

Medawar, P.B. *The Future of Man*. New York: Basic Books, 1959.

———. *The Hope of Progress*. Garden City, N.Y.: Doubleday, 1974.

———, and J.S. Medawar. *The Life Science*. New York: Harper & Row, 1977.

Mehra, Jagdish, ed. *The Physicist's Conception of Nature*. Boston: Kluwer, 1973.

———, ed. *The Quantum Principle: Its Interpretation and Epistemology*. Boston: Reidel, 1974.

Mehta, Ved. *Fly and the Fly-Bottle*. London: Penguin, 1965. Conversations with English philosophers.

Mellor, D.H. *Real Time*. London: Cambridge University Press, 1981. Philosophical study of time and tense.

Merleu-Ponty, Jacques, and Bruno Morando. *The Rebirth of Cosmology*, trans. Helen Weaver. New York: Knopf, 1976.

Michanowsky, George. *The Once and Future Star*. New York: Hawthorn, 1977. Theories of the Vela supernova.

Michel, Paul-Henri. *The Cosmology of Giordano Bruno*. Ithaca, N.Y.: Cornell University Press, 1973. Emphasizes Bruno's predecessors and influences.

Michelmore, Peter. *Einstein, Profile of the Man*. New York: Dodd, Mead, 1962.

Mihalas, Dimitri, and James Binney. *Galactic Astronomy*. San Francisco: Freeman, 1981. Textbook.

Miller, Arthur. *Albert Einstein's Special Theory of Relativity*. Reading, Mass.: Addison-Wesley, 1981. The development and early reception of the special theory; includes a new translation of Einstein's 1905 paper.

———. *Imagery in Scientific Thought*. Boston: Birkhäuser, 1984. Study of contributors to the theory of relativity.

Miller, John William. *The Paradox of Cause and Other Essays*. New York: Norton, 1978.

Milton, John. *Paradise Lost, Areopagitica* . . . Chicago: University of Chicago Press, 1952.

Minkoff Eli C. *Evolutionary Biology*. Reading, Mass.: Addison-Wesley, 1983. Textbook.

Mintz, Leigh. *Historical Geology*. Columbus, Ohio: Merrill, 1972.

Misner, Charles W., Kip S. Thorne, and John Archibald Wheeler. *Gravitation*. San Francisco: Freeman, 1973. Graduate-level textbook.

Mitton, Simon, ed. *The Cambridge Encyclopaedia of Astronomy*. New York: Crown, 1978.

———. *Exploring the Galaxies*. New York: Scribner's, 1976.

Moore, James R. *The Post-Darwinian Controversies*. London: Cambridge University Press, 1981. Study of Protestant reaction to Darwin's theory in the English-speaking world.

Moore, Patrick. *The Astronomy of Birr Castle*. London: Mitchell, 1971.

Moore, Ruth. *Niels Bohr*. Cambridge, Mass.: MIT Press, 1985. Accessible biography of the leading quantum physicist.

More, L.T. *Isaac Newton: A Biography*. New York: Scribner's, 1934.

Morison, Samuel Eliot. *Admiral of the Ocean Sea: A Life of Christopher Columbus*. Boston: Little, Brown, 1942; Boston: Northeastern University Press, 1983.

———, ed. and trans. *Journals and Other Documents on the Life and Voyages of Christopher Columbus*. New York: Limited Editions Club, 1963.

Moriyasu, K. *An Elementary Primary for Gauge Theory*. Singapore: World Scientific, 1981.

Morris, Richard. *Time's Arrows: Scientific Attitudes Toward Time*. New York: Simon & Schuster, 1984.

Morrison, Philip, and Phylis Morrison. *Powers of Ten*. New York: Freeman, 1982. Illustrated guide to natural structure on a wide spectrum of scales.

Moss, Norman. *Men Who Play God: The Story of the H-Bomb and How the World Came to Live With It*. New York: Harper & Row, 1968.

Motz, Lloyd, and Anneta Duveen. *Essentials of Astronomy*. New York: Columbia University Press, 1977. Undergraduate textbook, intended primarily for science majors.

Mourelatos, Alexander P.D., ed. *The Pre-Socratics*. Garden City, N.Y.: Doubleday, 1974.

Muller, H. J. *Out of the Night: A Biologist's View of the Future*. New York: Vanguard, 1935.

Mulvey, J.H. *The Nature of Matter*. London: Oxford University Press, 1981.

Mumby, F.A. *Publishing and Bookselling*. London: Cape, 1956.

Mumford, Lewis. *The Myth of the Machine*. New York: Harcourt, Brace, 1966.

Munitz, Milton K. *Cosmic Understanding: Philosophy and Science of the Universe*. Princeton, N.J.: Princeton University Press, 1986.

———. *Space, Time and Creation: Philosophical Aspects of Scientific Cosmology*. New York: Dover, 1981.

———. *Theories of the Universe*. New York: Free Press, 1957. Classic anthology.

Nagel, Ernest. *The Structure of Science*. New York and Indianapolis: Hackett, 1979.

Nagel, Thomas. *The View From Nowhere*. London: Oxford University Press, 1986. Philosophical study of objectivity.

Narlikar, Jayant. *Introduction to Cosmology*. Boston: Jones & Bartlett, 1983.

———. *The Structure of the Universe*. London: Oxford University Press, 1977. Semitechnical introduction to cosmology and stellar evolution.

Nasr, Seyyed Hossein. *An Introduction to Islamic Cosmological Doctrines*. London: Dent, 1932; Cambridge, Mass.: Harvard University Press, 1964.

Nathan, Otto, and Heinz Norden, ed. *Einstein on Peace*. New York: Avenel, 1960. Einstein's writings and speeches on world peace.

Needham, Joseph. *Order and Life*. Cambridge, Mass.: MIT Press, 1936, 1968.

———. *Science and Civilization in China*. London: Cambridge University Press, 1954–1984.

———. *Science in Traditional China*. Cambridge, Mass.: Harvard University Press, 1971.

Ne'eman, Yuval, and Yoram Kirsh. *The Particle Hunters*. London: Cambridge University Press, 1986. Nontechnical account of contemporary accelerator physics.

Neugebauer, Otto. *Astronomy and History: Selected Essays*. New York: Springer-Verlag, 1983.

———. *The Exact Sciences in Antiquity*. Chicago: University of Chicago Press, 1969.

Neumann, John von. *The Computer and the Brain*. New Haven, Conn.: Yale University Press, 1958.

Neurath, Otto, et al. *Foundations of the Unity of Science*. Chicago: University of Chicago Press, 1938.

Newby, Eric. *The Rand McNally World Atlas of Exploration*. New York: 1975.

Newman, James R., ed. *The World of Mathematics*. 4 vols. New York: Simon & Schuster, 1956. Standard reference work.

Newton, Isaac. *The Correspondence of Isaac Newton*, ed. A. Rupert Hall and Laura Tilling. London: Cambridge University Press, 1959.

———. *Mathematical Principles of Natural Philosophy and His System of the World*, trans. Florian Cajori and Andrew Motte. Berkeley: University of California Press, 1934.

———. *Philosophiae Naturalis Principia Mathematica*. Cambridge, Mass.: Harvard University Press, 1972. Reprint of 1726 edition.

Newton, Robert. *The Crime of Claudius Ptolemy*. Baltimore: Johns Hopkins University Press, 1977. How Ptolemy fudged his data.

Nicolson, Majorie Hope. *The Breaking of the Circle: Studies in the Effect of the New Science upon Seventeenth Century Poetry*. Evanston, Ill.: Northwestern University Press, 1950.

———. *Newton Demands the Muse: Newton's Optiks and the Eighteenth Century Poets*. Princeton, N.J.: Princeton University Press, 1963.

———. *Science and Imagination*. Ithaca, N.Y.: Cornell University Press, 1956.

Niebuhr, Reinhold. *The Nature and Destiny of Man*. New York: Scribner's, 1945.

Nisbet, Robert. *History of the Idea of Progress*. New York: Basic Books, 1980.

North, J.D. *Isaac Newton*. London: Oxford University Press, 1967.

———. *The Measure of the Universe: A History of Modern Cosmology*. London: Oxford University Press, 1965.

Nunis, Doyce, ed. *The 1769 Transit of Venus*. Los Angeles: Natural History Museum of Los Angeles County, 1982.

Ockham, William of. *Philosophical Writings*, trans. Philotheus Boehner. Indianapolis: Bobbs-Merrill, 1964.

Ogburn, Charlton, Jr. *The Forging of Our Continent*. New York: Van Nostrand, 1968. Introduction to the geology of North America, with historical background emphasizing establishment of the geological time scale.

Olby, Robert. *Origins of Mendelism*. Chicago: University of Chicago Press, 1985. Mendel and hybridization.

Oparin, A.I., ed. *The Origin of Life on the Earth*. London: Oxford University Press, 1959.

Oppenheimer, J. Robert. *Letters and Recollections*, ed. Alica Kimball Smith and Charles Weiner. Cambridge, Mass.: Harvard University Press, 1980.

———. *Uncommon Sense*, ed. N. Metropolis, Gian-Carlo Rota, and David Sharp. Boston: Birkhäuser, 1984. Collection of Oppenheimer's nontechnical essays.

Osborn, Henry Fairfield. *From the Greeks to Darwin*. New York: Macmillan, 1894. Survey history of pre-Darwinian evolutionary biology.

Packard, A.S. *Lamarck: The Founder of Evolution*. London: Longmans, 1901.

Pagels, Heinz. *The Cosmic Code*. New York: Bantam, 1982. Introduction to quantum physics, for general readers.

———. *Perfect Symmetry*. New York: Bantam, 1986. Comprehensive popularization of unified field theory and early-universe cosmology.

Pais, Abraham. *Inward Bound*. New York: Oxford University Press, 1986. History of twentieth-century physics, with an emphasis on the period preceding the advent of gauge theory.

———. *Subtle Is the Lord . . . The Science and the Life of Albert Einstein*. London: Oxford University Press, 1982. The standard scientific biography.

———. *A Tribute to Niels Bohr*. Geneva: CERN, 1985.

Pannekoek, Antonie. *A History of Astronomy*. New York: Interscience, 1961.

Panofsky, Erwin. *Renaissance and Renascences in Western Art*. New York: Harper & Row, 1969.

Papagiannis, M., ed. *The Search for Extraterrestrial Life: Recent Developments*. Boston: Reidel, 1985.

Park, David. *The Image of Eternity: Roots of Time in the Physical World*. Amherst: University of Massachusetts Press, 1980.

Parker, Barry. *Einstein's Dream: The Search for a Unified Theory of the Universe*. New York: De Capo/Plenum, 1986. Popular account of contemporary unified theories.

Parr, Charles McKew. *Ferdinand Magellan: Circumnavigator*. New York: Crowell, 1964.

Parry, J.H. *The Age of Reconnaissance*. Berkeley: University of California Press, 1963.

Parton, James. *The Life and Times of Benjamin Franklin*. Boston: Houghton Mifflin, 1882.

Pasachoff, Jay M. *Astronomy: From the Earth to the Universe*. New York: Saunders, 1983. Undergraduate textbook for nonscience majors.

———, and Marc L. Kutner. *Invitation to Physics*. New York: Norton, 1981. College textbook for nonscience majors.

Patterson, Colin. *Evolution*. London: British Museum, 1978. Brief, popular introduction emphasizing the empirical evidence for Darwinism.

Pauli, Wolfgang. *Theory of Relativity*. New York: Pergamon, 1958.

Pauling, Linus. *The Nature of the Chemical Bond*. London: Oxford University Press, 1960.

————, and E. Bright Wilson, Jr. *Introduction to Quantum Mechanics, with Applications to Chemistry*. New York and London: McGraw-Hill, 1935.

Payne-Gaposchkin, Cecilia. *Stars and Clusters*. Cambridge, Mass.: Harvard University Press, 1979. Popular exposition of stellar evolution, by a distinguished astrophysicist.

Peacocke, Arthur. *Creation and the World of Science*. London: Oxford University Press, 1979.

————. *Intimations of Reality: Critical Realism in Science and Religion*. Notre Dame, Ind.: University of Notre Dame Press, 1984.

Pedoe, Dan. *Geometry and the Visual Arts*. New York: Dover, 1976.

Peirce, Charles. *Philosophical Writings of Peirce*, ed. Justus Buchler. New York: Dover, 1955.

Perkins, Donald H. *Introduction to High Energy Physics*. Reading, Mass.: Addison-Wesley, 1982. Textbook.

Philip, J.A. *Pythagoras and Early Pythagoreanism*. Toronto: University of Toronto Press, 1966.

Philipson, Morris, ed. *Leonardo da Vinci: Aspects of the Renaissance Genius*. New York: Braziller, 1966.

Philo. *On the Creation*, trans. F.H. Colson and G.H. Whitaker. Cambridge, Mass.: Harvard University Press, 1981.

Pickering, Andrew. *Constructing Quarks*. Edinburgh: Edinburgh University Press, 1984. History of contemporary nucleon physics, disguised as a sociological study.

Pigafetta, Antonio. *Magellan's Voyage: A Narrative Account of the First Circumnavigation*. New Haven, Conn.: Yale University Press, 1969.

Pilbeam, David. *The Ascent of Man: An Introduction to Human Evolution*. New York, Macmillan: 1972.

Planck, Max. *The Philosophy of Physics*. New York: Norton, 1936.

Plato. *The Collected Dialogues, Including the Letters*, ed. Edith Hamilton and Huntington Cairns. Princeton, N.J.: Princeton University Press, 1969.

————. *Republic*, ed. and trans. Francis MacDonald Cornford. New York: Oxford University Press, 1961.

Playfair, John. *Illustrations of the Huttonian Theory of the Earth*. New York: Dover, 1964. Photocopy reproduction of the book that first popularized Hutton's ideas.

Plotinus. *The Essential Plotinus*, trans. Elmer O'Brien. Indianapolis: Hackett, 1984.

————. *The Six Enneads*, trans. Stephen MacKenna and B.S. Page. Chicago: University of Chicago Press, 1952.

Plumb, J.H. *The Renaissance*. New York: American Heritage, 1961.

Plutarch. *Lives*, trans. John Dryden, revised by A.H. Clough. 5 vols. New York: Harper, 1911.

————. *Moralia*, Vol. XI, trans. Lionel Pearson and F.H. Sandbach; Vol. XII, trans. Harold Cherniss and William C. Helmbold. Cambridge, Mass.: Harvard University Press, 1965, 1968. Contains Plutarch's essay "On the Face in the Orb of the Moon."

Poincaré, Henri. *Science and Hypothesis*. New York: Dover, 1952.

————. *Science and Method*, trans. F. Maitland. New York: Dover, 1952. Reprint of the 1908 edition.

————. *The Value of Science*, trans. G.B. Halstead. New York: Dover, 1958.

Polanyi, Michael. *Personal Knowledge*. Chicago: University of Chicago Press, 1958.

————. *Science, Faith and Society*. Chicago: University of Chicago Press, 1946.

————. *The Study of Man*. Chicago: University of Chicago Press, 1958.

Polkinghorne, J.C. *The Particle Play: An Account of the Ultimate Constituents of Matter*. San Francisco: Freeman, 1981.

Polo, Marco. *The Book of Ser Marco Polo the Venetian Concerning the Kingdoms and Marvels of the East*, trans. Henry Yule. London: Murray 1929.

————. *The Description of the World*, trans. A.C. Moule and Paul Pelliot. New York: AMS Press, 1976.

Ponnamperuma, Cyril. *The Origins of Life*. New York: Dutton, 1972.

————. ed. *Exobiology*. Amsterdam: North-Holland, 1972.

————, and A.G.W. Cameron, eds. *Interstellar Communication: Scientific Perspectives*. Boston: Houghton Mifflin, 1974.

Popper, Karl. *Conjectures and Refutations*. New York: Basic Books, 1968.

————. *The Logic of Scientific Discovery*. New York: Harper & Row, 1968.

————. *Objective Knowledge: An Evolutionary Approach*. London: Oxford University Press, 1972.

Press, Frank, and Raymond Siever. *Earth*. San Francisco: Freeman, 1978. Geology textbook.

Prestage, E. *The Portuguese Pioneers*. New York: Barnes & Noble, 1967.

Price, Lucien. *Dialogues of Alfred North Whitehead*. New York: Mentor, 1956.

Priestley, J.B. *Man and Time*. London: Allen & Unwin, 1964.

Pringele-Pattison, A.S. *The Idea of God in the Light of Recent Philosophy*. London: Oxford University Press, 1917.

Prigogine, Ilya. *From Being to Becoming*. San Francisco: Freeman, 1980. On the origin of complex structures in nature.

Pritchard, James B. *Ancient Near Eastern Texts Relating to the Old Testament*. 2 vols. Princeton, N.J.: Princeton University Press, 1955. Includes creation myths.

Ptolemy. *Almagest*, trans. R. Catesby Taliaferro. Chicago: University of Chicago Press, 1948.

Quine, Willard. V. *Elementary Logic*. Cambridge, Mass.: Harvard University Press, 1980.

————, and J.S. Ullian. *The Web of Belief*. New York: Random House, 1970. Brief study of faith in relation to observation, hypothesis, and explanation.

Rabel, Gabrielle. *Kant*. London: Oxford University Press, 1963.

Rabi, I.I. *Science: The Center of Culture*. New York: World, 1970.

Rae, Alastair I.M. *Quantum Physics, Illusion or Reality?* London: Cambridge University Press, 1986.

Randall, John Herman, Jr. *The Career of Philosophy*. 2 vols. New York: Columbia University Press, 1965.

——. *Philosophy After Darwin*. New York: Columbia University Press, 1977.

Raven, Charles. *John Ray, Naturalist*. London: Cambridge University Press, 1942.

Ray, John. *Wisdom of God Manifested in the Works of Creation*, New York: Garland, 1979. Discussion of the plurality of worlds.

Redondi, Pietro. *Galileo: Heretic*, trans. Raymond Rosenthal. Princeton, N.J.: Princeton University Press, 1987. Argues that Galileo was persecuted more for advocating atomism than Copernicanism.

Reeves, Hubert. *Atoms of Silence*. Cambridge, Mass.: MIT Press, 1984. Philosophical reflections by a nuclear physicist.

Reichenbach, Hans. *The Philosophy of Space and Time*, trans. Maria Reichenbach and John Freund. New York: Dover, 1958.

Reines, Frederick, ed. *Cosmology, Fusion and Other Matters*. Boulder: Colorado University Associated Press, 1972. Essays in memory of George Gamow.

Rhys, H.H., ed. *Seventeenth Century Science and the Arts*. Princeton, N.J.: Princeton University Press, 1961.

Rist, John M., ed. *The Stoics*. Berkeley: University of California Press, 1978.

Roe, Derek. *Prehistory*. Berkeley: University of California Press, 1972.

Ronchi, Vasco. *The Nature of Light: An Historical Survey*. Cambridge, Mass.: Harvard University Press, 1970.

Rood, Robert T., and James S. Trefil. *Are We Alone?: The Possibility of Extraterrestrial Civilizations*. New York: Scribner's, 1981.

Rosen, Edward. *Copernicus and the Scientific Revolution*, Malabar, Fla.: Krieger, 1984. Rosen's commentaries on Copernicus's *On the Revolutions*, along with short related documents.

——. *Kepler's Somnium*. Madison: University of Wisconsin Press, 1967.

——. *Three Copernican Treatises*. New York: Dover, 1959, New York, Octagon, 1971.

Rosenberg, David, ed. *The Realm of Science*. 21 vols. Louisville, Ky.: Touchstone, 1972.

Rosenthal-Schneider, Ilse. *Reality and Scientific Truth*. Detroit: Wayne State University Press, 1980. Reminiscences of discussions with Einstein, Max von Laue, and Planck.

Rossi, Paolo. *The Dark Abyss of Time: The History of the Earth and the History of Nations from Hooke to Vico*, trans. Lydia Cochrane. Chicago: University of Chicago Press, 1984.

——. *Francis Bacon: From Magic to Science*, trans. Sacha Rabinovitch. Chicago: University of Chicago Press, 1968.

Rothenberg, Jerome, ed. *Technicians of the Sacred*. New York: Anchor, 1969. Selected poetry from the "primitive" peoples of Africa, America, Asia, and Oceania.

Rowan-Robinson, Michael. *The Cosmological Distance Ladder*. San Francisco: Freeman, 1985.

Rowe, W.L. *The Cosmological Argument*. Princeton, N.J.: Princeton University Press, 1975.

Rozer, H.F. *History of Ancient Geography*. London: Cambridge University Press, 1935.

Rucker, Rudy. *The Fourth Dimension and How to Get There*. Boston: Houghton Mifflin, 1984.

———. *Infinity and the Mind*. Boston: Birkhäuser, 1982.

Rudnicki, Jozef. *Nicholas Copernicus*. London: Copernicus Quatercentenary Celebration Committee, 1943.

Ruse, Michael. *The Darwinian Revolution*. Chicago: University of Chicago Press, 1979. Darwin's work in the context of the history of science.

Russell, Bertrand. *The ABC of Relativity*. New York: Signet, 1958. Nontechnical introduction.

———. *The Foundations of Geometry*. New York: Dover, 1956.

———. *A History of Western Philosophy*. New York: Simon & Schuster, 1945.

———. *Human Knowledge*. London: Allen & Unwin, 1948.

———. *Mysticism and Logic*. New York: Doubleday, 1917.

———. *The Problems of Philosophy*. London: Oxford University Press, 1959.

———. *The Scientific Outlook*. New York: Norton, 1959.

Rutherford, Ernest. *The Collected Papers of Lord Rutherford of Nelson*, ed. James Chadwick. 3 vols. New York: Interscience, 1962. Includes reprints of Rutherford's accounts of radioactivity, radiochronology, and the structure of the atom.

———. *The Newer Alchemy*. London: Cambridge University Press, 1937. Explication of the transmutation of elements under nucleon bombardment.

———. *Radio-activity*. London: Cambridge University Press, 1904.

Ryder, Lewis. *Elementary Particles and Symmetries*. New York: Gordon & Breach, 1975.

———. *Quantum Field Theory*. London: Cambridge University Press, 1985. Introduction to field theory for students of particle physics.

Sagan, Carl, ed., *Communication with Extraterrestrial Intelligence*. Cambridge, Mass.: MIT Press, 1973. Based on a 1971 conference held in Armenia.

———, et al. *Murmurs of Earth*. New York: Random House, 1978. Details production and contents of the Voyager phonograph record.

Sainte-Beuve, Charles. *Portraits of Men*. Freeport, N.Y.: Books for Libraries, 1972. Reprint of 1891 edition. Contains a biographical sketch of Fontenelle.

Salmon, Wesley C. *Scientific Explanation and the Causal Structure of the World*. Princeton, N.J.: Princeton University Press, 1984.

———. *Space, Time, and Motion: A Philosophical Introduction*. Minneapolis: University of Minnesota Press, 1980.

Sanceau, Elaine. *Henry the Navigator*. New York: Norton, 1947; Hamden, Conn.: Archon Books, 1969.

Sandage, Allan. *The Hubble Atlas of Galaxies*. Washington, D.C.: Carnegie Institution, 1961.

———. *A Revised Shapley-Ames Catalog of Bright Galaxies*, Washington, D.C.: Carnegie Institution, 1981.

———, Mary Sandage, and Jerome Cristian, eds. *Galaxies and the Universe*. Chicago: University of Chicago Press, 1975.

Santillana, Giorgio. *The Origins of Scientific Thought*. New York: New American Library, 1970.

———, and Hertha von Dechend. *Hamlet's Mill: An Essay on Myth and the Frame of Time*. New York, Macmillan, 1969. The case for an ancient appreciation of the fact that the earth's axis precesses over a period of twenty-six thousand years.

Sarton, George. *The Appreciation of Ancient and Medieval Science During the Renaissance (1400–1600)*. Philadelphia: University of Pennsylvania Press, 1955.

———. *A History of Science*. New York: Norton, 1970.

Saunders, N.K. *Poems of Heaven and Hell from Ancient Mesopotamia*. Harmondsworth, Eng.: Penguin, 1972.

Sayen, Jamie. *Einstein in America*. New York: Crown, 1985.

Scharfstein, Ben-Ami. *The Philosophers: Their Lives and the Nature of Their Thought*. London: Oxford University Press, 1980.

Schilpp, Paul Arthur. *Albert Einstein: Philosopher-Scientist*. La Salle, Ill.: Open Court, 1969. Standard reference work; contains Einstein's "Autobiographical Notes."

Schneer, Cecil J. *The Evolution of Physical Science*. New York: University Press of America, 1984.

———, ed. *Toward a History of Geology*. Cambridge, Mass.: MIT Press, 1969. Proceedings of the New Hampshire Inter-Disciplinary Conference on the History of Geology, 1967.

Schrödinger, Erwin. *Space-Time Structure*. London: Cambridge University Press, 1985.

———. *What Is Life?* London: Cambridge University Press, 1946. Philosophy of biophysics.

Schwartz, John, ed. *Superstrings: The First 15 Years of Superstring Theory*. Singapore and Philadelphia: World Scientific, 1985. Collection of technical papers important to the origin and early progress of superstring theory.

Schwarzschild, Martin. *Structure and Evolution of the Stars*. Princeton, N.J.: Princeton University Press, 1958; New York: Dover, 1965.

Schwinger, Julian, ed. *Selected Papers on Quantum Electrodynamics*. New York: Dover, 1958.

Sciama, D.W. *Modern Cosmology*. London: Cambridge University Press, 1975. Nontechnical survey.

Searles, Herbert L. *Logic and Scientific Methods*, New York: Ronald, 1968.

Seelig, Carl. *Albert Einstein: A Documentary Biography*, trans. Mervyn Savill. London: Staples Press, 1956.

Segrè, Emilio. *Enrico Fermi: Physicist*. Chicago: University of Chicago Press, 1970.

———, *From X-Rays to Quarks*. Berkeley: University of California Press, 1980. Historical account of twentieth-century physics, by one of its participants.

———. *Nuclei and Particles*. Reading, Mass.: Benjamin, 1977. Textbook.

Seneca. *Letters from a Stoic*, trans. Robin Campbell. London: Penguin, 1969.

———. *Naturales Quaestiones*, trans. Thomas Corcoran. 2 vols. Cambridge, Mass.: Harvard University Press, 1971.

Setti, G., and L. Van Hove, eds. *Large-Scale Structure of the Universe, Cosmology and Fundamental Physics*. Geneva: CERN, 1984. Proceedings of a 1983 conference on cosmology and particle physics.

Sextus Empiricus. *Against the Physicists*, trans. R.G. Bury. Cambridge, Mass.: Harvard University Press, 1968.

Shapere, Dudley. *Galileo: A Philosophical Study*. Chicago: University of Chicago Press, 1974.

Shapley, Harlow. *Galaxies*. Cambridge, Mass.: Harvard University Press, 1972.

———. *Through Rugged Ways to the Stars*. New York: Scribner's 1969. Autobiography.

———, ed. *A Source Book in Astronomy*. New York: McGraw-Hill, 1929.

———, ed. *Source Book in Astronomy 1900–1950*. Cambridge, Mass.: Harvard University Press, 1960. Compendium of major scientific papers of the period.

Shirley, John. W., and F. David Hoeniger, eds. *Science and the Arts in the Renaissance*. London: Associated University Presses, 1985.

Shklovskii, Iosif S. *Stars: Their Birth, Life, and Death*. San Francisco: Freeman, 1978.

———, and Carl Sagan. *Intelligent Life in the Universe*. New York: Delta, 1966. Ground-breaking SETI study.

Shotwell, James T. *The History of History*. New York: Columbia University Press, 1939.

Shu, Frank. *The Physical Universe*. Mill Valley, Calif.: University Science Books, 1982. Leading astrophysics textbook.

Silk, Joseph. *The Big Bang: The Creation and Evolution of the Universe*. San Francisco: Freeman, 1980. Cosmology and particle physics for general readers.

Simpson, George Gaylord. *The Life of the Past*. New Haven, Conn.: Yale University Press, 1953.

———. *The Meaning of Evolution*. New Haven, Conn.: Yale University Press, 1950.

———. *This View of Life*. New York: Harcourt, Brace, 1964.

Singer, Charles. *A Short History of Scientific Ideas to 1900*. London: Oxford University Press, 1968.

———. *Studies in The History and Method of Science*, Vol. 1. London: Oxford University Press, 1917.

———, et al. *A History of Technology*. 7 vols. London: Oxford University Press, 1978–1979.

Singer, Dorothea Waley. *Giordano Bruno: His Life and Thought*. New York: Schuman, 1950.

Singh, Jagjit. *Great Ideas and Theories of Modern Cosmology*. New York: Dover, 1961. Nontechnical.

———. *Great Ideas of Modern Mathematics*. New York: Dover, 1959.

Singh, Thakur Jaideva. *Philosophy of Evolution: Western and Indian*. Prasaranga, India: University of Mysore Press, 1970.

Sklar, Lawrence. *Space, Time, and Spacetime*. Berkeley: University of California Press, 1977.

Smart, J.J.C. *Philosophy and Scientific Realism*. London: Routledge & Kegan Paul, 1963.

————, ed. *Problems of Space and Time*. New York: Macmillan, 1964, 1979. Anthology of philosophical and historical essays.

Smith, Cyril Stanley. *A Search for Structure: Selected Essays on Science, Art, and History*. Cambridge, Mass.: MIT Press, 1981.

Smith, D.E. *History of Mathematics*. 2 vols. New York: Dover, 1953.

Smith, Robert. *The Expanding Universe*. London: Cambridge University Press, 1982. Account of the "great debate" over the island universe theory during the first three decades of the twentieth century.

Snow, C.P. *The Physicists*. Boston: Little, Brown, 1981. Personal reminiscences, derived from a draft written just before Snow's death and flawed by a few errors.

————. *Variety of Men*. New York: Scribner's, 1967.

Snyder, Paul. *Toward One Science*. New York: St. Martin's, 1978.

Sober, Elliott. *The Nature of Selection: Evolutionary Theory in Philosophical Focus*. Cambridge, Mass.: MIT Press, 1984. Examination of philosophical questions about the nature of Darwinian selection.

Spencer, Herbert. *Essays*, Vol. 1. Appleton: New York, 1901. Includes Spencer on the "nebular hypothesis."

Spinoza, Benedict. *Chief Works*, trans. R.H.M. Elwes, New York: Dover, 1951.

————. *The Correspondence of Spinoza*, ed. and trans. A. Wolf. London: Allen & Unwin, 1928.

————. *Philosophy of Benedict de Spinoza*, trans. R.H.M. Elwes. New York: Tudor, 1936.

Spitzer, Lyman. *Searching Between the Stars*. New Haven, Conn.: Yale University Press, 1982. On interstellar matter.

Sprague, Rosamond Kent. *The Older Sophists*. Columbia: University of South Carolina Press, 1972. Translation of the fragments in Hermann Diels's *Die Fragmente der Vorsokratiker*, a compendium of primary source material on the Sophists.

Stachel, John, ed. *The Collected Papers of Albert Einstein*, Vol. 1. Princeton, N.J.: Princeton University Press, 1987.

Stanley, Steven M. *The New Evolutionary Timetable*. New York: Basic Books, 1981. Illustrated, nontechnical.

Starr, Chester G., et al. *A History of the World*. Chicago: Rand McNally, 1960.

Steen, Lynn Arthur, ed. *Mathematics Today*. New York: Springer-Verlag, 1978. Twelve semitechnical essays.

Stefansson, Vilhjalmur. *Ultima Thule*. New York: Macmillan, 1940. The "golden age" of navigation, circa 3000–1500 B.C.

Steinberg, S.H. *Five Hundred Years of Printing*. New York: Criterion Books, 1959.

Steinmetz, Charles Proteus. *Four Lectures on Relativity and Space*. New York: Dover, 1967. A minimum of mathematics.

Stern, Curt, and E.R. Sherwood, eds. *The Origin of Genetics: A Mendel Science Book*. San Francisco: Freeman, 1966.

Strachey, James, ed. *The Complete Psychological Works of Sigmund Freud*. London: Hogarth, 1955.

Straus, M. *Modern Physics and Its Philosophy*. Dordrecht: Reidel, 1972. Collection of papers and talks in which the author argues that the evolution of physics has been due more to the action of general laws than to the creativity of individual scientists.

Struve, Otto, and Velta Zebergs. *Astronomy of the Twentieth Century*. New York: Macmillan, 1962.

Sullivan, Walter. *Black Holes*. Garden City, N.Y.: Anchor, 1979.

Su Tung-p'o. *Selections from a Sung Dynasty Poet*, trans. Burton Watson, New York: Columbia University Press, 1965.

Sweeney, L. *Infinity in the Presocratics*. Boston: Kluwer, 1972.

Swinburne, Richard. *Space and Time*. New York: Macmillan, 1968.

Szilard, Leo. *Leo Szilard: His Version of the Facts*, ed. Spencer R. Weart and Gertrud Weiss Szilard. Cambridge, Mass.: MIT Press, 1978. Recollections and correspondence by a pioneer in nuclear fusion.

Taton, Rene. *History of Science*, trans. A.J. Pomerans. London: Basic Books, 1963.

————. *Reason and Chance in Scientific Discovery*, trans. A.J. Pomerans. New York: Science Editions, 1962.

Taube, Mieczyslaw. *Evolution of Matter and Energy in Cosmic and Planetary Scale*. Killwanger Switz.: self-published, 1982.

Taubes, Gary. *Nobel Dreams*. New York: Random House, 1986. Account of Carlo Rubbia's quest to identify the W and Z particles predicted by electroweak theory.

Taylor, A.E. *Aristotle*. New York: Dover, 1955. Survey of Aristotle's thought.

————. *Plato: The Man and His Work*. New York: Meridian, 1960. Standard reference, with commentary on each of the dialogues.

Taylor, F. Sherwood. *A Short History of Science and Scientific Thought*. London: Oxford University Press, 1963.

Taylor, J.C. *Gauge Theories of Weak Interactions*. London: Cambridge University Press, 1979.

Tedlock, Dennis, ed. *Popol vuh: The Definitive Edition of the Mayan Book of the Dawn of Life and the Glories of Gods and Kings*. New York: Simon & Schuster, 1985.

Teggart, Frederick J., and George Hildebrand, eds. *The Idea of Progress: A Collection of Readings*. Berkeley: University of California Press, 1949.

Temple, G. *General Principles of Quantum Theory*. London: Methuen, 1934.

Terzian, Yervant, and Elizabeth M. Bilson, eds. *Cosmology and Astrophysics*. Ithaca, N.Y.: Cornell University Press, 1982. Essays in honor of physicist Thomas Gold.

Thayer, H.S. *Newton's Philosophy of Nature*. New York: Macmillan, 1953. Selections from Newton's writings.

Thiebault, Dieudonne. *Original Anecdotes of Frederick the Great, King of Prussia*. London: Johnson, 1805; New York: Riley, 1806.

Thomas, Keith. *Man and the Natural World*. New York: Pantheon, 1983. Changing attitudes toward nature, 1500–1800.

Thomas, Lewis. *The Lives of a Cell*. New York: Viking, 1974.

————. *The Medusa and the Snail*. New York: Viking, 1979.

Thompson, D'Arcy W. *Science and the Classics*. London: Oxford University Press, 1940.